Ion/Molecule Attachment Reactions: Mass Spectrometry

Toshihiro Fujii
Editor

Ion/Molecule Attachment Reactions: Mass Spectrometry

Springer

Editor
Toshihiro Fujii
C & V Technix Co. Ltd.
Tokyo
Japan

ISBN 978-1-4899-7587-4 ISBN 978-1-4899-7588-1 (eBook)
DOI 10.1007/978-1-4899-7588-1

Library of Congress Control Number: 2015933647

Springer New York Heidelberg Dordrecht London

Printed on acid-free paper

Springer Science+Business Media LLC New York is part of Springer Science+Business Media (www.springer.com)

Preface

Ion/molecule reaction mass spectrometry has proved to be one of the most powerful, sensitive, and versatile research techniques and has been applied to analysis in a great variety of qualitative and quantitative studies. As a consequence, mass spectrometers have undergone endless cycles of development of ionization methods for a wide variety of applications. Under these circumstances, the novel ionization method appears that must make difference in science, especially in the field of chemical and environmental analysis, plasma diagnostics, process monitoring, biological, and clinical analysis.

The study of ion–molecule reactions started in the early 1950s with active researches in conventional mass spectrometers with modified ion sources operated at high pressure. Its unique feature is that the initial reaction step is driven by ion-permanent dipole or ion-induced dipole attraction between the reactants and it often occurs without any activation energy. The field has grown rapidly since that time. Pioneering works go back to experiments of Talrose and Melton and the technique was developed to be analytically useful by Munson and Field in the mid-1960s. Kinetic and thermodynamic studies have also grown rapidly during that time, with the development of mass spectrometry (MS) itself. Indeed, the majority of the thermochemical parameters in the gas phase for chemical compounds were determined using various MS techniques.

The ion–molecule association reaction is one of the various reaction types observed in ion–molecule reactions. It is a unique process similar to the combination of two atoms or two free radicals. The energy accumulated in the reaction complex often must be liberated by collision with a third body for the complex to survive. Studies of complex ions are relevant to phenomena such as solvation, clustering, plasma chemistry, radiation chemistry, flame and combustion, and even atmospheric and interstellar processes. The research area has been extended to the subject of a clustered ion formation in bridging the gap between the gas and condensed phase.

The characterization of reactive intermediate, free radical has been one of the most important challenges for the scientist in the history of chemistry. In early times, MS has appeared as a direct method to study radical intermediates due to its immediate response in the analysis. Among the various original methods developed in the purpose of detecting and characterizing free radicals in dynamic chemical systems such as plasmas and combustion, the ion attachment mass spectrometry

(IAMS) method to exploit alkali ion/molecule association reaction (cationization) appears to be an alternative to low-energy electron impact ionization methods. A real challenge for IAMS is to extend its capability to detect intermediary free radical species in various dynamic chemical systems. Recently this type of mass spectrometer is available commercially in a complete form.

Transition to the Gaseous to the Condensed Phase A variety of ionization techniques in the condensed phase are now available which enable MS to be used for the direct analysis of nonvolatile or/and thermally labile compounds. These include field desorption (FD), fast atom bombardment (FAB), plasma desorption (PD), secondary ion mass spectrometry (SIMS), electrospray ionization (ESI), and laser desorption (MALDI). Most methods can be classified into desorption ionization (DI). Cationization reactions are ubiquitous in DI MS, where traces of alkali salts are often sufficient to produce intense cationized species. Cationization appears to hold great promise for the analysis of nonvolatile and thermally labile compounds. Studies have been extensively done to understand the mechanism of this form of ionization in order to make a wider application of the technique. Cationization MS has already been regarded as the method of choice for the analysis of nonvolatile and thermally labile substances. Thus, the questions arise: (1) what are the distinct features of cationization MS, (2) what are problems of the technique and (3) what is its advantage for the analysis of biological substances, drugs and drug metabolites, medicinal and environmental chemicals. Gas-phase ion/molecule reaction chemistry can potentially lead to the understanding of the ion complexation (cationization) mechanism in the condensed phase. To our best knowledge, so far, no comprehensive monograph has been published for the basic fundamentals of ion/molecule association reaction and its mass spectrometric applications (conditions) for a wide diversity of analytical study.

Characteristics of This Book This monograph will cover primarily thermal alkali-metal ion association reactions of the general type: $A^{+}+M+N \rightleftarrows (A+M)^{+}+N$, where A denotes a positively charged alkali metal ion, M is a neutral species and N works as a third body. In this volume, nine chapters are presented. They include tutorials on important principles as well as reviews on special topics, covering fundamental operating principles, instrumentation, and representative applications for all areas of chemical dynamics, environmental chemistry, molecular medicine, and biological science. Cationization is defined (used) in this book as metal ion association, both in the gas-phase and in the condensed-phase. Attachment and association are used interchangeable. Other synonyms are listed elsewhere. Authors try to prepare the manuscript for reader to find a guide for tomorrow by taking lessons from the past. In other words, this book includes rather old but excellent studies, which seems to be obsolete. Individual chapters will be available for purchase from the publisher's website. In order for these chapters to be self-completed, it is inevitable to have few replications. Consequently, each chapter contains references to research articles and reviews.

This book is divided into nine parts:

1. Introduction
2. Ion chemistry (principle)
3. A theoretical approach to ion-molecule complexation
4. Experimental methods and instrumentation
5. Applications of association reactions in the gas-phase
6. Hybrid system with ion attachment techniques
7. Cationization MS for the condensed-phase samples
8. Direct analysis mass spectrometry
9. Future prospects

Chapter 1 provides a historical viewpoint (perspective) on the study of ion/molecule association (cationization) MS as well as explanation on the evolution of developments of the instrumental methods. In addition to serving as an introduction for the subject of cationization MS as it pertains to ion chemistry, this chapter briefs thermochemistry and chemical dynamics (and analytical application) of metal ion association reaction. The fundamentals for ion/molecule association reaction are described in *Chapter 2*, providing a basic introduction to the mechanism and dynamics of termolecular association reaction, dissociation and fragmentation reaction of associated ion and ion/molecule association mechanism in the condensed-phase.

Chapter 3 briefly describes the basic concepts behind ion-molecule interactions, such as ion-dipole, ion-induced dipole, and ion-π interactions. Fundamental theories explaining ion-molecule association/dissociation reaction dynamics are also outlined in this chapter. Computational methodologies to study the ion-molecule interactions are discussed. Finally, theoretical studies (using both *ab initio* and density functional theories) on ion interactions with organic molecules, biomolecules, and materials are given in detail with a lot of important and fundamental examples.

Chapter 4 discusses a variety of experimental methods and introduces instrumentation for the study of gas-phase ions and ion-molecule chemistry in general and of ion attachment processes in particular, with an emphasis on mass spectrometry (MS) methods. The section starts with a discussion on common methods to generate gas-phase alkali metal ions, to be applied in various ion-attachment experiments. Subsequently, (tandem) MS methods are discussed that enable the investigation of the structure of ions and the results of ion-molecule reactions. Separate sections are presented for beam instruments and ion-trapping instruments. In some cases, these instruments can be used to perform ion-molecule reactions as part of the measurement protocol. Emphasis is put on commercial available MS and MS–MS instruments. In subsequent sections, various other tools are discussed, that may be combined with MS and allow the study of gas-phase reactions of ions. These tools comprise (a) flowing-afterglow methods (FA-MS), including derived methods like selected-ion flow tubes (SIFT-MS) and proton-transfer reaction devices (PTR-MS), (b) drift tubes, ion-mobility spectrometry (IMS) and IMS–MS, and (c) high-pressure MS instruments. The text provides ample references for further reading.

Chapter 5 explores the application of alkali-metal ion/molecule association in the gas-phase to MS. This chapter: (1) surveys the fundamental basis to build the instrument, (2) describes the instrumentation to the measurement of mass spectrum, and (3) summarizes mass spectrometric applications to measurements on chemical compounds, which include intermediary free radical, interstellar, environmentally important, and unfamiliar or unstable species. The subject of application of the restriction of hazardous substances (RoHS) detective is also covered.

Chapter 6 describes the concept of hybrid mass spectrometric system with ion attachment technique as ionization method. A combined (hypernated) MS represents time-of-flight (TOF), ion trap quadrupole, ion mobility spectroscopy, ion cyclotron resonance (ICR) or aerosol MS, while descriptions of specially designed inlet system include chromatographic introduction (inlets), and various pyrolysis probes for evolved gas analysis. Some applications of each technology are presented, together with representative and/or illustrative examples. In addition, development of portable IAMS is provided along with explanations and spectral applications.

Chapter 7 highlights analytical mass spectrometry with use of Alkali metal ion reaction (association) in the condensed-phase. Methodology for cationization MS is emphasized in this chapter, which also describes role of M+ adduct of biologically important nucleosides, peptide or protein, sugars, as well as synthetic polymers. The approach and procedure for the structural characterization of biologically important substances from CID, BIRD, ECD, ISD and PSD MS/MS data, are described in detail, being reviewed by hundreds of current topics.

Chapter 8: Recently, we have seen a rapid development in direct introduction mass spectrometry (DIMS). Among the DIMS instruments mentioned are: atmospheric-pressure chemical ionization (APCI), proton-transfer-reaction mass spectrometry (PTR-MS), selected ion-flow-tube mass spectrometry (SIFT-MS), ion-molecule reaction mass spectrometry (IMR-MS), direct analysis in real time (DART) and desorption electrospray ionization (DESI). These technologies are also compared for applications to ambient mass spectrometry. Finally, selected examples of applications for the technologies are given, including research monitoring of volatile organic compounds (VOCs) in the applications to food-quality control, flavor release, process monitoring, environmental and medical sciences.

Chapter 9 discusses IAMS from the application point of view, highlighting important contributions IAMS can make in the study of dynamic systems like flames, discharges and plasmas, in pharmaceutical, biochemical and biotechnological, environmental, food safety, and polymer application areas.

Acknowledgments

This book embodies efforts with prominent colleagues, S. Arulmozhiraja, W. M. A. Niessen, and M. McEwan.

Among others whose contributions are greatly acknowledged are Drs. M. Nakamura, A. Maruyama, and K. Hino, editor's associates at Canon Anelva Corporation. They provided great support, especially on mass spectral data of IAMS derived from experiments.

Special thanks also go to Keiko Tatewaki, Mika Hayashi, Machiko Fujii, Dr. Hisao Kawamura, and Dr. Seiji Takahashi, who endured the final complications with submission to Springer.

This work was supported partially by a grant from the France–Japan Sasakawa Foundation. Thanks also go to kind librarians, who generously provided a various kind of helps for TF at the Ohme campus library of the Meisei University during the whole stages of this book project.

C & V Technix Co. Ltd.
Tokyo
Japan

Toshihiro Fujii

Contents

List of Authors

Sundaram Arulmozhiraja Catalysis Research Center, Hokkaido University, Sapporo, Japan

Toshihiro Fujii C & V Technix Co. Ltd., Akishima, Tokyo, Japan

Murray J McEwan Department of Chemistry, University of Canterbury, Christchurch, New Zealand

Wilfried M.A. Niessen Hyphen MassSpec, Leiden, The Netherlands

Abbreviations

ADDA	3-Amino-9-methoxy-2,6,8-trimethyl-10-phenyldeca-4,6-dienoic acid
AE	Appearance energy
amu	Atomic mass unit
AP	Appearance potential
APCI	Atmospheric pressure chemical ionization
API	Atmospheric pressure ionization
APPI	Atmospheric pressure photoionization
Arg	Arginine
ASAP	Atmospheric-pressure solids analysis probe
Asp	Aspartic acid
atm	Atmosphere (760 Torr)
B	Magnetic field sector
BDE	Bond dissociation energy
BDE	Bond dissociation energy (or enthalpy)
BIRD	Blackbody infrared radiative dissociation
Boc	*t*-Butyloxycarbonyl
CA	Collisional activation
CAD	Collisionally activated dissociation
cal	Calorie
cat	Cation
$C_{12}N$	Dodecylamine
CAPB	Cocamidopropylbetaine
CHCA	α-Cyano-4-hydroxycinnamic acid
CE	Capillary electrophoresis
CEMA	Channel electron multiplier array
CI	Chemical ionization
CID	Collision-induced dissociation
CIS	Coordination electrospray ionization
CRF	Charge-remote fragmentation
CRM	Certified reference material
CREMS	Charge reduction electrospray mass spectrometry
CVD	Chemical vapor deposition

CVTST	Canonical variational transition state theory
CZE	Capillary zone electrophoresis
Da	Dalton (unit of mass on the atomic scale)
DAC	Digital-to-analog converter
DAPCI	Direct atmospheric pressure chemical ionization
DART	Direct analysis in real time
DC	Direct current voltage
DCI	Desorption (direct) chemical ionization
DESI	Desorption electrospray ionization
DHAP	2,6-Dihydroxyacetophenone
DHB	2,5-Dihydroxybenzoic acid
DNA	Deoxyribonucleic acid
DSSI	Desorption sonic spray ionization
DI	Desorption/ionization
DIOS	Desorption ionization from silicon
DIP	Direct insertion probe, direct inlet probe
DT	Drift tube
e	Electron
E	Electric field sector
EA	Electron affinity
ECD	Electron capture dissociation
EE	Even electron
EE^+	Even-electron ions
EHI	Electrohydrodynamic ionization
EGA	Evolved gas analysis
EI	Electron ionization
ELS	Electrostatic lens system
EM	Electron multiplier
EPI	Enhanced product-ion analysis
ESA	Electrostatic analyzer
ESI	Electrospray ionization
ETD	Electron transfer dissociation
eV	Electron volt
FA	Flowing afterglow
FAPA	Flowing atmospheric pressure afterglow
FAB	Fast atom bombardment
F20TPP	Meso-tetrakis(pentafluorophenyl)porphyrin
FAIMS	High-field asymmetric waveform ion-mobility spectrometry
FD	Field desorption
FDT	Flow-drift tube
FI	Field ionization
FT	Fourier transform
FTMS	Fourier transform mass spectrometry (usually ICR)
Fuc	Fucose
FWHM	Full width half maximum

Gal	Galactose
GalNAc	*N*-Acetyl-galactosamine
GC-MS, GC/MS	Gas chromatography/mass spectrometry
GIB	Guilded ion beam
Glc	Glucose
GlcNAc	*N*-Acetyl-glucosamine
Glu	Glutamic acid
GluA	Glucuronic acid
GPA	Glycerophosphatidic acid
GPCho	Glycerophosphocholine
GPEtn	Glycerophosphoethanolamine
GPGro	Glycerophosphoglycerol
GPIno	Glycerophosphoinositol
GPSer	Glycerophosphoserine
H	Magnetic field
HCD	Higher energy collisional dissociation
HC(s)	Hydrocarbon(s)
H/D	Hydrogen/deuterium
HDE	Hydrogen/deuterium exchange
HFIP	hexafluoroisopropanol
HILIC	Hydrophilic interaction chromatography
His	Histidine
3-HPA	3-Hydroxypicolinic acid
HPAEC	High-performance anion-exchange chromatography
HPMS	High pressure source mass spectrometry
Hz	Hertz (cycles sec^{-1})
IA	Ion attachment
IAR	Ion attachment reaction
ICP	Inductively coupled plasma
ICR	Ion cyclotron resonance
IE	Ionization energy
IMR	Ion molecule reaction
IMS	Ion mobility spectrometry
IP	Ionization potential
IRMPD	Infrared multiphoton dissociation
ITMS	Ion trap mass spectrometer
J	Joule
KE	Kinetic energy
L	Liter
LC-MS, LC/MS	Liquid chromatography/mass spectrometry
LD	Laser-induced dissociation
LIFDI	Liquid injection field desorption ionization
LINAC	Linear-acceleration high-pressure collision cell
LIT	Linear quadrupole ion trap

LOD	Limit of detection
LSIMS	Liquid secondary ion mass spectrometry
LOQ	Limit of quantification
LPD	Laser photodissociation
Lys	Lysine
M^{+*} or M^{-*}	Molecular ion
MH^+	Protonated molecule
[M-H]−	Deprotonated molecule
MALDI	Matrix-assisted laser desorption/ionization
Man	Mannose
MDA (mda)	Minimum detectable amo
MID	Multi-ion detection
mmu	Millimass unit, 0.001 u, a millidalton
MS	Mass spectrometry
MS1	First-stage mass analysis in MS–MS system
MS2	Second-stage mass analysis in MS–MS system
MS/MS, MS-MS	Tandem mass spectrometry
MSn	Multiple stages of MS/MS
MPD	Multiphoton dissociation
MPI	Multiphoton ionization
m/z	Mass-to-charge ratio
NBS	National Bureau of Standards
nESI	Nano-electrospray ionization
Neu5Ac	*N*-acetyl-neuraminic acid or sialic acid
NIST	National Institute of Standards and Technology (formerly NBS)
NRMS	Neutralization-reionization mass spectrometry
oaTOF	Orthogonal injection TOF
OE	Odd electron
$OE^{+\bullet}$	Odd-electron ions
Pa	Pascal, SI unit of pressure
PA	Proton affinity
PCI	Positive-ion Cl
PD	Photodissociation
PDI	^{252}Cf-plasma desorption ionization
PEG	Polyethylene glycol
PES	Photoelectron spectroscopy
PFI	Pulsed field ionization
PIMS	Photoionization mass spectrometry
PQ-MS	Penta-quadrupole MS^3 instrument
PSD	Post-source decay
PST	Phase space theory
PTM	Post-translational modification
PTR	Proton-transfer reaction
Py/MS, Py-MS,	Pyrolysis mass spectrometry
q	RF-only Q collision cell in a tandem MS

Q	A quadrupole m/z analyzer in a tandem MS
QET	Quasi-equilibrium theory
QIT	Quadrupole ion trap
QMS	Quadrupole mass filter, quadrupole mass spectrometer
Q-TOF	Tandem MS with a Q followed by a TOF
r	Radius of electric or magnetic sector
R	Resolving power
reISD	Insource decay in reflectron mode
REMPI	Resonance-enhanced multiphoton ionization
reTOF	Reflectron TOF
RF	Radio frequency
RGA	Residual gas analyzer
RI	Relative intensity
RNA	Ribonucleic acid
RPLC	Reversed-phase liquid chromatography
RRK	Rice–Ramsperger–Kassel
RoHS	EU-Directive on the Restriction of Hazardous Substances
RSD	Relative standard deviation
RRKM	Rice–Ramsperger–Kassel–Marcus
SA	Stationary afterglow
s/b	Signal-to-background ratio
sccm	Standard cubic centimeter per minute
sec	Second
SELDI	Surface-enhanced laser desorption ionization
SFC	Supercritical fluid chromatography
SFDT	Selected flow drift flow tube
SFE	Supercritical fluid extraction
SID	Surface-induced dissociation
SIFT	Selected ion flow tube
SIM	Selected ion monitoring, single ion monitoring
s/n	Signal-to-noise ratio
SORI	Sustained off-resonance irradiation
SPE	Solid-phase extraction
SPME	Solid-phase microextraction
SRI-MS	Selective reagent ionization mass spectrometry
SRM	Selected reaction monitoring
STP	Standard temperature and pressure
TCDDs	Tetrachlorodibenzo-*p*-dioxins
TCID	Threshold collision-induced dissociation
TG	Thermogravimetric
THAP	2,4,6-Trihydroxyacetophenone
TIC	Total ion current
TIM	Total ion monitoring
TOF	Time-of-flight
Torr	Pressure unit equaling to 133 Pa

TQ	Tandem (or triple) quadrupole mass analyzer
TQS	Triple-quadrupole mass spectrometer
TSP	Thermospray
TS	Transition state
TST	Transition state theory
TMS	Tandem mass spectrometry
TRPI	Time resolved photoionization
U	Unified atomic mass unit. SI symbol = 1/12 mass of ^{12}C
V	volt
VOC	Volatile organic compound
VTST	Variational transition state theory
z	Number of charges on an ion

Lists of Synonym

Attachment and association are used interchangeable.
Other synonyms are:

AE (appearance energy)	AP (appearance potential)
Attachment	Association
Bond strength	Bonding energy, binding energy, and bond dissociation energy (BDE)
CVTST (canonical variational transition state theory)	VTST
Direct insertion probe	Direct inlet probe
Cation	Positive ion
Dissociation	Decomposition
Dynamics	Kinetics
FTMS	ICRMS
Hyphen (-)	Slash (/), for instance; ion/molecule and Ion–molecule
Ionization energy (IE)	Ionization potential (IP)
Ion attachment	Ion association and cationization
ITMS	QIT
Non-covalent bonding	Ionic bonding
Proton affinity	Basicity
Pseudomolecular ion	Quasimolecular ion
Pyrogram	Thermogram
Quadrupole mass spectrometer	Quadrupole mass filter
SIM (single ion monitoring)	MID (multi-ion detection)
Tandem mass spectrometry	MS/MS
TSQ	QQQ, Triple QMS

Chapter 1
Introduction

Toshihiro Fujii

1.1 Ion–Molecule Association Reactions

Thermal alkali-metal ion association reactions of the general type are:

$$A^{+} + M + N \rightleftharpoons (A{+}M)^{+*} + N, \tag{1.1}$$

where A denotes a positively charged alkali-metal ion, M is neutral species and N works as a third body. As indicated, most association reactions are reversible, though not necessarily appreciably so for all the systems of ionic and neutral molecular reactants. The bonding between ion and neutral species is generally found to be weak relative to normal chemical bond dissociation energy (BDE); BDE (A-M) is typically 200 kJ mol^{-1} or less, often much less. The A-M bond derives primarily from electrostatic forces such as ion-dipole attraction [1].

An ion–molecule association reaction is a process similar to the combination of two atoms or two free radicals. The initial association step takes place by ion-permanent dipole or ion-induced dipole attraction between the reactants and often occurs without any activation energy. The energy accumulated in the reaction complex $[(A+M)^{+*}]$ often must be liberated by collision with a third body, N, for the complex to survive. In some exceptional cases, however, a radiative association reaction or an elimination reaction involves an ion–molecule association complex (see Chap. 2). Termolecular reactions are second order in the condition of saturated termolecular kinetics and their rate depends upon a third body density. Typically, termolecular ion–molecule reactions (IMRs) have rate coefficients in the range 10^{-9}–10^{-10} $cm^3\ s^{-1}$ (10^{10}–10^{11} L mole s^{-1}). In general, they have larger rate coefficients in the range of the Langevin rate [2].

Considerable effort has been devoted to the ab initio molecular orbital (MO) calculation of three important aspects for ion–molecule association reactions: (i) the

T. Fujii (✉)
C & V Technix Co. Ltd., 3-6-1 Higashi, Akishima, Tokyo 196-0033, Japan
e-mail: fujii.toshihiro@c-vtechnix.com

T. Fujii (ed.), *Ion/Molecule Attachment Reactions: Mass Spectrometry*,
DOI 10.1007/978-1-4899-7588-1_1

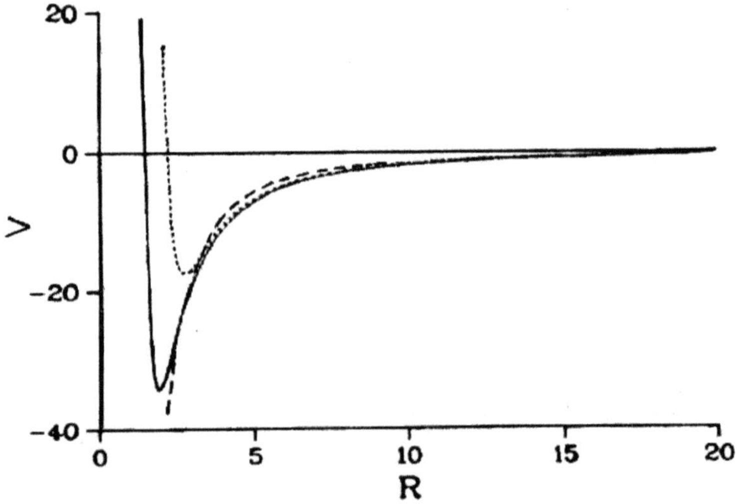

Fig. 1.1 Plot of potential energy along the minimum energy path with H_2O constrained in its equilibrium geometry. *R* is the distance between alkali ion and the H_2O center of mass. $Li^+ + H_2O$ *(line)* and $K^+ + H_2O$ *(dots)*. Alkali ion-water potential surface *(dashes)* obtained by collisional theories of ion-molecule kinetics: $V(r, \theta) = -(\alpha q_2/2r_4) - (q\mu_D/r_2)\cos\theta$, where μ_D is the dipole moment of the neutral and θ is the angle formed by the dipole and *r* (Reprinted with permission from [5]. ©1982, American Institute of Physics)

properties of the transition states, (ii) the accuracy of the canonical variational transition state theory (CVTST) rate constant, and (iii) the potential energy surface [3, 4]. Many of these studies have utilized the ion-permanent dipole and ion-induced dipole potential. These studies are very illustrative to understand the fundamentals of ion–molecule association reactions. The Li^+ and $K^+(H_2O)$ equilibrium complex is shown in Fig. 1.1, along the minimum energy path. This complex has C_{2v} symmetry and a classical binding energy of 33.78 kcal/mol with respect to completely dissociated $Li^+ + H_2O$. The Li^+, O, and H atoms from Li^+ are in the same plane. The potential minimum occurs at R = 1.865 A when R is the distance between Li^+ and the H_2O center of mass [5].

1.2 Origin

1.2.1 Flowing Afterglow

A study of IMRs has its roots in the National Oceanic and Atmospheric Administration (NOAA) flowing afterglow (FA) apparatus. In the early 1960s, Ferguson et al. [6, 7] developed the FA method for investigating the kinetics of IMRs which were important in the chemistry of the upper atmosphere. FA has since emerged as one of the most versatile methods for studies of gas-phase ion chemistry. In its basic form, a FA apparatus consists of an ion source, a relatively high-pressure flow-tube reactor, an ion sampling system, and a quadrupole mass spectrometer. The early FA studies included metal-ion chemistry, which was directed to metals such as lithium, magnesium, sodium, and so on. These measurements established fundamental features for the chemical kinetics of metal ions. Termolecular kinetics was established for the association reactions of Li^+ with N_2, CO_2, SO_2, halogenated organics under the conditions of a pressure range from 0.3 to 1.0 Torr [8].

1.2.2 *Drift-Tube*

The conventional type of drift-tube (DT) experiment, in which ions drift under the influence of a weak electric field, has been widely used in the measurement of ion mobilities, equilibrium constants, and IMR rate constants as a function of ion kinetic energy from 0.05 to about 5 eV [9]. The capabilities of this technique have been extended soon by incorporating an ion drift-tube section into the gas-flowing region of the basic FA device [10]. These FA and DT instruments (namely, flow-tube mass spectrometry) and their many derivative techniques such as the powerful selected ion flow tube (SIFT) method, variable temperature FA, and guided ion beam tandem mass spectrometry have been employed in many laboratories worldwide.

1.2.3 *High-Pressure Mass Spectrometry*

The thermodynamics of alkali-ion metal–molecule reactions began to be uncovered in the mid-1960s [11, 12]. Work on the thermodynamics of the solvation of ions may be considered as a root in the studies of alkali-metal IMRs. In the 1970s, several research groups studied the gas phase solvation of alkali-metal ions by water, ammonia, and various solvent molecules. The thermochemistry for clustering of water molecules around alkali ions was established [13]. The dynamics (kinetics) are also important for understanding the ion–molecule association. In this regard, the most important early development was that of the high-pressure mass spectrometer (HPMS) technique which enabled ion thermalization and the establishment of equilibrium cluster ion distributions [14]. This method has grown to be one of the most valuable in deriving thermochemical data of ion–molecule association reaction.

1.2.4 *Present*

Over the last 50 years, all three basic elements of the flow-tube mass spectrometry, the ion source, reactor, and detector, have been continuously modified and improved, giving rise to a variety of useful new instruments (see Chaps. 4, 5, 6 and 8).

Those are, for instance, known as active chemical ionization mass spectrometry (ACIMS) for atmospheric trace-gas measurements in the chemosphere, proton transfer reaction mass spectrometry (PRT-MS), selected ion flow tube mass spectrometry (SIFT-MS), ion–molecule reaction mass spectrometry (IMR-MS), ion attachment mass spectrometry (IAMS), and ion mobility spectroscopy (IMS). Those have now become part of the extensive resources of experimental techniques dedicated to all the scientific fields, especially such as fundamentals of gas-phase ion kinetics and thermochemistry, physical organic chemistry, atmospheric chemistry, interstellar chemistry, plasma, and combustion chemistry, radiation chemistry, analytical chemistry, and ultimately even medicine and biology.

1.3 Importance and Interest

Apart from their intrinsic interest as a chemical process, such ion–molecule association reaction can provide a great deal of information on the nature of ion clustering [14–17]. Equation 1.1 is regarded as representing merely the first step in a series of reactions leading to the fully clustered ion. It has also been found that ion–molecule association is a particularly important process in the atmosphere, ionosphere, interstellar environments, and plasma, flame, combustion, and radiation chemistry.

The formation of ion clusters in the gas phase proceeds via a sequence of association reactions between ions and the molecules.

$$A^+M_{n-1} + M + N \rightleftharpoons A^+ M_n + N, \tag{1.2}$$

where A^+ designates the ion about which molecule M is clustered. N is the third body which serves to stabilize the cluster during the association process. These reactions occur for a wide variety of molecules and are observed in most cases to be highly exothermic with a collision rate constant and little or no activation energy. Alkali halide clusters $[A^+(AX)_n]$ have been studied extensively for abundance distribution and thermodynamics [17, 18]. The study of the properties of ion clusters is very interesting since they are concerned with the phase transitions, the development of surfaces, clustering, and formation of the condensed state [15].

Experimental results of ion–molecule association reactions also provide insight into a wide range of atmospheric issues. For instance, studies of gas phase molecular aggregates of acid molecules can contribute to an understanding of acid rain, smog, and the chemistry of the upper atmosphere [19]. Another application to atmospheric issues deal with a wide range of reactions occurring in the ionosphere through all altitudes up to 90 km, and the chemistry taking place on polar stratospheric clouds (PSCs) surface [20].

Recently many propositions have been made in the field of astronomical science. Herbst et al. proposed plausible mechanisms for converting molecular hydrogen and atomic carbon in interstellar clouds into hydrocarbon molecules via sequences of ion-neutral reactions [21]. These mechanisms have been incorporated into IMR models. The chemistry of the early universe includes several reaction routes like non-reactive collisions, IMRs, and two-body association reactions. The possible presence of LiH^+ was first postulated by Dalgarno and Lepp, who suggested it to be formed by efficient ion–atom radiative association [22]. It was suggested at the time of the original discovery of C_{60} [23] that polycyclic and polyaromatic hydrocarbon (PAH) molecules and ions, and possibly larger fullerene molecules and ions, are present in interstellar and circumstellar environments. These propositions definitely enhance the fundamental studies of association reactions of ions with molecules [24].

The importance of the ion chemistry in plasma [25], flame [26], combustion [27], and radiation [28, 29] is well recognized. Attention is drawn to dynamical and accurate measurements of IMRs which are required to improve our understanding of the ion chemistry proceeding in such extremely complex environments. For instance, in analysis of flames and related combustion processes, difficulties are associated with the fact that a flame is a complex reacting chemical system which contains many interfering species with steep gradients of reactive species [30]. This needs measurement of spatial composition with a precise position in the system. The study of ions, however, provides an ideal position for the dynamical and spatial study of reaction dynamics because of the ease with which ions are handled by mass spectrometry (MS) due to their charges.

1.4 Ion Thermochemistry

1.4.1 Importance

Ion thermochemistry is of great practical importance to MS. Mass spectral patterns are intimately related to ion energy. In chemical ionization (CI)MS, the absence of appreciably endoergic IMRs results in analytical selectivity. Schemes for analyzing and sequencing peptides and other biomolecules by MS involve site-specific protonation or cationization. The site specificity is determined by the relative proton affinities or metal ion affinities of various functional groups. The dynamics of the reactions is important too, but thermodynamics ultimately controls the favored processes [31].

A variety of MS and ion spectroscopy techniques have been developed to measure ion thermochemical quantities [32]. MS allows a gas-phase ion of known mass to be isolated and manipulated. The most precise and accurate measurements of molecular energetic are made from spectroscopic measurements, where transitions between initial and final quantum states are individually resolved and assigned. Aside from spectroscopic determinations, IMR equilibrium measurements can provide the most accurate ion thermochemistry values. Ion concentrations are sampled directly from an equilibrium system using MS. For nonequilibrium systems or where not all the reactants and products are stable or easily produced, unimolecular and bimolecular kinetics methods have been developed by ion chemists to obtain thermochemical information. Theoretical calculations using ab initio molecular orbital (MO) theory or density functional theory (DFT) have become extremely important for the experimental chemist in general and for ion chemists in particular [33]. Modern theory rivals the accuracy of experimental thermochemistry in some cases and is indispensable for the interpretation of experiments in others. Thermochemical applications of theory, as well as comprehensive compilations of ion and neutral thermochemical quantities are available [33–36].

1.4.2 Thermodynamics

The thermodynamical studies of alkali-ion metal–molecule reactions in a HPMS began in the mid-1960s. Kebarle reported the gas-phase reaction involving the clustering of water molecules around the potassium ion [37]. They established the thermodynamics on the solvation of ions [38]. The study of dynamics (reaction rate) is also important in ion–molecule association, which is detailed in the Chaps. 2 and 3. These two different areas of interest can in practice be investigated using MS and computational techniques.

An ion/molecule equilibrium:

$$A^{+} + B \rightleftharpoons C^{+} + D \tag{1.3}$$

is established in a FA, drift-tube, and HPMS, or ion cyclotron resonance spectrometer. The equilibrium constant is determined by observing the relative abundances of the two ions, A^{+} and C^{+} after an equilibrium is established:

$$K_{eq} = [C^{+}][D]/[A^{+}][B]. \tag{1.4}$$

Here, A^{+} and C^{+} represent the respective measured ion intensities. The neutral reactants, B and D, are present in great abundance compared to the ionic reactants, and therefore, the ratio [D]/[B] can be assumed as unchanged after equilibrium is established.

As can be seen from the following equation:

$$-RT \ln K_{eq} = \Delta G^{\circ} = \Delta H^{\circ} - T\Delta S^{\circ}, \tag{1.5}$$

a single measurement leads to a value for the Gibbs change in free energy of reaction at the temperature of the measurement. Such measurements only generate relative thermodynamics scales of ΔH° and ΔS°. Practically, many studies have been published in which measurements were made at a single temperature. In order to determine the absolute values for thermochemical quantities of these reactions, values for the entropy changes of reaction must be obtained, through measurements of the equilibrium constant at different temperatures (Van't Hoff plot) or through statistical mechanical estimations.

1.4.3 Bond Dissociation Energy (Metal Ion Affinity)

The study of the binding of metal cation with various chemical species has been an area of significant interest over the past 50 years. Especially the interaction between organic compounds or biologically active molecules and alkali-metal cations in the gas phase has attracted much attention. The same type of fundamental reactions, protonation and ion-solvent molecules association, occur also in the gas phase [39]. Such interactions relate to chemical and biological processes occurring in the heterogeneous regions; for example, ion solvation, catalysis, transport through membranes, affinity of active compounds toward receptors, and antibiotic activity. Furthermore, the corresponding complexes have useful applications in analytical MS.

Understanding these phenomena requires knowledge of the corresponding interactive energetic in the gas phase, a solvent-free environment. Such an intrinsic ion-neutral interaction can be combined with data on ion solvation for better understanding of the condensed phase phenomena. Metal cation affinities are a good basis for an analysis and a modeling of such interactions in complexes systems. Experimental studies have focused primarily on the binding of alkali-metal ions and transition-metal ions. Much experimental bonding energy can be found in the exhaustive reviews [34–36, 40]. Most of the reported data were obtained by HPMS and ion cyclotron resonance (ICR).

The BDEs of ion–molecule association complexes are afforded by enthalpy changes of the reaction

$$MA^{+} \rightarrow M + A^{+} \quad \Delta H = D\left(M - A^{+}\right), \tag{1.6}$$

where M and A^+ represent a molecule and a metal cation, respectively. Many techniques have been used to investigate such BDEs ranging from equilibrium measurements, bracketing experiments, photodissociation, and threshold collision induced dissociation. On the other hand, the relationship between the enthalpy of formation of MA^+ and its neutral counterpart, M, is also defined in terms of a quantity called the ion affinities. This is the negative of the enthalpy change of the hypothetical ion association (cationization) reaction:

$$M + A^{+} \rightarrow MA^{+} \; - \Delta H = \text{ion affinities}. \tag{1.7}$$

The term "ion affinities" is a quantity defined at a finite temperature, usually 298 K. The principal instruments that have produced information on the bonding in ion–molecule association complexes include HPMS, FA, ICR, and so on.

Most ion/molecule equilibrium studies have been devoted also to the derivation of extensive scales of BDEs. From the temperature dependence of the equilibrium constants (Van't Hoff equation), enthalpy changes have been obtained. The BDEs are defined as the standard enthalpy changes at 0 K for the reaction (Eq. 1.6). These can be calculated using Kirchhoff's equation:

$$BDE = H^{\circ}{}_{T} - RT - \Delta C_{v}T. \tag{1.8}$$

ΔH°_T is the enthalpy change at a given temperature of T. ΔC_v is the difference in the heat capacity between products and reactants. However, the enthalpy change, ΔH°_T, is assumed to approximate the ion–molecule BDE. This assumption is reasonable since the enthalpy change is expected to be only weakly temperature dependent for ion–molecule association reaction.

Developments in ab initio MO quantum chemistry have resulted in theoretical models [41]. These are able to predict the properties of neutral molecules and ions within so-called chemical accuracy (about 2 kcal/mol). This makes these procedures extremely useful in ion chemistry, especially in thermochemical studies. Another approach is DFT that has attracted intense interest. DFT offers the promise of less drastic scaling with the size of the system than traditional ab initio methods. In general, hybrid density functions frequently provide the best overall agreement and surpass MP2 result of ab initio theory. There have been a number of theoretical studies of the structures and binding energies of alkali ions to many bases, including from free radical, biradical, amino acid, and crown ether to DNA bases (see details in Chap. 3).

1.5 Cationization Mass Spectrometry; Application to Analytical Mass Spectrometry

1.5.1 Alkali Metal Ion in the Gas Phase

Methodology leading to ion attachment (cationization) of chemical compounds as a method of ionization is extremely developing area of current MS. Cationization of molecules is an important analytical method because of its simplicity, its extreme sensitivity, and its ability to provide real time measurements as a consequence of fast response time of ion attachment reaction. Cationization of organic molecules with alkali ions to provide molecular weight information was first demonstrated with the reaction of Li^+, produced by field ionization and 1-hexanol [42].

The reactivity of metal-ion attachment to organic molecules has stimulated further interest in alkali-metal ion CI. Metal ions have been used as special reagents in analytical MS to provide new or improved information for analytical chemistry [43]. The adduct ions can be used to determine relative molecular weight of unknown species and to find numerous applications in the analysis of mixtures. The target species includes intermediary free radical and other species [44], which are sometimes not found under ordinary conditions (see Chap. 5).

1.5.2 Transition to the Gaseous to the Condensed Phase

Nonvolatile Sample Molecules For nonvolatile or thermally labile sample molecules, a variety of ionization methods have been developed over the past 50 years.

These methods include field desorption (FD), fast atom bombardment (FAB), plasma desorption (PD), (static) secondary ion mass spectrometry (SIMS), laser desorption (LD), and electrospray ionization (ESI). Most methods can be classified into desorption ionization (DI) or spray ionization methods. Both DI and spray methods sample directly from the condensed phase, either liquid or solid. In DI, the processes of volatilization and ionization are not distinct and associated, while in spray ionization, a nebulization is used to separate sample molecules and/or ions from the solvent liquid. One of the earliest applications to produce cationized adducts was done by using FD and FAB ionization. Furthermore, the direct sampling and following ionization from the condensed phases is important in a variety of sciences. These ionization methods significantly extended the range of compounds amenable to MS analysis. These days MS permits the application of these methods to molecular systems in condensed phases chemistry, biology, and medicine (see Chap. 7).

Significance of Cationized (Associated) Ions The metal ion adducts were found to have considerable potential for identifying unknown components of organic compounds, since cationization of molecules with alkali-metal ions is a universal way of generating secondary ions and little fragmentation of the $[M+Cat]^+$ ions occurred. This approach is most often applied to nonvolatile, thermally labile materials using various DI methods. For instance, cationization reactions are ubiquitous in field desorption mass spectrometry (FDMS), where the use of inorganic alkali salts is often sufficient to produce intense cationized species and to simplify the spectra. Consequently, the chemistry of metal ion adducts frequently excludes the necessity for chemical derivatization of the analyte. Cationization has also been used to characterize materials that are unstable toward protonation. Compounds such as sugars and glycosides do not form abundant or stable $[M+H]^+$ ions. These DI and spray ionization methods combined with tandem mass spectrometry (MS/MS) are particularly useful for structure elucidation because collision-induced dissociation (CID of $[M+Cat]^+$ adducts), which are unlike decompositions of $[M+H]^+$ ions, are increasingly being used for structure determination of multifunctional biomolecules. The use of alkali-metal ions as reagents has also become important in DI mass spectrometry. In biological systems alkali-metal ions, especially Na^+ and K^+, associate with a variety of biologically important peptides, proteins, sugars, and synthetic polymers.

Alkali metal ion adducts reported in DI and ESI mass spectra, however, were considered nuisances in some cases [45–47]. Cationization takes place unfavorably, especially in matrix-assisted laser desorption/ionisation (MALDI) and ESI. The unavoidable presence of sodium and potassium salts as impurities produces cation adducts and often comes up with a considerable decrease in sensitivity in DI methods. Furthermore, the presence of protonated species together with alkali ion adducts can appreciably complicate the identification of the components of a complex mixture.

$[M+Cat]^+$ Adducts in DI and Spray Ionization MS The next part of this section will describe briefly on how useful metal ion adducts ($[M+Cat]^+$) ions can be formed in various DI methods to facilitate the analysis of samples (see details in

Sect. 4.2). PD, FAB, liquid secondary-ion mass spectrometry (liquid SIMS) and FDMS are (arguably) almost obsolescent ionization techniques that revolutionized the analyses of large biological molecules in the 1980s. Such ionization method shows significant similarities to each other. These ionization methods have been used to form metal ion adducts to confirm relative molecular mass, using MS and MS/MS techniques [48–54]. In most cases, alkali ion adducts of the type $(M+Li)^+$, $(M+Na)^+$ and $(M+K)^+$ are formed and observed for analytes (M). Alkali-ion attachment is probed to yield excellent sensitivity for a wide variety of compounds, including carbohydrates, amino acids, moderately large peptides, nucleotides, and natural products. The metal-ion complexes have the advantage over the protonated ions or molecular ions, in some cases, that their fragmentation can yield more structural information.

Owing to its compatibility with solution samples, ESI is preferred over other ionization methods in many MS fields. Applications of metal ion adducts have been reported for ESI [55–57]. For example, ESI can be used to produce alkali-metal adducts of antibiotics that do not form abundant $[M+H]^+$ ions. Informative adducts between alkali-metal ions and peptides have been observed under a variety of conditions of electrospray ionization mass spectrometry (ESI-MS). It should be noted, however, that the presence of salt ion adducts cause the signal suppression and interference with the interpretation of the mass spectra, particularly in analytical MS of proteins and other biological molecules.

In MALDI, quasi-molecular ions ($[M+alkali]^+$ formation) can be formed by attachment of alkali ions, if alkali-metal salts are present in the sample [58, 59]. Cations generally do not need to be added to the sample: ubiquitous alkali-metal impurities are sufficient to give strong alkali-cationized signals. It is reported that for polysaccharides, alkalization occurred much more frequently than protonation. Imaging of MALDI samples shows that the positions of analyte molecules and alkali atoms are highly correlated. In some cases, cationization takes place unfavorably, also in MALDI [46].

1.6 Review, Tutorial, and Web Articles

1.6.1 Reviews and Web Sites

The last 70 years have been marked by a dramatic increase in research work on the formation of gas-phase and condensed-phase ion–molecule complexes. As a result of this interest, a large amount of research literatures, review and tutorial articles, and monographs have appeared. The present section is intended to summarize several early reviews, which contain the tutorial description, tutorial articles, and World Wide Web URLs. The literature [60] reported by K. Murray should be noted. This paper provides a more complete URLs listing and discussion of on-line resources available for MS. World Wide Web sites are tabulated for search pages, index pages for MS research groups and companies, MS societies, local discussion groups, journals, and software.

Early general review by Good [1] explores the mechanism of third-order ion–molecule clustering reactions alkali-metal ion–molecule association reaction and survey the instrumental basis to study its kinetic, covering the associated experimental results. Other related review on ionic clusters is available in the Castleman and Keesee's literature [15, 35, 61]. More recently Wu and McMahon [62] review structures, energetics, and dynamics of gas phase ions studied by Fourier transform ion cyclotron resonance (FTICR) and HPMS. While thermochemistry has been discussed in early reviews by Kebarle [38], advances in flow reactor techniques for the mechanism and dynamic study of gas-phase ion chemistry is focused by Grad and Squires [63, 64].

Several reviews on experimental techniques and methods have been written. Ervin's review [31] discusses a wide variety of experimental methods available for examination of thermochemical and spectroscopic properties of gas-phase ions. This review has detailed theoretical analysis of experimental results, especially employing statistical rate models for activated decomposition processes and accounting for ion internal energies, is critically important for obtaining accurate thermochemical results and is required for precise experimental work. Eberlin has reviewed [65] how studies of IMR can be done using a triple-stage pentaquadrupole.The general concepts, advantages, and applications of pentaquadrupole mass spectrometry and of a variety of pentaquadrupole multidimensional triple-stage (MS^3) scans as applied to the study kinetic, of gas-phase IMRs are detailed. Gronert reviews [66, 67] the general features and limitations of quadrupole ion traps, discussing by a description of representative ion trap studies focused on fundamental organic reaction mechanisms. There is a wide variety of commercial and customized equipment for performing gas-phase IMRs. Ryzhov [68] discuss the recent advances in instrumentation development involving IMR as an analytical tool.

Many reviews encompass a wide range of analytical applications. The some materials focus on how IMRs can be used for structural studies. Lebrilla [69] reviews how proton transfer reactions and H/D exchange reactions can be used as probes of gas phase structures of peptide and protein. Wesdemiotis et al. [70] reviews the fragmentation pathways of synthetic polymer ions that have been energized to decompose via collisionally activated dissociation (CAD), the most widely used activation method in polymer analysis. Brodbelt [71] reviews applications of IMRs for solving increasingly complex analytical problems. It is suggested that product distributions from IMRs provide key diagnostic information for structure identification. The Fujii's review is concerned [72] primarily with thermal alkali-metal ion association reactions. It includes reaction mechanism, instrumentation, and application to MS, together with reaction rates and alkali ion affinities of the classified compounds. The review literature [73] covered by Teesch and Adams focuses on how metal ions have been used as reagents in various MS methods since the early 1970s and how important the use of metal ions as reagents in DI of molecules is.

1.6.2 Tutorial Articles

Other publications give us tutorial into ion–molecule association reactions: cationization MS. These will be a useful starter for those who are newcomers to the unfamiliar field and wish to learn the various subjects in more details. The tutorial material prepared by Busch [74] and Hanley et al. [75] focused on general aspects of DI-MS and its ionization mechanism. DI methods tend to produce even-electron ions such cationized molecules such as $[M+Na]^+$; these stable ions undergo only a minimum amount of fragmentation. While kinetic energy releases in unimolecular fragmentation have been discussed in tutorial by Lifshitz [76], internal energy effects in MS are given by Vékey [77].

Several tutorial literatures encompass a wide range of MS hardware for the study of noncovalent complexes, including MS/MS [78], quadrupole ion trap MS [79], time-of-flight (TOF) [80], linear ion traps combined with quadrupole [81], and FTICR [82], and quadrupole-TOF [83]. The monograph *Ion Cyclotron Resonance Spectrometry II* edited by Hartmann and Wanczek [2], covers all the fields studied with ion cyclotron resonance. It includes tutorial reviews on systematic studies of ion chemistry as well as theories of ion–molecule association reaction. Several approaches to ion activation in MS/MS have been developed in recent years for use in ion trapping instruments. The Volmer's tutorial [84] presents the most common ion activation techniques employed in MS/MS. McLuckey's tutorial [85] discusses how slow heating methods can be used as ion activation in MS/MS. Pramanik's tutorial [86] discusses how noncovalent complexes in the mass range 19,000–34,000 Da, can be detected using ESI-MS.

References

1. Good A. Third-order ion-molecule clustering reactions. Chem Rev. 1975;75:561–83.
2. Bass L, Michael T, Bowers MT. Ion-molecule association reactions, lecture notes in Chemistry. In Hartmann H, Wanczek K-P, editors. Ion cyclotron resonance spectrometry II. Springer: Berlin; 1982. pp. 432–63.
3. Linde SV, Hase WL. Dynamics of ion-molecule recombination IV. $Li^+ + (CH_3)_2O$ association. Comp Phys Commun. 1988;51:17–34.
4. Mondro SL, Linde SV, Hase WL. Reaction path and variational transition state theory rate constant for $Li^+ + H_2O \rightarrow Li^+ (H_2O)$ association. J Chem Phys. 1986;84:3783–7.
5. Swamy KN, Hase WL. Dynamics of ion–molecule recombination. II. An alkali ion and a water molecule. J Chem Phys. 1982;77:3011–21.
6. Fehsenfeld FC, Goldan PD, Schmeltekopf AL, Schiff HI, Ferguson EE. Thermal energy ion-neutral reaction rates, I. J Chem Phys. 1966;44:4087–94.
7. Fehsenfeld FC. Diagnostics of the flowing afterglow. Int J Mass Spectrom Ion Phys. 1975;16:151–66.
8. Spears KG, Ferguson EE. Termolecular and saturated termolecular kinetics for Li^+ and F^-. J Chem Phys. 1973;59:4174–83.
9. McFarland M, Albritton DL, Fehsenfeld FC, Ferguson EE, Schmeltekopf AL. Flow-drift technique for ion mobility and ion-molecule reaction rate constant measurements. I. Apparatus and mobility measurements. J Chem Phys. 1973;59:6610–9.

10. Bohme DK. Experimental studies of positive ion chemistry with flow-tube mass spectrometry: birth, evolution, and achievements in the 20th century. Int J Mass Spectrom Ion Phys. 2000;200:97–136.
11. Kebarle P, Searles SK, Zolla A, Scarborough J, Arshadi M. The solvation of the hydrogen ion by water molecules in the gas phase. Heats and entropies of salvation of individual reactions: $H^{+}(H_2O)_{n-1} + H_2O = H^{+}(H_2O)_n$. J Am Chem Soc. 1967;89:6393–9.
12. Searles SK, Kebarle P. Hydration of the potassium ion in the gas phase: enthalpies and entropies of hydration reactions $K^{+}(H_2O)_{n-1} + H_2O = K^{+}(H_2O)_n$ for $n = 1$ to $n = 6$. Canadian J Chem. 1969;47:2619–27.
13. Davidson WR, Kebarle P. Binding energies and stabilities of potassium ion complexes from studies of the gas phase ion equilibria $K^{+} + M = K^{+}M$. J Am Chem Soc. 1976;29:6133–8.
14. Castleman AW Jr, Bowen KH Jr. Clusters: structure, energetics, and dynamics of intermediate states of matter. J Phys Chem. 1996;100:12911–44.
15. Castleman AW Jr, Keesee RG. Clusters: bridging the gas and condensed phases. Acc Chem Res. 1986;19:413–9.
16. Garvey JF, Peifer WR, Coolbaugh MT. Some novel ion-molecule chemistry within van der waals clusters. Acc Chem Res. 1991;24:48–54.
17. Wang G, Cole RB. Solvation energy and gas-phase stability influences on alkali metal cluster ion formation in electrospray ionization mass spectrometry. Anal Chem. 1998;70:873–81.
18. Bloomfield LA, Conover CWS, Yang YA, Twu YJ, Phillips NG. Experimental and theoretical studies of the structure of alkali halide clusters. Nucl Instrum Methods Phys Res B. 2012;273:102–4.
19. MacTaylor RS, Castleman AW Jr. Cluster ion reactions: insights into processes of atmospheric significance. J Atmos Chem. 2000;36:23–63.
20. MacTaylor RS, Gilligan JJ, Moody DJ, Castleman AW Jr. Molecular activation by surface coordination: a new model for HCl reactivity on water-ice polar stratospheric clouds. J Phys Chem A. 1999;103:4196–201.
21. McEwan MJ, Scott GBI, Adams NG, Babcock LM, Terzieva R, Herbst E. New H and H_2 reactions with small hydrocarbon ions and their roles in benzene synthesis in dense interstellar clouds, Astrophys J. 1999;513:287–93.
22. Bovino S, Stoecklin T, Gianturco FA. The ionic pathways of lithium chemistry in the early universe: quantum calculations for LiH^{+} reacting with H. Astrophys J. 2010;708:1560–4.
23. Sellgren, K. Aromatic hydrocarbons, diamonds, and fullerenes in interstellar space: puzzles to be solved by laboratory and theoretical astrochemistry. Spectrochim Acta A. 2001;57:627–42.
24. Waite JH, Young DT, Cravens TE, Coates AJ, Crary FJ, Magee B, Westlake J. The process of Tholin formation in Titan's upper atmosphere. Science. 2007;316:870–5.
25. Marotta E, Paradisi CA. Mass spectrometry study of alkanes in air plasma at atmospheric pressure. J Am Soc Mass Spectrom. 2009;20:697–707.
26. Bohme DK. Chemical ionization in flames. In: Kinetics of ion-molecule reactions. NATO Advanced Study Institutes Series. Vol. 40. New York: Springer; 1979. pp. 323–43.
27. Semo NM, Koski WS. Some ion-molecule reactions pertinent to combustion. J Phys Chem. 1984;88:5320–4.
28. Willis C. Ion-molecule reactions in gas phase radiation chemistry. J Chem Educ. 1981;58:88–92.
29. Castleman AW Jr, Holland PM, Keesee RG. Ion association processes and ion clustering: elucidating transitions from the gaseous to the condensed phase. Radia Phys Chem. 1982;20:57–74.
30. Sablier M, Fujii T. Mass spectrometry of free radicals: a methodological overview. In: Webb G, editor. Progress in chemistry, Sect. C (Phys Chem). Cambridge: Royal Society of Chemistry; 2005. pp. 53–99.
31. Ervin KM. Experimental techniques in gas-phase ion thermochemistry. Chem Rev. 2001;101:391–444.

32. Kebarle P. Gas phase ion thermochemistry based on ion-equilibria from the ionosphere to the reactive centers of enzymes. Int J Mass Spectrom. 2000;200:313–30.
33. Irikura KK, Frurip DJ. Computational thermochemistry. Prediction and estimation of molecular thermodynamics. Washington, DC: American Chemical Society; 1998. p. 480.
34. Armentrout PB, Rodgers MT. Thermochemistry of non-covalent ion–molecule interactions. In: Mass spectrometry (Tokyo), Vol. 2 Special Issue: Proceedings of 19th International Mass Spectrometry Conference, 2013; S0005
35. Keesee RG, Castleman AW Jr. Thermochemical data on Gas-phase ion-molecule association and clustering reactions. J Phys Chem Ref Data. 1986;15:1011–71.
36. Lias SG, Bartmess JE, Liebman JF, Holmes JL, Levin RD, Mallard WG. Gas-phase ion and neutral thermochemistry. J Phys Chem Ref Data. 1988;16(1):861.
37. Dzidic I, Kebarle P. Hydration of the alkali ions in the gas phase. Enthalpies and entropies of reactions $M^+(H_2O)_{n-1} + H_2O = M^+(H_2O)_n$. J Phys Chem. 1970;74:1466–74.
38. Kebarle P. Ion thermochemistry and solvation from gas phase ion equilibria. Annu Rev Phys Chem. 1977;28:445–46.
39. Hunter EP, Lias SG. Evaluated gas phase basicities and proton affinities of molecules. J Phys Chem Ref Data. 1998;27:413.
40. Rodgers MT, Armentrout PB. Noncovalent metal-ligand bond energies as studied by threshold collision-induced dissociation. Mass Spectrom Rev. 2000;19:215–47.
41. Hoyau S, Norrman K, McMahon TB, Ohanessian G. A quantitative basis for a scale of Na^+ affinities of organic and small biological molecules in the gas phase. J Am Chem Soc. 1999;121:8864–75.
42. Rollgen FW, Schulten HR. Molecular weight determination by cationization. Org Mass Spectrom. 1975;10:660–8.
43. Hodges RV, Beauchamp JL. Application of alkali ions in chemical ionization mass spectrometry. Anal Chem. 1976;48:825–8.
44. Fujii T. A novel method for detection of radical species in the gas phase: usage of lithium ion attachment to chemical species. Chem Phys Lett. 1992;191:162–8.
45. North S, Okafo G, Birrell H, Haskins N, Camilleri P. Minimizing cationization effects in the analysis of complex mixtures of oligosaccharides. Rapid Commun Mass Spectrom. 1997;11:1635–42.
46. Mohr MD, Bornsen KO, Widmer HM. Matrix-assisted laser desorption ionization mass spectrometry: improved matrix for oligosaccharides. Rapid Commun Mass Spectrom. 1995;9:809–14.
47. Lavarone AT, Udekwu OA, Williams ER. Buffer loading for counteracting metal salt-induced signal suppression in electrospray ionization. Anal Chem. 2004;76:3944–50.
48. Reynolds WD. Field desorption mass spectrometry. Anal Chem. 1979;51:283A–293A.
49. Prome JC, Puzo G. Field desorption mass spectrometry of oligosaccharides in the presence of metallic. Org Mass Spectrom. 1977;12:28–32.
50. Keough T. Cationization of organic molecules using fast atom bombardment mass spectrometry. Anal Chem. 1985;57:2028–34.
51. Macfarlane RD, Torgerson DF. Californium-252 plasma desorption mass spectroscopy. Science. 1976;191:920–5.
52. Demirev P, Fenyö D, Håkansson P, Sundqvist BUR. Mechanism of formation of alkali metal cation adducts in plasma desorption mass spectrometry of biomolecules. Org Mass Spectrom. 1991;26:471–5.
53. Pachuta SJ, Cooks RG. Mechanisms in molecular SIMS. Chem Rev. 1987;87:647–69.
54. Hagenhoff B. Optimisation methods: cationization, TOF-SIMS. In: Vickerman JC, Briggs D, editors. Surface analysis by mass spectrometry. Manchester: IM Publications and Surface Spectra; 2001. pp. 285–307.
55. Pilosof D, Kim HY, Dyckes DF, Vestal ML. Determination of nonderivatized peptides by thermospray liquid chromatography/mass spectrometry. Anal Chem. 1984;56:1236–40.

56. Kempen EC, Brodbelt JS, Bartsch RA, Jang Y, Kim JS. Investigation of alkali metal cation selectivities of lariat ethers by electrospray ionization mass spectrometry. Anal Chem. 1999;71:5493–500.
57. Harvey DJ. Electrospray mass spectrometry and fragmentation of n-linked carbohydrates derivatized at the reducing terminus. J Am Soc Mass Spectrom. 2000;11:900–95.
58. Zenobi R, Knochenmuss R. Ion formation in MALDI mass spectrometry. Mass Spec Rev. 1998;17:337–66.
59. Cancilla MT, Penn SG, Carroll JA, Lebrilla CB. Coordination of alkali metals to oligosaccharides dictates fragmentation behavior in matrix assisted laser desorption ionization/fourier transform mass spectrometry. J Am Chem Soc. 1996;118:6736–45.
60. Murray KK. Internet resources for mass spectrometry. J Mass Spectrom. 1999;34:1–9.
61. Castleman AW Jr, Keesee RG. Ionic clusters. Chem Rev. 1986;86:589–818.
62. Wu R, McMahon TB. Structures, energetics, and dynamics of gas phase ions studied by FTICR and HPMS. Mass Spectrom Rev. 2009;28:546–85.
63. Grad ST, Squires RR. Advances in flow reactor techniques for the study of gas-phase ion chemistry. Mass Spectrom Rev. 1988;7:263–358.
64. Squires RR. Advances in flowing afterglow and selected-ion flow tube techniques. Int J Mass Spectrom Ion Proc. 1992;118/119:503–18.
65. Eberlin MN. Triple-stage pentaquadrupole (QqQqQ) mass spectrometry and ion/molecule reactions. Mass Spectrom Rev. 1997;16:113–44.
66. Gronert S. Mass spectrometric studies of organic ion/molecule reactions. Chem Rev. 2001;101:329–60.
67. Gronert S. Quadrupole ion trap studies of fundamental organic reactions. Mass Spectrom Rev. 2005;24:100–20
68. Osburn S, Ryzhov V. Ion–molecule reactions: analytical and structural tool. Anal Chem. 2013;85:769–78.
69. Green MK, Lebrilla CB. Ion molecule reactions as probes of gas-phase structures of peptides and proteins. Mass Spectrom Rev. 1997;16:53–71.
70. Wesdemiotis C, Solak N, Polce MJ, Dabney DE, Chaicharoen K, Katzenmeyer BC. Fragmentation pathways of polymer ions. Mass Spectrom Rev. 2011;30:523–59.
71. Brodbelt JS. Analytical applications of ion-molecule reactions. Mass Spectrom Rev. 1997;16:91–110.
72. Fujii T. Alkali-metal ion/molecule association reactions and their applications to mass spectrometry. Mass Spectrom Rev. 2000;19:111–38.
73. Teesch LM, Adams J. Metal Ions as special reagents in analytical mass spectrometry. Org Mass Spectrom. 1992;27:931–43.
74. Busch KL. Desorption ionization mass spectrometry. J Mass Spectrom. 1995;30:233–40.
75. Hanley L, Kornienko O, Ada ET, Fuoco E, Trevor JL. Surface mass spectrometry of molecular species. J Mass Spectrom. 1999;34:705–23.
76. Laskin J, Lifshitz C. Kinetic energy release distributions in mass spectrometry. J Mass Spectrom. 2001;36:459–78.
77. Vékey K. Internal energy effects in mass spectrometry. J Mass Spectrom. 1996;31:445–63.
78. Hoffmann E. Tandem mass spectrometry: a primer. J Mass Spectrom. 1996;31:445–63.
79. March RE. An introduction to quadrupole ion trap mass spectrometry. J Mass Spectrom. 1997;32:351–69.
80. Guilhaus M. Principles and instrumentation in time-of-flight mass spectrometry, physical and instrumental concepts. J Mass Spectrom. 1995;30:1519–32.
81. Hopfgartner G, Varesio E, Tschäppät V, Grivet C, Bourgogne E, Leuthold LA. Triple quadrupole linear ion trap mass spectrometer for the analysis of small molecules and macromolecules. J Mass Spectrom. 2004;39:845–55.
82. Guan S, Marshall AG. Ion traps for Fourier transform ion cyclotron resonance mass spectrometry: principles and design of geometric and electric configurations. J Mass Spectrom Ion Proc. 1995;146/147:261–96.

83. Chernushevich IV, Loboda AV, Thomson BA. An introduction to quadrupole–time-of-flight mass spectrometry. J Mass Spectrom. 2001;36:849–65.
84. Sleno L, Volmer DA. Ion activation methods for tandem mass spectrometry. J Mass Spectrom. 2004;39:1091–112.
85. McLuckey SA, Goeringer DE. Slow heating methods in tandem mass spectrometry. J Mass Spectrom. 1997;32:461–74.
86. Pramanik BN, Bartner PL, UMirza UA, Liu Y, Ganguly AK. Electrospray ionization mass spectrometry for the study of non-covalent complexes: an emerging technology. J Mass Spectrom. 1998;33:911–20.

Chapter 2
Fundamentals of Ion Chemistry

Toshihiro Fujii

2.1 Termolecular Association Reactions

2.1.1 Association Reaction Mechanism

It was experimentally found that the overall process of ion–molecule association reactions can more descriptively be schemed [1] as follows: collisional stabilization reaction (Eq. 2.1), radiative stabilization reaction (Eq. 2.2), and elimination reaction (Eq. 2.3) from ion–molecule complexes.

$$A^{+} + M \rightleftharpoons (A+M)^{+*} \rightarrow (A+M)^{+} \tag{2.1}$$

$$A^{+} + M \rightleftharpoons (A+M)^{+*} \rightarrow (A+M)^{+} + h\nu \tag{2.2}$$

$$A^{+} + M \rightleftharpoons (A+M)^{+*} \rightarrow \left[A(M-RX)\right]^{+} + RX, \tag{2.3}$$

where $(A+M)^{+*}$ is an intermediate complex of cation A^{+} and molecule M, which has enough internal energy to become stabilized with respect to dissociation, hν is the radiation, and RX is the component species of M.

Ion–molecule association reactions by collisional stabilization are of interest in all the fields of chemistry. The process is typically written as

$$A^{+} + M \xrightarrow{kc} (A+M)^{+*} \xrightarrow{ks[N]} (AM)^{+}. \tag{2.4}$$

T. Fujii (✉)
C & V Technix Co., Ltd., 3-6-1 Higashi, Akishima, Tokyo 196-0033, Japan
e-mail: fujii.toshihiro@c-vtechnix.com

T. Fujii (ed.), *Ion/Molecule Attachment Reactions: Mass Spectrometry*,
DOI 10.1007/978-1-4899-7588-1_2

The collision rate constants, k_c and k_s, are given by Langevin theory [2, 3] nicely. It is detailed later in this section. It has already been pointed out (see Sect. 1.1) that this type of reaction is totally analogous to neutral radical combination reactions.

The ion–molecule reactions show a characteristic behavior; they present a very small or even no activation barrier. This behavior can be understood by considering a Lindemann type formalism [4, 5]. According to this model, an intermediate activated molecular complex is formed through a bimolecular collision. A subsequent collision of the intermediate activated complex $(A+M)^{+*}$ with another molecule (third body, N) is necessary in order to release the excited energies. Furthermore, it is well known that ion–molecule complexes stick together by ion/dipole and ion/induced dipole forces, and these classical electrostatic forces between the components of an ion/molecule complex and the stability of this complex increase with the dipole moment and the polarizability of the neutral component, which varies with its size and the presence of functional groups.

Radiative Association Reactions The study of radiative association reactions, (Eq. 2.2), has been of considerable interest [6–8] in chemical kinetics, planetary and interstellar chemistry, flames, and a variety of other areas. The kinetic study makes it possible to model the formation of complex molecular species in the interstellar science. At the very low molecular number densities in interstellar environments, the probability of formation of the products of association reactions by collisional stabilization is very low. Therefore, the radiative association process becomes an extremely important one for the production of the complex molecular species observed by astronomical physicist. The methodology is either flowing afterglow (FA) or Fourier transform ion cyclotron resonance (FT-ICR) mass spectrometry. For the study of the apparent bimolecular rate constant for formation of association products as a function of pressure of a third body (N), the pressure should be set up to be sufficiently high in order to release the energy in the associated complex. Under the high pressure conditions collisional stabilization has competed with and usually dominated over radiative association. As a result, the radiative association rate was then extrapolated from the intercept of a plot of apparent rate constant versus pressure of a third body, N.

Woodin and Beauchamp have pioneered the study for rapid radiative association of Li^+ with a variety of carbonyl compounds at low pressure in a trapped-ion ion cyclotron resonance experiment [6]. More recently, Dearden et al. [9] have presented convincing evidence for very rapid radiative association of laser-desorbed various alkali ions with thermally vaporized valinomycin. These reactions are observed so readily, since they are exothermic processes with no third body present to release the excess energy. Collisions with neutral samples at the pressures of FT-ICR trapping cell are too infrequent to account for collisional stabilization of the complexes, so these reactions are most likely stabilized by emitting the excess energy as infrared (IR) photons.

Elimination Reactions Possible elimination reactions of the type (Eq. 2.5) have been investigated, using ion cyclotron resonance (ICR) [10, 11].

$$A^{+} + M \rightarrow [A(M-RX)]^{+} + RX, \tag{2.5}$$

where RX is either alkali halide or organic halide. The occurrence of such a process implies that the reaction is exothermic and hence BDE(R-X) > BDE[A-(M-RX)] if the reactants are at thermal energies. To investigate such elimination reactions, Bueachamp et al. [10] have performed trap ion experiments using the thermionic source, which was mounted inside the source region of an ICR cell. For instance, association of an alkali ion and organic halide leads to the formation of an activated species which eliminate RX to give lithiated alkene, with the alkali ion remaining bound to alkene species. Interestingly, they also reported that intermediates dissociate to an alkali halide and carbonium ion. These observations indicate the possibility of determining the relative binding energies of alkali ions to various bases.

2.1.2 Termolecular Reaction Kinetics

The kinetics of termolecular ion association reaction is usually approximated by a two-step mechanism and is described as the general form:

$$A^{+} + M \rightleftarrows (A+M)^{+}* \quad (\text{combination}, k_1/\text{dissociation}, k_2) \tag{2.6}$$

$$(A+M)^{+}* + N \rightleftarrows (A+M)^{+} + N \quad (\text{stabilization}, k_3/\text{reverse}, k_4), \tag{2.7}$$

where A denotes a positively charged ion, M and N are neutral species, and k_1, k_2, k_3, and k_4 represent the reaction rate constants for the combination (or association), decomposition, stabilization, and reverse reaction, respectively [12, 13]. As indicated, most association reactions are reversible. It has been pointed out that an ion–molecule association reaction is a process very similar to the combination of two atoms or two free radicals. The association process is considered to proceed via an intermediate complex $(A+M)^{+*}$, which has a characteristic lifetime, τ ($=1/k_2$), against unimolecular decomposition back to the reactants A^+ and M. An initial combination step proceeds by ion-permanent dipole or ion-induced dipole attraction between the reactants. The energy thus accumulated in the reaction complex could then be liberated by collision with a third body, N. It is expected that k_1 contains no activation energy term and that it has thus no temperature dependence to a first approximation. It was experimentally found that ion–molecule reactions have negative temperature dependence, deviating from an Arrhenius plot. On the contrary, stabilization rate constants, k_3, should depend on the identity of N; k_3 may have some slight temperature dependence, since the excess energy to be lost may be distributed according to Maxwell–Boltzmann statistics. The bonding between ion and neutral species is generally found to be weak relative to normal chemical bond strengths.

Employing the steady-state approximation for the activated intermediate complex $(A+M)^{+*}$, the expression for the overall second-order reaction rate constant, $k(2)$, is given by:

$$k(2)_{\text{overall}} = k_1 k_3 [N] / (k_2 + k_3 [N]). \tag{2.8}$$

If the time interval between the formation of the activated intermediate complex and the collision for stabilization is long compared to the lifetime (τ, $1/k_2$) of the intermediate product, $(A+M)^{+*}$, the activated complex will decompose back to the initial reactants. Since the formation of the association products depends on the collision with the third-body gas molecules (N), the entire process is a competition between intermolecular collisional energy transfer and intramolecular vibrational energy redistribution of the excess energy of the activated intermediate complex among its degrees of freedom.

Two cases should be considered as a function of the buffer gas concentration $[N]$. In the high-pressure region, the stabilization process occurs much faster than the dissociation of the activated intermediate complex. In this case, $k_3[N] \gg k_2$ and the reaction rate becomes:

$$k(2)_{\text{overall}} = k_1. \tag{2.9}$$

From Eq. 2.9, it can be seen that in this regime, the overall second-order reaction rate is equal to ion–molecule association reaction rate, k_1, and does not depend on the buffer gas pressure. In the low-pressure region, the condition $k_3[N] \ll k_2$ is fulfilled, and the rate determining step is represented by the collisional energy transfer to the buffer gas. The reaction rate in this case is given by:

$$k(2)_{\text{overall}} = k_1 k_3 [N] / k_2. \tag{2.10}$$

These considerations clearly explain the experimental finding that the order of the reaction rate constant changes from second order in the high-pressure regime to third order in the low-pressure limits. The expression for $k(2)_{\text{overall}}$ can be determined when k_1, k_2, and k_3 are known. k_1 and k_2 can be calculated by using the Langevin collision theory. For the calculation of the dissociation rate constant, k_2, statistical models like Rice–Ramsperger–Kassel–Marcus (RRKM) theory [14, 15] are used. The Langevin and RRKM theories will be presented briefly in the next section and detailed in Chap. 3.

Langevin Theory The theoretical treatment of ion–molecule reactions is presented. This is a useful aid in understanding the collision dynamics of two species. The collision rate for an ion and a polarizable molecule having no permanent dipole moment is given by the Langevin theory [2, 3] as follows. For an ion and a molecule approaching each other with a relative velocity v and impact parameter b (Fig. 2.1), the Langevin theory describes the molecular interaction potential between an ion

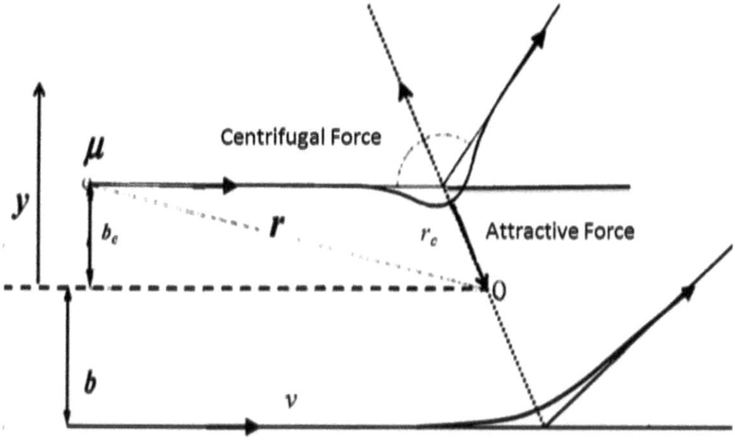

Fig. 2.1 Schematics of ion molecule collisional behavior for critical impact parameter b_e. The particles orbit the scattering center with r_e. For the impact parameter b greater than b_e, the particles are simply scattered at large values of the relative intermolecular distance r

and a nonpolar molecule. According to this, the effective interaction potential is given at an ion/molecule separation r by the following expression:

$$V_{\text{eff}}(r) = U(r) + L^2/2\mu r^2 = -1/2(ae^2/r^4) + L^2/2\mu r^2, \tag{2.11}$$

where α represents the polarizability of the neutral molecule, e denotes the elementary charge, μ is the reduced mass, and L represents the classical angular momentum. The angular momentum has the form $L = \mu vb$. Equation 2.11 contains an attractive ion/induced dipole term ($\sim r^{-4}$) from standard electrostatic forces and is referred to as the central potential term. The second term describes the repulsive term ($\sim r^{-2}$) between the two particles and is referred to as the centrifugal term.

When $E_{\text{trans}}(r)$ is defined as the translational energy along the line of the center of the collision, the total relative energy of the system is given as follows,

$$E(r) = E_{\text{trans}} + V_{\text{eff}}(r). \tag{2.12}$$

The variation of $V_{\text{eff}}(r)$ with r, depends on the value of the impact parameter b. In the special case ($b = b_e$) where the centrifugal term equals to $E(r)$, namely, $V_{\text{eff}}(r) = E(r)$, the particles will orbit the scattering center with a constant ion/molecule separation r_e. For impact parameter values $b \leq b_e$, the capture collision probability is $p = 1$ while for $b > b_e$, the capture reaction does not occur and $p = 0$.

Similar to the hard-sphere collision model, the capture collision cross section or the Langevin cross section is defined as:

$$\sigma(v) = \pi b_e^{\,2}. \tag{2.13}$$

From the conditions that a $\delta V_{\text{eff}}(r)/\delta r=0$ and $V_{\text{eff}}(r)=E(r)$, at $r=r_c$, a capture collision cross section and collision rate constant are deduced by

$$\sigma_c(v)=\pi b_c^{\,2}=(2\pi q/v)(\alpha/\mu)^{1/2} \tag{2.14}$$

$$k_c=v\sigma_c(v)=(2\pi q)(\alpha/\mu)^{1/2}. \tag{2.15}$$

As can be seen from Eq. 2.14, classical theory predicts the capture cross section should vary inversely as the relative velocity of the colliding pair and hence the Langevin rate constant, should be independent of the relative velocity and the temperature. Equation 2.15 predicts reasonably well the rate constants for ion–molecule reactions involving only ion-induced dipole interactions. This indicates that a reaction occurs on every collision for many ion/molecule pairs; there can be no activation energy for the reaction. The rate constant predicted is of the order of 1×10^{-9} cm^3 molecule^{1}s^{-1}.

2.1.3 *Kinetic Data Base*

The techniques for measuring rate constants and product distribution (branching ratios) for ion–molecule reactions are varied, but the majority of the data have been determined using the FA, drift-tube (DT), selected ion flow tube (SIFT), high-pressure mass spectrometric techniques, or ICR. These methods are detailed in Chap. 4. A number of surveys of all classes of ion–molecule reactions have appeared in the literature; Ferguson [16], Sieck et al. [17], Albritton [18], Anicich et al. [19], Ikezoe et al. [20], some of which include termolecular reactions or are limited to selected methods. Anicich listed [21] an index to the literature for gas phase bimolecular positive ion–molecule reactions as a comprehensive survey of ion–molecule reaction kinetics and product distribution of the reactions. Over 2300 references are cited. This index covers the literature from 1936 to 2003. It was limited to selected reactions, listed by reactant ion, that were important for chemical modeling ionic processes in planetary atmospheres, cometary comas, and interstellar clouds.

Compared to the enormous amount of study devoted to the subject of bimolecular ion–molecule reactions, the subject of termolecular ion reactions is comparatively small. Consequently, kinetic data on the ion association are relatively sparse. The measurement has been largely confined to systems of planetary or interstellar atmospheres with experiments like SIFT operated at high pressure. An extensive study of the Li ion attachment has been carried out by Ferguson et al. [22]. Some reaction rates at termolecular kinetics are very large and near the Langevin collision rate. The association of Li^+ with the water, rare gases, N_2, and some organics has also been reported in drift tubes and ICR. Passarella et al. [23] reported rate coefficients for the association of HCl, CO_2, NH_3, ND_3, SO_2, and CH_3OH with Na^+. At ambient temperature three-body rates in helium are found to range from 1.6×10^{-30}

to 6.3×10^{-28} cm^6/s, increasing in magnitude from ligand to ligand. Ikezoe et al. listed the termolecular reactions whose rate constants are in the range of 7×10^{-26} to 2×10^{-32} cm^6/s. Termolecular reactions that have reaction rate constants faster than 10^{-26} cm^6/s appear to be competitive with bimolecular reactions. These are called then the pseudo-bimolecular or saturated termolecular reactions.

2.2 Dissociation Reaction

2.2.1 Unimolecular Dissociation

The basic theory of unimolecular dissociations of the ionized molecules under the condition of mass spectrometry are discussed in this section. Useful monograph of the development of the theory as applied to unimolecular reactions have been presented by Forst and by Baer and Hase [24, 25]. Historically, the unimolecular dissociative properties of gaseous precursor ions were studied by fragmentations occurring either in the ion source by electron ionization (EI) or by metastable ion decompositions during an ion's flight from the ion source to the detector. This is followed by tandem mass spectrometry (MS-MS), which involves the activation of a precursor ion formed in the ion source and the mass analysis of the precursor and product ions. The ion activation method determines what types of products result. Several ion activation techniques have been developed over the past 50 years. The advent of soft ionization techniques, such as fast atom bombardment (FAB), electrospray ionization (ESI), and matrix-assisted laser desorption/ionization (MALDI), extended the range of application of mass spectrometry to biological, thermally labile compounds. However, these ionization techniques primarily yield quasimolecular species with little or no fragmentation and limit the structural information available in a single-stage mass spectrum. Consequently, collision-induced dissociation (CID) was introduced in most MS-MS applications. While CID remains the most common dissociation technique, several new methods have become increasingly useful for specific applications, complementing conventional CID mass spectra. These include surface-induced dissociation (SID), electron-capture dissociation (ECD), electron-transfer dissociation (ETD), and infrared multiphoton dissociation (IRMPD). Blackbody infrared radiative dissociation (BIRD) can thermally activate ions in quadrupole ion trap and FT-ICR instruments. These activation techniques can be used to determine the bond energies (ion affinities) of metal ion/molecule complexes and to obtain the structural information in chemical and biological applications. Lastly, in-source decay (ISD) and post-source decay (PSD) in MALDI is presented with a focus on significance of the structural elucidation. A brief description of each of these methods is also presented in this section.

Quasi-Equilibrium Theory (QET) Ionization process in an EI source is very fast, taking $< 10^{-16}$ s, which forms molecular ions in ground and excited states. Ionization occurs by a vertical transition, and therefore the interatomic distances remain

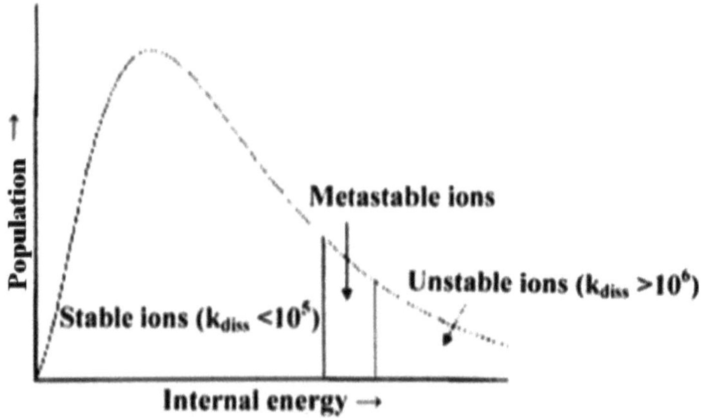

Fig. 2.2 Hypothetical distribution of internal energies of ions formed in a typical beam-type instrument. Ions with dissociation rate constants $<10^5\ s^{-1}$ remain intact on the timescale of the mass spectrometer (10^{-5} s). Those with rate constants $>10^6\ s^{-1}$ are unstable ions and fragment in the ion source, while the intermediate situation ($10^6\ s^{-1} > k_{diss} > 10^5\ s^{-1}$) represents metastable ions that fragment between the source and the detector. (Reprinted with permission from [26]. ©2004, John Wiley and Sons)

largely unchanged (Franck–Condon principle). Excited molecular ions redistribute their internal energy (E) over the degrees of freedom in a statistical manner. It is postulated that all quantum states have equal statistical probability and all the degrees of freedom participate with the same probability in the energy distribution. After the accumulation of internal energy into appropriate modes, unimolecular dissociation (fragmentation) of the ion occurs. The whole system is described with a three-dimensional potential energy surface (PES) over the reaction coordinate, indicating the possible dissociation channels. Observed fragmentations in EI result from unimolecular dissociations. The QET or, equivalently, the RRKM theory can be used to explain unimolecular dissociations in the gas phase in high vacuum conditions.

The rate constant $k(E)$ for unimolecular dissociation defines the resulting EI mass spectra [26]. Ions with dissociation rate constants $k(E) < 10^5\ s^{-1}$ can reach the detector as molecular ions since the total time in mass spectrometers $\sim 10^{-5}$ s. Those with rate constants $k(E) > 10^6\ s^{-1}$ are unstable ions and fragment in the ion source, while the intermediate situation ($10^6\ s^{-1} > k(E) > 10^5\ s^{-1}$) represents metastable ions (see below) that fragment between the ion source and the detector. Figure 2.2 illustrates the relative portions of stable, metastable, and unstable ions in an internal energy distribution curve.

Metastable Ion Decomposition Molecular ions formed in the ion source to dissociate (fragment) spontaneously in a field-free region between the source and the detector are defined as metastable ions (see Fig. 2.2). These fragments are unimolecular dissociation products decomposed in the field-free regions of a mass spectrometer. Beynon et al., in the early 1970s, used a reverse geometry sector instrument, in which the magnetic analyzer (B) is placed in front of the electrostatic analyzer (E) [27]. Metastable ions (precursor ions) are chosen with a constant B values in the magnetic sector and then scanned by E for detecting metastable ion decompositions in sector instruments. This technique is called mass-analysed ion kinetic energy spectroscopy (MIKES). A MIKES experiment allows not only the study of fragmentation pathways, but also a direct measurement of this kinetic energy released during the fragmentation between the two sectors by measuring the width of the resulting peak.

The MIKES scans can also be applied very nicely with a collision cell placed between the two sectors for CID. Consequently, several scan types are developed for recording the products of metastable or CID. The linked scan with holding the electric field (E) and the magnetic field (B) of an analyser constant (B/E-linked scan) and the B^2/E constant scan are very common among them. The B/E-linked scan is useful in both EB and BE geometry instruments and the resolution is enhanced as compared to the MIKES experiment. The B^2/E, constant scan, is useful for identifying a family of compounds present in a mixture. Another technique, the defocusing method, accelerating voltage scan, can be used to determine the precursor ions for a given fragment formed in the source of a single focusing instrument.

Even though a two-stage sector MS-MS experiment is rather obsolete and complex methods, it may be applied for specific applications [28]. There are a number of important data for fundamental studies of the intrinsic, gas-phase chemistry of specific types of products, such as complexes between alkali-metal ions and biological molecules [29, 30]. Furthermore, analytical applications of such metal ion association reactions have resulted in new approaches for structural elucidation of large biological molecules. These applications have basis in the early work of Rollgen et al., who first reported mass spectra of collision-induced decompositions (CID) of $(M+Cat)^+$ ions formed by Li^+, Na^+, and K^+ ion attachment under field desorption (FD) conditions, in which M = monosaccharide molecule and Cat = alkali-metal ion [31]. It is shown that the elimination of the alkali ions is determined by the alkali ion affinities of the molecules (M) and competes with a fragmentation of M which is almost independent of the alkali ion attached. Correspondingly, the alkali ion is predominantly retained in the fragment ions. The usefulness of this method is demonstrated by showing that CID of cationized complexes can be significantly different from CID of analogous protonated $(M+H)^+$ ions. The CID and structure determination of complexes between alkali-metal ions and more complicated biological molecules such as peptides are very interesting. These are detailed in the Chap. 7.

The unimolecular dissociation of ions has been analyzed with the aid of statistical theory, which is based on the assumption that excited states rapidly convert their energy into vibrational energy of the ground electronic state [24, 25, 32]. The statistical theory for unimolecular reactions is known as RRKM, QET, transition state theory (TST), variational transition state theory (VTST), transition state switching model, or phase space theory (PST) [15], although the basic assumptions in these versions are identical. The significance of the theory is the calculation of the rate constant for unimolecular dissociation from the energy barrier for dissociation and a given reactant state (see details in Sect. 3.2).

2.2.2 Collisional Dissociation

CID The CID process is assumed to occur by a two-step mechanism; the activation (excitation) of the precursor ions, $(A+B)^+$, and their fragmentations.

$$(A+B)^+ + N \rightarrow (A+B)^{+*} + N \tag{2.16}$$

$$(A+B)^{+*} \rightarrow A^+ + B. \tag{2.17}$$

The second step of this mechanism is a unimolecular dissociation of an excited ion. Therefore, the subsequent dissociation of activated ions is adequately explained by RRKM (or QET). Fragmentation of a precursor ion can occur if the collision energy is sufficiently high enough to be excited beyond its threshold for dissociation. Collisions between the precursor ion and a neutral target gas, N, are accompanied by an increase in internal energy, which induces dissociation with improved probability of fragmentation as compared with metastable ion decompositions (see the previous section). By varying the translational energy of the $(A+B)^+$ ions in activation reaction (Eq. 2.16) and detecting the appearance of the A^+ product ions, the threshold energy for CID can be determined and directly related to the bond dissociation energy, BDE, of $(A+B)^+$ [26, 33]. Since the powerful CID methods can potentially address bond dissociation energies (bond strengths) and conformational information between a metal ion and biologically important materials, such as a peptide or small protein, these have been applied to many systems [34].

SID As analogy to CID, SID method represents an activation procedure, which uses a solid surface as a collision target instead of an inert gas molecule [35, 36]. Kinetic energy of the projectile ion is transferred into potential energy upon collision with a solid surface, with resulting activation of the precursor ions and subsequent dissociation.

Threshold Collision-Induced Dissociation (TCID) In the TCID method, a tandem mass spectrometer is used to isolate the parent ion by mass, accelerate or decelerate the ions to a controlled kinetic energy, collide them with an inert target gas under single collision conditions, and then mass analyze the fragment ions [37]. The reaction process is given above by Eqs. (2.16) and (2.17). Xenon is preferred as the target gas, N, because it is heavy and polarizable, promoting efficient energy transfer. In TCID, the apparent cross section for formation of the fragment ion A^+ is measured as a function of the relative ion/target collision energy in the center-of-mass frame. The threshold energy for dissociation, E_0, is equal to the 0 K reaction endothermicity of the dissociation.

TCID experiments employ tandem mass spectrometers of various types. Guided ion beam (GIB) instruments use a radio frequency octopole in the interaction region for high-efficiency collection of scattered products and low-energy ions. These instruments typically use a magnetic sector or Wien filter for initial mass analysis and a quadrupole for post-reaction mass analysis, but configurations have been reported using tandem sector, quadrupole, or time-of-flight mass spectrometers (TOFMSs) with an octopole ion beam guide. Commercial triple quadrupole mass spectrometers have been used for TCID experiments as well as for analytic CID. Energy-resolved CID experiments have also been carried out in ICR or quadruple traps. In these experiments, the activation energy is varied either by changing the pressure (number of collisions) or by changing the ion kinetic energies.

Much of the available thermodynamic data, such as bond dissociation energies, gas-phase acidities and basicities, and heats of formation for ions and neutral target species, has been determined using TCID [37, 38]. These data include a variety of metals (alkalis, magnesium, aluminum, and first and second row transition metals), and many types of target molecules. For instance, Armentrout [39] studied an absolute cation affinity scale, thermochemistry of alkali-metal cation interactions with histidine,and host–guest interactions of crown ethers with alkali ions using TCID.

Low-Energy CID All CID processes may be classified into one of two categories based on the kinetic energy of the precursor ion [26]. Low-energy collision, occurred in the 1–100 eV range of collision energy, is mostly observed in triple quadrupoles (QqQ, see next paragraph), quadrupole ion traps (QIT), and FT-ICR instruments, while high-energy collisions, in the kiloelectron volt range, are seen in sector MS and TOF instruments. There are quite some differences in the resulting CID spectra under low and high collision energy conditions, since high-energy collisions involve electronic excitation, whereas low-energy collisions involve mostly vibrational excitation.

In a QqQ configuration, the collision cell is the second quadrupole (q) and is operated in r.f. mode to allow the precursor ions to be focused. The collision cell is filled with a neutral inert gas, usually N_2, Ar, or Xe, and ion activation is achieved by multiple collisions at high gas pressure. Several scan modes are available in the case of a $Q_1q_2Q_3$ instrument, just like sector tandem mass spectrometer. The most common scan is the product ion scan, where a specific precursor is selected in Q_1, fragmented in q_2, and the products are subsequently determined in Q_3. A second scan is the precursor ion scan, which is essentially the reverse of the product ion scan. The third quadrupole is set to select a specific product ion formed in q_2 and the first quadrupole is scanned for all precursor ions forming the chosen fragment.

In a QIT, the precursor ions are isolated and accelerated by on-resonance excitation. Multiple collisions can occur during the excitation period. Ion activation times are of the order of tens of milliseconds. Because of this relatively long timescale, interesting excitation techniques have been developed, such as "sustained off-resonance irradiation" (SORI) or BIRD with much longer excitation periods. Furthermore, multiple stages of CID can be applied to QIT instruments. Several hybrid geometries have recently been introduced, including the quadrupole-linear ion trap (QqLIT) and the quadrupole time of flight (QqTOF) instruments.

As is the case for QIT, in FTICR instruments isolation and excitation take place in the same confined space, where the ions are trapped for a specific time in a combined magnetic and electrostatic field [26]. On-resonance excitation in the FTICR cell can be achieved by using a short (hundreds of μs) high-amplitude a.c. signal at the natural cyclotron frequency of the precursor ion. This procedure excites the precursor ion rapidly via multiple collisions and deposits a large amount of energy in the ion. The short irradiation time is necessary to minimize precursor ion losses in the on-resonance experiment. Some commercial instruments can even achieve low-energy CID in an r.f.-only quadrupole with a hybrid Qq-FTICR instrument. Additionally, a recently introduced FTICR instrument employs a linear ion trap for multiple stages of front end CID.

High-Energy CID In considering high-energy collisions, collisional activation occurs in the kiloelectron volt range in sector MS and TOF instruments. A collision cell is placed between the two mass analyzers. The precursor ions, having a kinetic energy of a few kiloelectron volt, can enter the collision cell, absorbing energy by single collisions to dissociate before mass analysis of the product ions. Ion activation at high kinetic energies caused a vertical electronic transition (Franck–Condon principle) and hence ion excitation is mainly electronic. However, vibrational and rotational energies can also be involved in the excitation. As is the case for metastable ion decompositions, several linked scans can also be employed (see Sect. 2.2.1) for the analysis of CID products in sector instruments.

2.2.3 *Time-Resolved Photoactivated Dissociation, IRMPD, and BIRD*

Photoactivated dissociation processes [40–42] are briefed. Ion photo dissociation time-resolved measurements have been made for a long time. In time-resolved photo dissociation, the rate of dissociation reaction can be measured by monitoring photo fragmentation of ions as a function of time. MS-MS allows selection of the target ion to be dissociated and analysis of the product ions. FTICR or QIT are well suited for time-resolved photo dissociation experiments. The photoactivated dissociation is assumed to occur by a two-step mechanism; the photo activation (excitation) of the precursor ions, $(A+B)^+$, and their fragmentations, similar to CID.

$$(A+B)^{+} + nhv \rightarrow (A+B)^{+*} \tag{2.18}$$

$$(A+B)^{+*} \rightarrow A^{+} + B, \tag{2.19}$$

where n describes the number of absorbed photons and hv is the photon energy. The mechanism of activation is through the absorption of IR radiation, followed by the statistical internal rapid redistribution of excited energy over all the vibrational degrees of freedom. The second step of this mechanism is a unimolecular dissociation of an excited ion. Therefore, the subsequent dissociation of activated ions is adequately explained by RRKM (or QET). Fragmentation of a precursor ion can occur if the absorbed energy is sufficiently high enough to be excited beyond its threshold for dissociation.

Photo activation of gaseous ions in a mass spectrometer has been performed by using IR lasers. These lasers do not provide enough energy to initiate dissociation of precursor ions. In IRMPD experiments, multiphoton processes are consequently needed to excite ions sufficiently for efficient dissociation. IRMPD as an activation technique in MS-MS is suitable to QIT and FT-ICR mass spectrometers, because of their capability to store ions for long times. Over last decades, IRMPD spectroscopy has proven to be a powerful method [43–47] for obtaining structural information on protonated and alkali-metal ion complexes of small biological molecules including nucleic acid bases which make up the A:T base pair, amino acids (Adenine, proline, phenylalanine) and dipeptides, and medicines (therapeutically theophilline). Most work is done in combination with computational methods [46, 47].

In BIRD experiments of FTI-CR mass spectrometers [42], the IR radiation by heating from the ICR cell is employed for unimolecular dissociation rate constants. Heating makes possible for the trapped ions to absorb blackbody photons radiated from the walls of the ICR cell. There are two essential requirements for BIRD methods: (i) collision-free conditions at the low ambient pressure ($<10^{-8}$ Pa) and (ii) long reaction time of the dissociating molecules (of the order of seconds). FTICR mass spectrometers can achieve these conditions, with background pressures below 10^{-8} Pa and ion trapping times of milliseconds or longer. Dissociation occurs without collisional activation due to the extremely low pressures in this instrument. In BIRD conditions, the precursor ions reach equilibrium at the chamber temperature because (i) radiative photon absorption are much faster than dissociation and (ii) blackbody IR activation provides a thermal energy distribution. Under these conditions, the simple Arrhenius rate equation may be applied [48],

$$k(\mathrm{E}) = A\mathrm{e}^{-Ea/RT}, \tag{2.20}$$

where k represents the rate constant, A is the pre-exponential Arrhenius factor, Ea is the activation energy, R is the gas constant, and T is the absolute temperature of the ICR cell. Dissociation rate constants can be calculated from measurements of the dissociated ion yields at different activation times. If the experiment is performed at several temperatures, the activation energy and the pre-exponential Arrhenius factor are easily calculated for the dissociation of the precursor ions. Ea is assumed to be equal to the bond dissociation energy for unimolecular dissociation. Since BIRD has also proven to be a powerful tool for obtaining structural information as well as dissociation pathway, kinetics and Arrhenius parameters of various kinds of compounds including protonated and alkali-metal ion complexes [49, 50], the BIRD research done in less than a decade has extended to variety of applications and systems. Alkali ion complexes of biomolecules are attractive BIRD targets. These include amino acids, hydrated complexes of valine, leucine enkephalin cationized with alkali-metal ions, and so on.

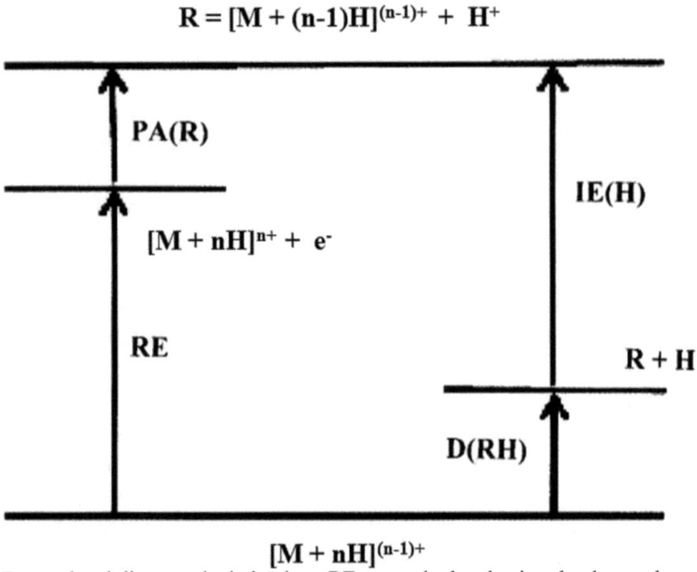

Fig. 2.3 Energy level diagram depicting how REs are calculated using the thermodynamic cycle. R = $[M + (n-1)H]^{(n-1)+}$, PA(R) = proton affinities of R (up to ~6 eV), D(RH) = bond dissociation energies, and IE(H) = ionization energies of H, 13.6 eV

2.2.4 Electron-Capture Dissociation and Electron-Transfer Dissociation

Fragmentation following gas-phase electron capture by positive ions is termed dissociative recombination. ECD is based on the dissociative recombination of multiply-charged cations $(M+nH)^{n+}$ with low-energy electrons (<1 eV), generating charge-reduced radical species $(M+nH)^{(n-1)+*}$ and product ions [51, 52].

$$[M+nH]^{n+}+e^{-}\rightarrow[M+nH]^{(n-1)+*}. \quad (2.21)$$

The recombination energy (RE) can be estimated from the thermodynamic cycle of ECD (Fig. 2.3).

$$RE = IE(13.6\ eV) - PA[(M+(n-1)H)^{(n-1)+}] + D[(M+(n-1)H)^{(n-1)+} - H], \quad (2.22)$$

where PA and $D[(M+(n-1)H)^{(n-1)+} -H]$ are the proton affinity of $[M+(n-1)H]^{(n-1)+}$ and bond dissociation energy between $(M+(n-1)H)^{(n-1)+}$ and hydrogen atom, respectively. The 5–7 eV energy released by recombination (neutralization) appears to induce cleavage. The ECD process is non-ergodic; bond cleavage occurs before energy randomization (intermolecular vibrational energy redistribution), and the electron forms radical species whose activation energies for dissociation should be lower.

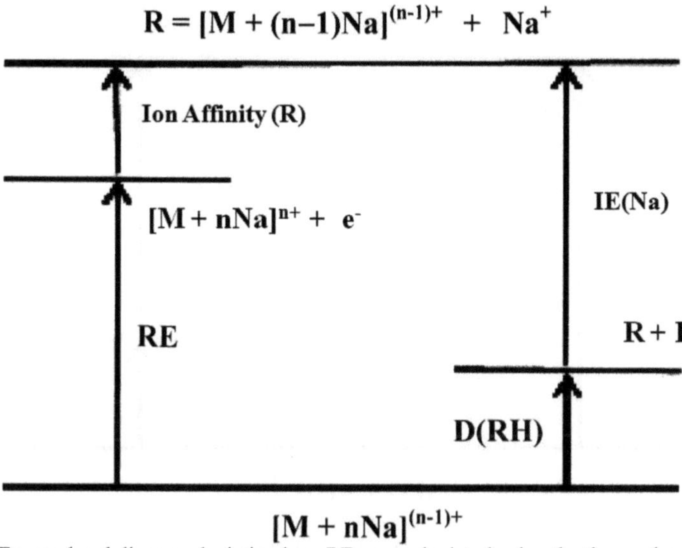

Fig. 2.4 Energy level diagram depicting how REs are calculated using the thermodynamic cycle. $R = [M + (n-1)Na]^{(n-1)+}$, IA(Na) = sodium ion affinities (up to ~2 eV), D(RH) = bond dissociation energies, and IE(Na) = ionization energies of Na, 5.14 eV

The advantage of ECD is (i) ECD spectra are characteristic of the molecular composition and structure, (ii) ECD causes dissociation in very large biomolecules, especially useful for protein sequencing applications, (iii) ECD cleaves homogeneously and effectively many more bonds than conventional CID, and its fragment ions come from single bond cleavage, and (iv) the labile bond as well as ionic (non-covalent) bond remain intact in ECD spectrum. Therefore, the fragments reflect the original structure. FTICR-MS is ideal for the application of ECD and, accordingly, all commercial FTICR instrument could be used for ECD experiments without hardware modification.

The ECD technique in FTICR-MS has been applied [53–55] to precursor cationized with metal ions. The application of ECD to oligosaccharides is made by employing alkali, alkaline earth, and transition metals as charge carriers in ESI. It is demonstrated that complementary structural information is obtained with ECD of metal ion adduct complexes as compared with other dissociation methods, and hence it is a valuable method for structural characterization of oligosaccharides and polyethers. Another finding in these studies is that ECD of multiply-sodiated precursor produces significantly less fragmentation than ECD of multiply-protonated one. It is consistent [52] with the much lower RE of sodiated species (see Fig. 2.4) compared with protonated species.

More recently, ETD, a dissociation technique very similar to ECD, was introduced for peptide/protein sequencing [56–58]. In ETD, multiply-charged cations react with an anion radical (A^{-*}, e.g., SO_2^-, SF_6^-, anthracene, fluoranthene), leading to electron transfer and the generation of both odd- and even-electron fragments and molecular species with lower charge states as follows:

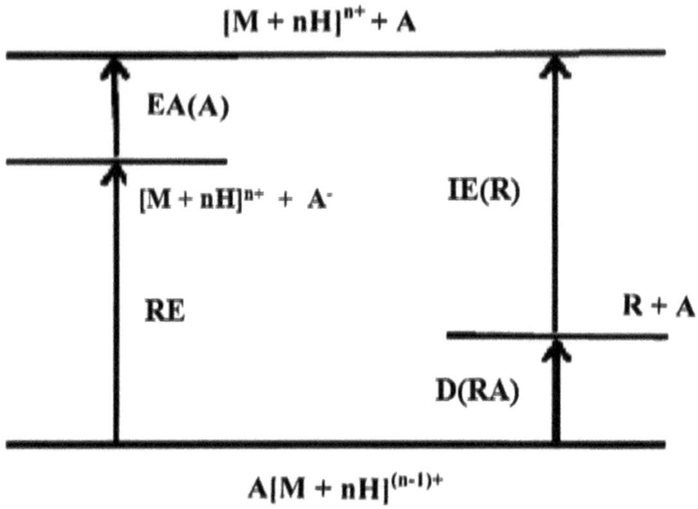

Fig. 2.5 Energy level diagram depicting how REs are calculated using the thermodynamic cycle. $R = [M + nH]^{(n-1)+}$, EA(A) = proton affinities of A, and D(RA) = bond dissociation energies

$$[M+nH]^{n+} + A^{-*} \rightarrow [M+nH]^{(n-1)+*} + A. \tag{2.23}$$

Equation 2.23 is less exothermic than Eq. 2.21 by the electron affinity (EA) of A (see Fig. 2.5). Similar to ECD, the electron-transfer reaction deposits excitation energy, only lowered by the EA of the radical anion (0.5–1.5 eV, depending on the radical anion), for hydrogen atom release. The next reaction (Eq. 2.24) is the proton transfer reaction which may be competing pathway to Eq. 2.23.

$$[M+nH]^{n+} + A^{-} \rightarrow [M+(n-1)H]^{(n-1)+} + AH \tag{2.24}$$

ETD induces a less-exothermic (i.e., more "gentle") electron-transfer reaction than ECD and its process is non-ergodic. ETD can be performed in quadrupole ion trap mass spectrometers, which are widely available to the mass spectrometry community. Several types of mass spectrometers have been configured to perform ETD experiments [59, 60]. These include a hybrid ion trap (3D, a linear multipole, or ring guide)-Fourier transform ion cyclotron resonance mass spectrometer (LTQ-FT), hybrid triple quadrupole-linear ion trap instrument (QqQ-LIT), and hybrid quadrupole time of flight (QTOF).

ETD and electron activated dissociation (ExD) are applied to structural identification of carbohydrates and results in some new and detailed information for glycan (a series of linear milk sugars) structural studies [61, 62]. Different metal ion adducts of Na^+, Li^+, Ca^{2+},Mg^{2+}, and Zn^{2+} were used as precursor ions to determine which metal adducts provided the most informative dissociation behavior under ETD conditions. It is demonstrated that ETD of metal ion complexes should contribute substantially to confident structural analysis of a wide variety of oligosaccharides [61]. ETD is also employed to analyze synthetic phospho- and sulfopeptides [63]. It is reported that informative fragmentation behavior may be observed by fragmenting peptide-alkali-metal adducts, such as sodiated or potassiated peptides.

2.2.5 *ISD and PSD in MALDI*

Similar to fragmentations occurring either in the ion source by EI, or by metastable ion decompositions during an ion's flight from the ion source to the detector, quite some degree of fragmentations is observed during the MALDI process. The rate of a fragmentation reaction is important in deciding whether or not that reaction will take place. As for peptide and protein ions, very fast reactions, such as loss of water or ammonia, occur within the ion source before ion extraction. On the other hand, some peptide bond cleavages require more time. Two types of fragmentation reactions can be distinguished and referred to in MALDI-MS as ISD, which is "prompt fragmentation" of ions occurring in the source region of the mass spectrometer and PSD, which involves the fragmentation of metastable ions occurring in the field-free drift region of a reflectron TOFMS [64–67].

PSD seems to be induced by converting the excess energy deposited on ions into vibrational energy that is distributed over the entire molecule. PSD mechanisms are the randomization of the internal energy among all internal degrees of freedom before the metastable ion decomposition. Instead, ISD is postulated to be induced by plenty of electrons as well as hot hydrogen atoms in the MALDI plume [68]. When a precursor ion enters the conventional TOFMS, metastable ions that decompose during their flight through the drift tube are detected at the precursor's m/z. These PSD fragment ions can be distinguished, however, by the reflectron TOFMS. Recently, ISD and PSD fragments can be observed in the same spectrum. In addition, PSD experiments make use of a collision cell placed between the ionization source and the drift tube to induce fragmentation by collision. The unimolecular decompositions of ions possessing excess energy in MALDI process, as well as collisional activation of analyte ions, are contributed to the observed ISD and PSD spectra.

PSD-MALDI techniques have been developed to obtain chemical structure information from MALDI. PSD-MALDI-MS has been applied to precursor cationized with metal ions. The reported PSD-MALDI-MS studies have dealt with metal ion adduct precursors of a wide variety of compounds. For instance, the application of PSD to oligosaccharides [69–71] and synthetic polymers [72, 73] is made by employing alkali-metal ions for precursor ions in MALDI. Recently ISD-MALDI-MS investigation has been reported on metal ion complexes of oligosaccharides [74].

2.3 Ion–molecule Association in the Condensed Phase

2.3.1 *Cationization in Desorption Ionization*

Observations of alkali-metal ion adducts of the type $[M+Li]^+$, $[M+Na]^+$, etc. are common in the desorption ionization (DI) mass spectra of a variety of polar molecules. In fact, alkali-metal ion association reactions are observed with FD ionization, FAB ionization, ^{252}Cf plasma desorption (PD), secondary ion mass spectrometry (SIMS), MALDI, and ESI. Ion yields can be greatly enhanced by addition of alkali-metal salts to the sample. Particularly for the MALDI analysis of synthetic polymers, metal cations are often intentionally added to enhance signals. A qualitative description of the current understanding of formation mechanism of alkali-metal ion complexes from the condensed phase was presented [75]. Knowledge of the ionization mechanisms is important and helpful from the perspective of increasing the analytical utility of the method.

Although the mechanism of production of cationized molecules is not fully understood and details of the mechanisms may be debated [76], a qualitative statement has been made by Busch [77]. He describes qualitatively the features of ion generation in a number of particle-induced emission techniques. It is suggested that ions may be formed in distinctive beam-induced processes. Finally, ions and molecules must survive passage through a so-called selvedge region. The selvedge is phenomenologically defined as the boundary above the surface beyond which no ion–molecule reactions occur, and past which only unimolecular dissociation reactions occur. In this region, the pressure drops from that of the condensed phase to that of the free vacuum of the mass spectrometer source and ions and molecules may undergo association and dissociation reactions both with each other and with other solution components as well as the matrix. DI methods tend to produce even-electron ions such as protonated molecules $[M+H]^+$ or cationized molecules such as $[M+Na]^+$; these stable ions undergo only a minimum amount of fragmentation. A metal cation can be something as ubiquitous as Na^+ or K^+, or transition metal cations such as Fe^+ or Ag^+.

2.3.2 *Cationization Mechanism in MALDI*

The factors determining the observed ion distribution in MALDI are of considerable interest. The question regarding metal cation adducts in MALDI is why they give the dominant signal in certain cases, although proton affinities are always much higher. There have been a number of qualitative mechanistic proposals [77, 78] which include a wide variety of rather different processes. Knochenmuss et al. have proposed that secondary reactions in the MALDI plume may in many cases be the dominant determinant of the final detected mass spectrum. This quantitative, thermodynamically based proposal was built on earlier qualitative indications that analyte ions are formed either predominantly or in part via secondary reactions with matrix or metal ions. The thermodynamics of possible plume cation transfer reactions is slowly becoming known, as is their role in MALDI. Cationization is found to be typically important for analytes with a low proton affinity, less than that of the matrix. The thermodynamic approach also extends earlier studies of systematic influences on MALDI spectra.

2.3.3 Cationization Mechanism in ESI

Cationization appears to hold great promise for the ESI analysis of nonvolatile and thermally labile compounds. The initial ESI droplets are positively charged due to the presence of excess ions that can include H^+, NH_4^+, Na^+, and K^+, although protons are often the main contributor to the net droplet charge. Studies in a number of laboratories are performed to understand the mechanism of this form of ionization in order to make a wider application of the technique. In recent year, some advances have been made in the understanding of the ESI mechanism [79–81]. Two competing mechanisms explain how macromolecules become multiply charged in ESI, the charged-residue mechanism (CRM) and the ion-emission mechanism (IEM). According to the CRM, the excess charges on ESI droplets are transferred to and remain on macromolecules enclosed within the droplets after solvent evaporation. The charge on a macromolecule is limited by the maximum charge (i.e., the Rayleigh limiting charge) that a droplet similar in size to the macromolecule can contain without fissioning. Conversely, according to IEM theory, analyte ions desorb directly from charged nanodroplets, driven by the large electric field at the droplet surface. The formation of low molecular weight, singly charged ions in ESI has been explained well by the IEM. Despite extensive study of macromolecular ESI, neither mechanism explains quantitatively the extent of macromolecular multiple charging.

References

1. Bass L, Michael T, Bowers MT. Ion-molecule association reactions, lecture notes in chemistry. In Hartmann H, Wanczek K-P, Editors. Ion cyclotron resonance spectrometry II. Berlin: Springer; 1982. p. 432–63.
2. Gioumousis G, Stevenson DP. Reactions of gaseous molecule ions with gaseous molecules. V Theor, J Chem Phys. 1958;29:294–9.
3. Vogt E, Wannier GH. Scattering of ions by polarization forces. Phys Rev. 1954;95:1190–8.
4. Good A. Third-order ion-molecule clustering reactions. Chem Rev. 1975;75:561–83.
5. Chang JS, Golden DM. Kinetics and thermodynamics for ion-molecule association reactions. J Am Chem Soc. 1981;103:496–500
6. Woodin RL, Beauchamp JL. Bimolecular infrared radiative association reactions. Attachment of Li^+ to carbonyl compounds in the gas phase. Chem Phys. 1979;41:1–9.
7. Gerlich D, Kaefer G. Ion trap studies of association processes in collisions of CH_3^+ and CD_3^+ with n-H_2, p-H_2, D_2, and He at 80 K. Astrophys J. 1989;347:849–54.

8. Zhang H, Dearden DV. The gas-phase macrocyclic effect: reaction rates for crown ethers and the corresponding glymes with alkali metal cations. J Am Chem Soc. 1992;114:2154–5.
9. Wong PSH, Antonio BJ, Dearden DV. Gas-phase studies of valinomycin-alkali metal cation complexes: attachment rates and cation affinities. J Am Soc Mass Spectrom. 1994;5:632–7.
10. Staley RH, Wieting RD, Beauchamp JL. Carbenium ion stabilities in the gas phase and solution. An ion cyclotron resonance study of bromide transfer reactions involving alkali ions, alkyl carbenium ions, acyl cations, and cyclic halonium ions. J Am Chem Soc. 1977;99:5964–72.
11. Allison J, Ridge DP. Reactions of atomic metal ions with alkyl halides and alcohols in the gas phase. J Am Chem Soc. 1979;101:4998–5009.
12. Meot-Ner M, Field FH. Decomposition rates of excited reaction complexes. Temperature and pressure effects in association reactions involving NH_4^+, $CH_3NH_3^+$, and $(CH_3)_2NH_2^+$. J Am Chem Soc. 1975;97:5339–45.
13. Neilson PV, Bowers MT, Chau M, Davidson WR, Aue DH. Energy transfer in excited ionic species. Rates and mechanism of dimerization of protonated amines with their neutral bases. J Am Chem Soc. 1978;100:3649–58.
14. Wieder GM, Marcus RA. Dissociation and isomerization of vibrationally excited species. II. Unimolecular reaction rate theory and its application. J Chem Phys. 1962;37:1835–52.
15. Bass L, Chesnavich WJ, Bowers MT. Gas-phase ion-molecule association reactions. A statistical phase space theory approach. J Am Chem Soc. 1979;101:5493–502.
16. Ferguson EE. Rate constants of thermal energy binary ion-molecule reactions of aeronomic interest. Atomic Data Nucl Data Tables. 1973;12:159~.
17. Sieck LW, Lias SG. Rate coefficients for ion-molecule reactions I. Ions containing C and H. J Phys Chem Ref Data. 1976;5:1123–46.
18. Albritton DL. Ion-neutral reaction rate constants measured in flow reactors through 1977. Atomic Data Nucl Data Tables. 1978;22:1~.
19. Anicich VG. Evaluated bimolecular ion-molecule gas phase kinetics of positive ions for use in modeling planetary atmospheres, cometary, comae, and interstellar clouds. Phys Chem Ref Data. 1993;22:1469–569.
20. Ikezoe Y, Matsuoka S, Takebe M, Viggiano A. Gas phase ion-molecule reaction rate constants through 1986. Tokyo: Maruzen Company; 1987. p. 244.
21. Anicich VG. An index of the literature for bimolecular gas phase cation–molecule reaction kinetics. JPL Publication: U.S. (03–19); 2003. p. 1172
22. Spears KG, Ferguson EE. Termolecular and saturated termolecular kinetics for Li^+ and F^-. J Chem Phys. 1973;59:4174–83.
23. Passarella R, Castleman AW Jr. Thermal energy ion-molecule association reactions involving sodium ions. J Phys Chem. 1989;93:5840–5.
24. Forst W. Unimolecular reactions: a concise introduction. London: Cambridge University Press; 2003. p. 332.
25. Baer T, Hase WL. Unimolecular reaction dynamics: theory and experiments (International series of monographs on chemistry). London: Oxford University Press; 1996. p. 438.
26. Sleno L, Volmer DA. Ion activation methods for tandem mass spectrometry. J Mass Spectrom. 2004;39:1091–112.
27. Beynon JH, Cooks RG, Amy JM, Baitinger WE, Ridley TY. Design and performance of a mass-analysed ion kinetic energy (MIKE) spectrometer. Anal Chem. 1973;45:1023A–31A.
28. Campana JE, Greed BN. Unicluster dissociation of large alkali iodide cluster ions? J Am Chem Soc. 1984;106:531–5.
29. Grese RP, Cerny RL, Gross ML. Metal ion-peptide interactions in the gas phase: a tandem mass spectrometry study of alkali metal cationized peptides. J Am Chem Soc. 1989;111:2835–42.
30. Tang X, Ens W, Standing KG, Westmore JB. Daughter ion mass spectra from cationized molecules of small oligopeptides in a reflecting time-of-flight mass spectrometer. Anal Chem. 1988;60:1791–9.
31. Rollgen FW, Borchers F, Giessmann U, Levsen K. Collisional activation of ions formed by $[Li]^+$ ion attachment. Org Mass Spectrom. 1977;2:541–3.

32. Bush MF, Oomens J, Saykally RJ, Williams ER. Alkali metal ion binding to glutamine and glutamine derivatives investigated by infrared action spectroscopy and theory. J Phys Chem A. 2008;112:8578–84.
33. Ervin KM. Experimental techniques in gas-phase ion thermochemistry. Chem Rev. 2001;101:391–444.
34. Wells JM, McLuckey SA. Collision-induced dissociation (CID) of peptides and proteins. Methods Enzymol. 2005;402:148–85.
35. Verena G. Collisions of ions with surfaces at chemically relevant energies: instrumentation and phenomena. Rev Sci Instrum. 2001;72:3149–79.
36. Mabud M, Dekrey MJ, Cooks RG. Surface-induced dissociation of molecular ions. Int J Mass Spectrom Ion Process. 1985;67:285–94.
37. Rodgers MT, Armentrout PB. Noncovalent metal-ligand bond energies as studied by threshold collision-induced dissociation. Mass Spectrom Rev. 2000;19:215–47.
38. Rogers MT, Armentrout PB. A thermodynamic "vocabulary" for metal ion interactions in biological systems. Acc Chem Res. 2004:37:989–98.
39. Armentrout PB. Cation-ether complexes in the gas phase: thermodynamic insight into molecular recognition. Int J Mass Spectrom. 1999;193:227–40.
40. Polfer NC. Infrared multiple photon dissociation spectroscopy of trapped ions. Chem Soc Rev. 2011;40:2211–21.
41. Polfer NC, Oomens J, Dunbar RC. Alkali metal complexes of the dipeptides PheAla and AlaPhe: IRMPD spectroscopy. ChemPhysChem. 2008;9:579–89.
42. Dunbar RC. BIRD (Blackbody infrared radiative dissociation): evolution, principles, and applications. Mass Spectrom Rev. 2004;23:127–58.
43. Drays MK, Armentrout PB, Oomens J, Schafer M. IR spectroscopy of cationized aliphatic amino acids: stability of charge-solvated structure increases with metal cation size. Int J Mass Spectrom. 2010;297:18–27.
44. Citir M, Stennett EMS, Oomens J, Steill JD, Rodgers MT, Armentrout PB. Infrared multiple photon dissociation spectroscopy of cationized cysteine: effects of metal cation size on gas-phase conformation. Int J Mass Spectrom. 2010;297:9–17.
45. Marta RA, Wu R, Eldridge KR, Martens JK, McMahon TB. The sodium cation-bound dimer of theophylline: IRMPD spectroscopy of a highly symmetric electrostatically bound species. Int J Mass Spectrom. 2010;297:76–84.
46. Rajabi K, Gillis EAL, Fridgen TD. Structures of alkali metal ion-adenine complexes and hydrated complexes by IRMPD spectroscopy and electronic structure calculations. J Phys Chem A. 2010;114:3449–56.
47. Dunbar RC, Steill JD, Jos Oomens J. Cationized phenylalanine conformations characterized by IRMPD and computation for singly and doubly charged ions. Phys Chem Chem Phys. 2010;12:13383–93.
48. Schnier PD, Price WD, Strittmatter EF, Williams ER. Dissociation energetics and mechanisms of leucine enkephalin $(M + H)^+$ and $(2M + X)^+$ ions (X = H, Li, Na, K, and Rb) measured by blackbody infrared radiative dissociation. J Am Soc Mass Spectrom. 1997;8:771–80.
49. Jockusch RA, Lemoff AS, Williams ER. Hydration of valine-cation complexes in the gas phase: on the number of water molecules necessary to form a zwitterion. J Phys Chem A. 2001;105:10929–42.
50. Jockusch RA, Lemoff AS, Williams ER. Effect of metal ion and water coordination on the structure of agas-phase amino acid. J Am Chem Soc. 2001;123:12255–65.
51. Zubarev RA, Haselmann KF, Budnik B, Kjeldsen F, Jensen F. Towards an understanding of the mechanism of electron-capture dissociation: a historical perspective and modern ideas. Eur J Mass Spectrom. 2002;8:337–49.
52. Cerda BA, Horn DH, Breuker K, Carpenter BK, McLafferty FW. Electron capture dissociation of multiply-charged oxygenasted cations. A nonergodic process. Eur J Mass Spectrom. 1999;5:335–8.
53. Haselmann KF, Budnik BA, Olsen JV, Nielsen ML, Reis CA, Clausen H, Johnsen AH, Zubarev RA. Advantages of external accumulation for electron capture dissociation in Fourier transform mass spectrometry. Anal Chem. 2001;73:2998–3005.

54. Adamson JT, Håkansson K. Electron capture dissociation of oligosaccharides ionized with alkali, alkaline earth, and transition metals. Anal Chem. 2007;79:2901–10.
55. Budnik BA, Haselmann KF, Elkin YN, Gorbach VI, Zubarev RA. Applications of electron-ion dissociation reactions for analysis of polycationic chitooligosaccharides in Fourier transform mass spectrometry. Anal Chem. 2003;75:5994–6001.
56. Coona JJ, Syka JEP, Schwartz JC, Shabanowitz J, Hunt DF. Anion dependence in the partitioning between proton and electron transfer in ion/ion reactions. Int J Mass Spectrom. 2004;236:33–42.
57. Anusiewicz I, Berdys-Kochanska J, Simons J. Electron Attachment step in electron capture dissociation (ECD) and electron transfer dissociation (ETD). J Phys Chem A. 2005;109:5801–13.
58. Wolff JJ, Leach III FE, Laremore TN, Kaplan DA, Easterling ML, Linhardt RJ, Amster JJ. Negative electron transfer dissociation of glycosaminoglycans. Anal Chem. 2010;82:3460–6.
59. Campbell JL, Hager JW. Creating an evanescent ion/ion reaction region within a low-pressure linear ion trap. Int J Mass Spectrom. 2012;323/324:14–20.
60. Zhang Y, Zhang H, Cui W, Chen H. Tandem MS analysis of selenamide-derivatized peptide ions. J Am Soc Mass Spectrom. 2011;22:1610–21.
61. Han L, Costello CE. Electron Transfer dissociation of milk oligosaccharides. J Am Soc Mass Spectrom. 2011;22:997–1013.
62. Yu X, Huang Y, Lin C, Costello CE. Energy-dependent electron activated dissociation of metal-adducted permethylated oligosaccharides. Anal Chem. 2012;84:7487–94.
63. Medzihradszky KF, Guan S, Maltby DA, Burlingame AL. Sulfopeptide fragmentation in electron-capture and electron-transfer dissociation. J Am Soc Mass Spectrom. 2007;18:1617–24.
64. Hillenkamp F, Peter-Katalinic J. MALDI MS. A practical guide to instrumentation, methods and applications. Weinheim: Wiley-Blackwell; 2013. p. 480.
65. Hoteling AJ, Owens KG. Improved PSD and CID on a MALDI TOFMS. J Am Soc Mass Spectrom. 2004;15:523–35.
66. Harvey DJ. Matrix-assisted laser desorption/ionization mass spectrometry of carbohydrates. Mass Spectrom Rev. 1999;18:349–451.
67. Gevaert K, Vandekerckhove J. Protein identification methods in proteomics. Electrophoresis. 2000;21:1145–54.
68. Kolcher T, Engstrom Å, Zubarev RA. Fragmentation of peptides in MALDI in-source decay mediated by hydrogen radicals. Anal Chem. 2005;77:172–7
69. Cancilla MT, Penn SG, Carroll JA, Lebrilla CB. Coordination of alkali metals to oligosaccharides dictates fragmentation behavior in matrix assisted laser desorption ionization/fourier transform mass spectrometry. J Am Chem Soc. 1996;118:6736–45.
70. Ngoka LC, Gal JF, Lebrllla CB. Effects of cations and charge types on the metastable decay rates of oligosaccharides. Anal Chem. 1994;66:692–8.
71. Keki S, Deak G, Zsuga M. Fragmentation study of rutin, a naturally occurring flavone glycoside cationized with different alkali metal ions, using post-source decay matrix-assisted laser desorption/ionization mass spectrometry. J Mass Spectrom. 2001;36:1312–6.
72. Muscat D, Henderickx H, Kwakkenbos G, Benthem R, de Koster CG, Fokkens R, Nibbering NMM. In-source decay of hyperbranched polyesteramides in matrix-assisted laser desorption/ionization time-of-flight mass spectrometry. J Am Soc Mass Spectrom. 2000;11:218–27.
73. Hantona SD, Pareesa DM, Owensb KG. MALDI PSD of low molecular weight ethoxylated polymers. Int J Mass Spectrom. 2004;238:257–64.
74. Yang H, Yu Y, Song F, Liu S. Structural characterization of neutral oligosaccharides by laser-enhanced in-source decay of MALDI-FTICR MS. J Am Soc Mass Spectrom. 2011;22:845–55.
75. Keough T. Cationization of organic molecules using fast atom bombardment mass spectrometry. Anal Chem. 1985;57:2027–34.
76. Mallis LM, Russell DH. General aspects of the chemistry of organo-alkali metal ions. An overview of recent work. Int J Mass Spectrom. 1987;78:147–78.
77. Busch KL. Desorption ionization mass spectrometry. J Mass Spectrom. 1995;30:233–40.

78. Knochenmuss R, Zenobi R. MALDI ionization: the role of in-plume processes. Chem Rev. 2003;103:441–52.
79. Hogan CJ Jr, Carroll JA, Rohrs HW, Biswas P, Gross ML. Combined charged residue-field emission model of macromolecular electrospray ionization. Anal Chem. 2009;81:369–77.
80. Marchese R, Grandori R, Carloni P, Raugei S. A computational model for protein ionization by electrospray based on gas-phase basicity. J Am Soc Mass Spectrom. 2012;23:1903–10.
81. Konermann L, Ahadi E, Rodriguez AD, Vahidi S. Unraveling the mechanism of electrospray ionization. Anal Chem. 2013;85:2–9.

Chapter 3
A Theoretical Approach to Ion–Molecule Complexation

Sundaram Arulmozhiraja

3.1 Basic Ion–Molecule Interactions

Non-covalent interactions play an important role not only in chemistry but also in molecular biology and material science. The well-known nonbonding electrostatic interactions mainly are dipole–dipole, ion–dipole, dipole–induced dipole, ion–induced dipole, and ion–π interactions. Electrostatic interactions are normally between cations and anions, which are species with formal charges. In fact, all molecular interactions are inherently electrostatic in nature. It might have been better to have called these interactions between cations and anions charge–charge interactions; however, by convention we use the term electrostatic to describe interactions between formally charged species.

Before going to see what a dipole–dipole interaction is, one should know about dipoles and electronegativities. In a molecule with unlike atoms, electrons are not shared equally. The tendency of any atom to pull electrons away from other atoms is characterized by a quantity called electronegativity. Oxygen has the highest electronegativity (approximately 3.5) and hydrogen has the least electronegativity (around 2.1). In a molecule containing atoms of different electronegativities, atoms with the smallest electronegativities hold partial positive charges, and the atoms with largest electronegativities hold partial negative charges (polar molecule). In a water molecule, the oxygen atom pulls electron density away from the hydrogen atoms. The oxygen atom then carries a partial negative charge whereas both the hydrogen atoms carry partial positive charge. This phenomenon of charge separation is called polarity. So, water is a polar molecule. The extent of charge separation within a molecule is characterized by the dipole moment μ, which is determined by the magnitudes of the partial charges (q) and by the distances (r) between them ($\mu = qr$).

In a dipole–dipole interaction, two dipoles have a kind of interaction (either attractive or repulsive), and the strength of a dipole–dipole interaction depends on the

S. Arulmozhiraja (✉)
Catalysis Research Center, Hokkaido University, Kita 21, Nishi 10, Kita-ku,
Sapporo 001-0021, Japan
e-mail: raja@cat.hokudai.ac.jp

T. Fujii (ed.), *Ion/Molecule Attachment Reactions: Mass Spectrometry*,
DOI 10.1007/978-1-4899-7588-1_3

size of each dipole and on their relative orientation. On the other hand, dipole–induced dipole interaction differs slightly: A molecule with a permanent dipole moment will induce a dipole moment in a nearby molecule. This phenomenon even can induce a temporary dipole in a nonpolar molecule (this inducing phenomenon is called polarization) and thus produces a net interaction between these two molecules. This is called dipole–induced dipole interaction. The strength of a dipole–induced dipole interaction depends on the size of the dipole moment of the first molecule and on the polarizability of the second molecule. Polarizability is a measure of the ease with which electrons are shifted by an external electronic field. Molecules with π-electrons, such as in benzene and phenylalanine, are more polarizable than molecules that lack π electrons. Dipole–induced dipole interactions are important even between molecules with permanent dipole because a permanent dipole is altered by the dipole of the other.

Ion–molecule complexes are formed mainly due to one of the following three interactions: (1) ion–dipole, (2) ion–induced dipole, and (3) ion–π interactions. According to Coulomb's law, the electrostatic force between two charged particles is defined as $f = \frac{q_1 q_2}{r}$, where q_1 and q_2 are the charges of the particles and the r is the distance between them. Accordingly, the electrostatic force between an ion and a neutral particle should be zero; however, this law assumes that the two particles are point charges having zero radii. A real particle, a molecule in this case, occupies a certain volume of space. Even in the case of a neutral molecule, it is possible that the spatial distribution of the electron density representing the most loosely bound electrons might be asymmetrical. This asymmetrical nature gives an electric dipole moment. There are two kinds of dipole moments as explained above: (1) permanent dipole moment that can arise when bonding occurs between dissimilar atoms of different electronegativities, and (2) induced dipole moment, which is created when an external electric field causes distortion of the electron cloud of a neutral molecule. Even if a molecule is neutral and possesses no permanent dipole moment, a nonpolar molecule, an electric field of a nearby ion can cause the centers of positive and negative charges of a neutral molecule to shift in opposite directions, which results in induced dipole moment. Normally, charged ions polarize nearby molecules more than a dipole of a polar molecule and induce favorable dipoles.

In an ion–dipole (or charge–dipole) interaction, a dipole, in a polar molecule, will orient itself to the nearby ion so that the end whose partial charge is opposite to the ion charge will point towards the ion that causes the interaction between the dipole and the ion. A simple example is a cation interaction with an H_2O molecule (Fig. 3.1 (a)). The strength of the ion–dipole interaction depends on the magnitude of the dipole moment of the polar molecule and the charge density of the ion. The later quantity is the charge of the ion divided by its volume. Hence, the ion–dipole interaction energy decreases with increase in the size of the ion. In the ion–induced dipole interaction, which can also be called as ion–polarization interaction, the induced dipole (of the nonpolar neutral molecule) interacts with the nearby ion that induced it (for example, cation interaction with a CH_4 molecule, Fig. 3.1 (b)). In the ion–induced dipole interaction case, the larger the ion the more readily will the electron cloud be distorted by an external field. Thus, unlike in an ion–dipole

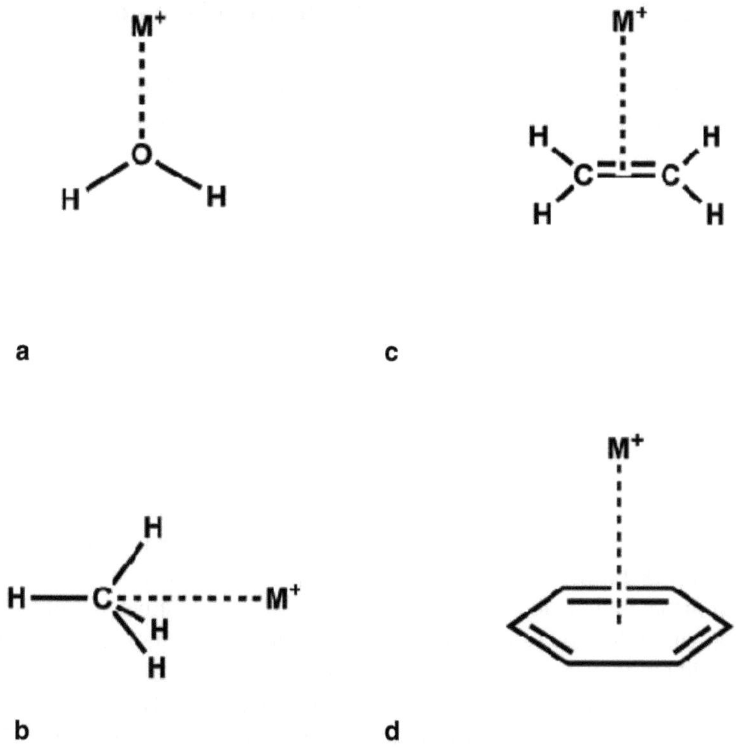

Fig. 3.1 Typical ion-molecule complexes. **a** ion-dipole, **b** ion-induced dipole, **c** & **d** ion-π interactions.

interaction, the ion–induced dipole interaction energy cannot decrease with the increase in size of the ion. It is obvious that a small to significant amount of ion–induced dipole interaction element also is involved in any ion–dipole interaction complex.

The last one, ion–π interaction, differs much from these two interactions. A π-system, like benzene, has π-electron density above and below the plane of the ring. A cation can interact favorably with this negatively charged electron density cloud when it comes close to the face of the π-system (see Figs. 3.1 (c) and (d)). In the most stable complexes, the cation is centered directly over the π-system and the cation–π interaction energy is roughly close to that of the cation–dipole interaction.

There are a few other important non-covalent interactions such as, (1) dispersive interaction (fluctuating electrons that cause molecules and atoms to behave like oscillating dipole. In molecules that are located near each other, the oscillating dipoles are coupled and the movement of the electrons in the adjacent molecules is correlated. These coupled fluctuating dipoles experience favorable electrostatic interaction known as dispersive interaction.), and (2) hydrogen bonding interactions (Interaction involving hydrogen atoms—an acceptor with a basic lone pair can interact with an acidic proton (hydrogen) bond to an electronegative atom). Since they are out of scope of this present topic, these are not detailed here.

3.2 Ion–Molecule Association/Dissociation Dynamics

Gas-phase ion–molecule reaction is an interesting class of association reactions without a potential energy barrier. The ion–molecule complex formation can be visualized to form through a sequence of reaction steps which are as follows:

$$M^{+} + L \underset{k_r}{\overset{k_f}{\rightleftharpoons}} (ML)^{+*}, \tag{1}$$

$$(ML)^{+*} + B \underset{k_a}{\overset{k_s}{\rightleftharpoons}} (ML)^{+} + B, \tag{2}$$

and the overall reaction is

$$M^{+} + L \underset{k_{ro}}{\overset{k_{fo}}{\rightleftharpoons}} (ML)^{+}_{s} \tag{3}$$

where M^+ is the metal ion and L refers to a neutral molecule. The ion–molecule complex formation, reaction 3, must be considered to proceed in two independent steps, complex formation, reaction 1, and stabilization, reaction 2. The bonding between ion and the ligand is generally found to be weak relative to normal chemical bond strengths; $\Delta E(\mathrm{L\text{-}M^+})$ is typically around 50 kcal/mol, often much less. It seems likely that the $\mathrm{L{-}M^+}$ complexation derives primarily from electrostatic forces such as ion–dipole or ion–induced dipole attraction. The ion–molecule association process is considered to proceed via an intermediate complex $(\mathrm{ML})^{+*}$. Collision with a third body (buffer gas), B, removes the excess energy from the ion–molecule complex and results in the formation of a stable ion–molecule complex.

Let us briefly see some of the important theories that explain reaction dynamics. These theories were developed to answer experimental questions.

3.2.1 Transition State Theory (TST) and Variational TST

TST is a semiclassical theory [1, 2], where the dynamics along the reaction coordinate is treated classically, while the perpendicular directions take into account the quantization of, for example, the vibrational energy. It also assumes an equilibrium energy distribution among all possible quantum states at all the points along the reaction coordinate. The probability of finding a molecule in a given quantum state is proportional to $e^{-\Delta E/kT}$, which is a Boltzmann distribution. Assuming that the molecules in the transition state (TS) are in equilibrium with the reactant, the rate constant (for the reaction (3), we consider this reaction is our main ion–molecule association reaction) can be expressed as in Eq. (4).

$$k_{\mathrm{fo}} = \frac{kT}{h} e^{-\Delta G^{*}/RT}, \tag{4}$$

where

$\Delta G^{*} = \Delta G_{TS} - \Delta G_{reactant}$.

Actually, the TST expression only holds if all molecules that pass from the reactant over the TS go on to the product, the ideal situation to the ion–molecule barrierless association reaction. The TST assumption is that no recrossing occurs for a given temperature, i.e., all molecules passing through the dividing surface will go on to form the product.

From the expression in Eq. (4), it is clear that if the free energy of the reactants and TS can be calculated, the reaction rate can also be calculated. Similarly, the equilibrium constant for the reaction can be calculated from the free energy difference between the reactants and the product.

$$K_{eq} = e^{-\Delta G_0/RT}. \tag{5}$$

Variational transition state theory (VTST) is useful when no TS can be explicitly identified (for Morse-like potential, e.g., direct bond dissociation), where there is no TS. It is based on the idea that there is a "bottleneck" in the phase space during the dissociation. This can be explained by the fact that during the dissociation process the molecule needs to reach at a certain point a very specific conformation, without which it cannot go further to disassociate. The Arrhenius equation can be written in terms of exponential of Gibbs free energy and exponential of entropy, which characterize the number of distinct states reachable with that amount of energy.

This variational version of TST can be used to calculate temperature-dependent rate constants for the barrierless reactions [3–6]. In VTST, the TS state is located at the free energy maximum along the reaction path, and the association rate constant is given by

$$k(T) = \frac{kT}{h}\exp\left(-\Delta G^{\dagger}/RT\right) \tag{6}$$

where $\Delta G^{\dagger}$ is the free energy difference between the variational transition state and the reactants. The temperature-dependent TST rate constant can also be found with the use of microcanonical VTST. The microcanonical variational transition state is dependent on total energy E and angular momentum J and is located at the minimum in the sum of the states $N^{\ddagger}(E,J)$ along the reaction path. The association rate constant is then

$$k(T) = \frac{1}{hQ_r}\int_0^{\infty}\sum_{J=0}^{\infty} N^{\ddagger}(E,J)\,e^{-E/k_BT}\,dE, \tag{7}$$

where Q_r is the reactant partition function and all the other parameters are described earlier.

3.2.2 Rice–Ramsperger–Kassel–Marcus Theory (RRKMT) and Quasi-equilibrium Theory (QET)

The canonical transition state theory assumes fast energy exchange with the surroundings, i.e., the reacting molecule is in thermal equilibrium with the environment. For unimolecular reactions in the gas phase, this assumption may not hold, especially not if the pressure is low. To avoid the shortcomings of both TST and RRM methods, Rice, Marcus, and others merge TST with RRK to bring a new theory, known as Rice–Ramsperger–Kassel–Marcus (RRKM theory) [7−10], considering individual vibrational frequencies of the activated complex. The fundamental assumption in the RRKM theory is that no recrossing occurs for a given total energy of the molecule. When combining with the RRK approximation, this means that the reaction rate coefficient can be calculated from the rate of reaction paths that cross the barrier by the total number of states in ensemble at the required energy (phase space). The other assumptions in RRKM are (1) the time required for dissociation is long, relative to the time required for ion formation/excitation/activation; (2) the time required for dissociation is long, relative to the time required for redistribution of energy over all internal modes of ion and the distribution of reactants is defined by a microcanonical ensemble; (3) the reaction is adiabatic; it takes place on a single potential energy surface, upon which the motion is classical, linking reactants to products and is separable from the other molecular motions.

The new reaction scheme is therefore

$$A + M \underset{k_{-1}}{\overset{k_1}{\rightleftharpoons}} A^* + M. \tag{8}$$

$$\mathrm{A}^* \xrightarrow{k_2} \mathrm{A}^\# \xrightarrow{k^\#} \mathrm{P}. \tag{9}$$

In this reaction, a molecule A acquires energy by collision with a molecule M to form an energized molecule A*, with the energy being distributed between the transition, rotation, and vibrational degrees of freedom. The vibrational energy can be transferred between different modes owing to vibrational anharmonicity, and if it is higher than the activation energy $E^\#$, it may at some point accumulate in a specific mode to reach an activated state $\mathrm{A}^\#$ (TS) leading to the reaction product P. Assuming that the decay rate constant $k^\#$ for the TS $\mathrm{A}^\#$ is much faster than k_2, the rate for production of P can be written in terms of the k_1, k_{-1}, and k_2 rate constants by making a steady state approximation for A^*, as shown in Eq. (10).

$$\frac{d[P]}{dt} = k_2\left[A^*\right] = \frac{k_1 k_2 [M][A]}{k_{-1}[M] + k_2} = k_{eff}\,[a], \tag{10}$$

k_{eff}, the effective rate constant, is thus a function of the concentration of M, i.e, the pressure of the gas. The energy transferred to A* by M will be a variable, and the rate constant for the activation and reaction (but not the deactivation) will depend on the energy, i.e., $k_1(E)$ and $k_2(E)$. The effective rate constant in a small energy interval around E is obtained by rearranging Eq. (10).

$$k_{eff}(E+dE)=\frac{\left(dk_1(E)/k_{-1}\right)k_2(E)}{1+k_2(E)/k_{-1}[M]}. \tag{11}$$

The ratio k_1/k_2 is the equilibrium constant for the reaction (Eq. 8), and $dk_1(E)/k_{-1}$ is the probability of A^* being in a state with energy E, $P(E)$. The k_{-1}[M] factor is the collision frequency for deactivation that is usually denoted by ω. The unimolecular reaction rate constant can be obtained by integrating the effective rate constant over all energies higher than the activation energy.

$$k_{uni}=\int_{E^{\#}}^{\infty}\frac{k_2(E)}{1+k_2(E)/\omega}P(E)\,dE, \tag{12}$$

$P(E)$, the probability factor, is given by a Boltzmann distribution for the reactant, while $k_2(E)$ is determined by the number of vibrational quantum states for the activated state $A^{\#}$. For further details readers can refer to the references 9 and 10, but the essence of Eq. (12) is that the rate constant can be evaluated from the geometries and vibrational frequencies of the reactant and activated complex. In a fast energy exchange limit (i.e., $\omega\rightarrow\infty$), the RRKM expression becomes equivalent to the TST expression in Eq. 4.

The QET was developed simultaneously with the RRKM theory by Rosenstock [11] and used in the mass spectrometry literature to explain breakdown curves whereas the RRKM theory was developed for neutral reaction kinetics, but they are basically identical.

3.2.3 *Phase Space Theory (PST)*

The RRKM theory is equivalent to TST, and it allows an arbitrarily detailed description of the TS. PST [12–19], however, assumes that the collisional rate coefficient k_1 and the rate coefficient k_{-1} for dissociation of the complex are governed by an orbiting or other type of loose TS, but unlike the usual formulation of RRKM theory, it rigorously conserves angular momentum. The statistical dissociation rate constant can be calculated from the point of view of the reverse reaction. Again, this procedure is limited to reaction without energy barrier.

It is considered that the disassociation reaction is governed by the phase space available under strict energy and angular momentum conservation. Considering equilibrium between reactants and products, one gets:

$$\frac{k_{uni}}{k_{bi}}=\frac{[A][B]}{[AB]}\Leftrightarrow k_{uni}=k_{bi}\frac{\rho(E-E_0,j_A)*\rho(E-E_0,j_B)}{\rho_{AB}(E,J)}, \tag{13}$$

for which there is no need for information on the TS.

There were many studies using PST to explain the dynamics of the ion–molecule reaction. For example, Base et al. [20] studied PST models of the reaction $CH_3^+ + HCN \rightarrow (CH_3^+HCN) + h\nu$ and analyzed the radiative stabilization of the complex. Bass coworkers [21] also applied statistical PST to clustering reactions of $CH_3OH_2^+$, $(CH_3)_3OH^+$, and $(CH_3OH)_2H^+$ with CH_3OH. They found good agreement between their calculated and experimental rate coefficients.

3.2.4 Quasi-classical Trajectory Theory (QCT)

The quasi-classical trajectory theory provides a procedure for calculating trajectories of the reaction in which the quantization of the reactants is taken into account, but the course of the reaction is treated classically. In the standard QCT method [22–25], the reagent molecules are given initial coordinates and momenta that correspond to the semiclassical eigenstates being studied, and then the classical equations of motion are numerically integrated to simulate collisions. For reactive trajectories, states of the product molecules are assigned by determining the good actions that characterize vibrational–rotational motion of the product molecules. These actions are the classical equivalent of the quantum numbers, and it is customary to round them off to the nearest integer multiple of $\hbar$ to define quantum states. One of the issues is that the determination of vibrational actions for polyatomic molecules involves a nonstandard calculation that must be done with some care, if accurate results are to be obtained. These actions (both vibrational and rotational) are well-defined only if the molecular motion is quasi-periodic, which means that trajectory motion follows the surface of a torus in phase space. In this case the action J_k is calculated from the following relation:

$$J_k = \frac{1}{2\pi}\int_{C_k} P \cdot dX, \tag{14}$$

where X and P are normal coordinates and momenta, and C_k specifies a closed contour in phase space that encircles the torus, and the index k labels the $3n-6$ independent contours that exist for a molecule with n atoms.

The evaluation of the integral in Eq. (14) may be done in a number of ways. Harmonic oscillator expressions for vibrational actions can be used for weakly anharmonic molecules. A more accurate rate method, which is most useful, is based on a Fourier series representation of the coordinates and momenta in Eq. (14). In this method, the normal coordinates and momenta are calculated as a function of time by integrating the molecule equations of motion by standard numerical integration methods. The X's and P's are represented in a Fourier series, and then the actions of Eq. (14) are evaluated from the Fourier coefficients:

$$X_k(t) = \frac{1}{T}\sum_n D_{kn} \exp\left[\frac{2\pi \mathrm{int}}{T}\right], \tag{15}$$

$$P_k(t) = \frac{1}{T}\sum_n F_{kn} \exp\left[\frac{2\pi \mathrm{int}}{T}\right], \tag{16}$$

where t is time, T is the time interval over which the coordinates and momenta have been determined by numerical integration, and D_{kn} and F_{kn} are coefficients that are obtained by fast Fourier transfer methods. By examining the Fourier coefficients, it is possible then to determine the fundamental frequencies Ω_k for each mode, and the actions are obtained from

$$J_k = \left(4\pi/\Omega_k T^3\right)\sum_n n\left[Re\left(D_{kn}\right)Im\left(F_{kn}\right) - Im\left(D_{kn}\right)Re\left(F_{kn}\right)\right]. \tag{17}$$

It is necessary to determine vibrational semiclassical eigenvalues to determine initial coordinates for a quasi-classical calculation. This requires finding molecular coordinates and momenta such that the resulting good actions are integer multiples of $\hbar$. Usually, the harmonic actions can be used to define initial conditions that are approximately correct, and then the ratio of desired to calculated actions is used to scale the coordinates and momenta until the calculated actions are equal to the desired actions within some tolerance. Once the semiclassical eigenvalue has been determined, it is necessary to calculate molecular coordinates and momenta that can be used as initial conditions for collision simulations. These can be determined from the Fourier representation, or one can save coordinates and momenta from the trajectory that is used to determine the semiclassical good actions.

Ion–molecule association is seemingly well suited for the application of the QCT method. Since there is no potential barrier and the centrifugal potential is broad, quantum mechanical tunneling, which is an issue with the QCT method, is typically unimportant. Energy transfer from relative translational to vibrational and/or rotational motions of the ion–molecule complex should be reasonably classical because of the large density of states involved. Additionally, since the variational transition state has an early location along the reaction path, quantization of reactant vibrational motions should result in a reasonably correct treatment of these motions at the TS.

Hase and coworkers [26–32] made a number of classical trajectory studies on the association reactions: $M^{+}+H_2O$ and $M^{+}+D_2O$ with M=Li, Na, K, $Li^{+}(H_2O)+H_2O$, $Li^{+}+(CH_3)_2O$, and $Cl^{-}+CH_3Cl$. In their studies, the occurrence of multiple inner turning points in the time dependence of the association of radical coordinate was taken as the criterion for complex formation. Comparison of association probabilities from various studies leads to the conclusion that softer and/or floppier ions and molecules that have low-frequency vibrations typically recombine most efficiently. Thus, it has been found that $Li^{+}+(CH_3)_2O$ association is more likely to occur than $Li^{+}+H_2O$ association, and similarly H_2O association with $Li(H_2O)^{+}$ is more likely to occur than with the bare Li^{+} ion. They found a non-monotonic dependence of association probability on the assumed H_2O bending frequency and also a dependence on the impact parameter, the rotational temperature, and the orientation of the H_2O dipole during the collision.

Although the RRK theory is sometimes used, the more advanced RRKM theory and PST give more insight because of their closer connection to the true molecular dynamics. QCT calculations have provided important information about ion–molecule reactions. In addition to giving accurate rate constants, the trajectory calculations provide valuable microscopic details about the recombination dynamics [28].

3.3 Overview of Computational Chemistry

The term theoretical chemistry may be defined as the mathematical description of chemistry, and computational chemistry uses well-developed mathematical methods of chemistry to generate numerical data using computers. Computational chemistry is the application of chemical, mathematical, and computational skills to the solution of interesting chemical problems. It uses computers to generate information such as structure and properties of molecules. Computational theoretical chemistry is mainly concerned with the numerical computation of molecular electronic structures and molecular interactions. Quantum mechanics, classical mechanics, statistical physics, and thermodynamics are the foundation for most of the computational theoretical chemistry. There are a lot of computational chemistry programs (software) available in the market; some are free and can be downloaded from the Web. The following are some of the popular computational chemistry programs:

- Gaussian [33]
- GAMESS [34]
- Q-Chem [35]
- NWChem [36]
- Molpro [36]
- Molcas [38, 39]
- ACES II [40]
- Turbomole [41]
- HyperChem [42]

Theoretical studies have provided an excellent complement to the experimental, gas-phase ion–molecule interactions. In fact, modern theoretical approaches with the vast improvement of computational hardware potentials have become a way to prove experimental results. In addition to that, though hundreds of ion–molecule complexes have been studied by using various methodologies through experiments, most of these studies have mainly focused to study the interaction energies and stabilities of these ion–molecule complexes, but the geometries of these complexes have been obtained very rarely. This might be because of the lack/limitations of sophisticated techniques.

Computational chemistry methodology has become a powerful tool for predicting a number of molecular properties, including structure and energetics of ion–molecule complexes. High-level theoretical studies have become handy in studying the ion–molecule complex structures as well as accurate ion–molecule interaction energies, reliably.

The following are the available computational chemistry methodologies to study the chemistry of molecules and molecular complexes:

- Ab initio [43] (Latin word for "from scratch") uses various quantum mechanical concepts and methods to solve the Schrodinger equation to find the wave function, structure, and other physical and chemical properties of molecules. It does not use any experimental values other than fundamental constants.

- Density functional theory (DFT) [44, 45] is also based upon the foundations of quantum mechanics to solve the energy equation. Within this theory, the properties of molecular systems can be determined by using functionals, i.e., functions of another function, which in this case is the spatially dependent electron density.
- Semiempirical methods—these methods use approximations from empirical (certain number of experimental results) data throughout the calculation to provide the input for the mathematical models. For example, bond lengths and bond angles of specific types will have fixed values independent of the molecule. This dramatically speeds up the calculations, but in general, the results are not accurate. Usually, semiempirical methods are used for very large systems. Some of the examples of semiempirical methods are MINDO3 [46], AM1 [47], and PM3 [48].
- Molecular mechanics, which uses classical mechanics and predetermined force fields to explain the behavior of atoms in molecules. They consider atoms as spheres and bonds as springs. They use an algebraic classical mechanics equation for the energy calculation. The constants in the equation are obtained from experimental data or other calculations and are stored in a data library. The combination of constants and equations is called a force field. These calculations are simple and can be done with many software programs such as AMBER [49] and CHARMM [50]. There are lot of force fields available and some of the popular force fields are UFF [51], MM4 [52], and MMFF94 [53].

Among these methods, ab initio and DFT methodologies are important and they play central roles in understanding the molecular structure and other physical and chemical properties of molecular systems. The starting point of these quantum mechanics-based computational chemistry is the well-known Schrodinger equations [54]. These theories consider Born–Oppenheimer approximation [55], through which the position of the nuclei is considered as fixed and electrons move in a field of fixed nuclei, and variational principle to find the solution for the Schrodinger equation. When more than one electron is present in a system, solving the Schrodinger equation is impossible because of the interelectronic term. To produce a solvable Schrodinger equation, some kind of approximation and assumptions are made.

According to the ab initio molecular orbital theory methodology, atomic orbitals (set of functions, also called basis sets) combine in a way to form molecular orbitals that surround the molecule. The molecular orbital theory considers the molecular wave function as an antisymmetrized product of orthonormal spatial molecular orbitals. Then they are constructed as a Slater determinant [56]. Essentially, the calculations initially use a basis set, atomic wave functions [57, 58], to construct the molecular orbitals. The first and basic ab initio molecular orbital theory approach to solve the Schrodinger equation is the Hartree–Fock (HF) method [59, 60]. Almost all the ab initio methodologies have the same basic numerical approach but they differ in mathematical approximations. As it is clear that finding the exact solution for the Schrodinger equation, for a molecular system, is not possible, various approaches and approximations are used to find the reliable to close-to-accurate solutions [61–68].

Wave function, the backbone of the ab initio-based molecular orbitals theory, depends on three spatial and one spin coordinates of each electron, assuming that electrons move in a fixed nuclear potential. However, the DFT is based purely on electron density, ρ [69–72], and the ground-state electronic energy is determined completely by electron density. In other words, the ground-state energy from the Schrodinger equation is a unique functional of the electron density. The significance of this is that when a wave function for an N-electron system contains $3N$ spatial and a spin coordinate, the electron density, which is a square of the wave function, integrated over $N-1$ electron coordinates, only depends on three coordinates, independent of the number of electrons. The advantage is that while the complexity of a wave function increases with the increase in number of electrons, the electron density has the same number of variables. The disadvantage is that although it has been proved that each different density yields a different ground-state energy, the exact functional connecting these two quantities is not known. Because of this, considerable research effort has gone into finding the more accurate density functionals. Since detailing these methods are beyond the scope of this book, readers are advised to check basic and advanced ab initio molecular orbital theory books and reviews for theoretical details [43, 73–75].

3.4 Computational Methodologies

All these theoretical procedures try to find the minimum energy structure of a molecular system, which is normally called "optimization." The procedure calculates the wave function (or electron density) and the energy at a starting geometry, and then proceeds to search the new geometry of a lower energy. This is repeated until the lowest energy geometry is found. The procedure calculates the force on each atom by evaluating the gradient (first derivative of the energy) of the energy with respect to atomic positions. Sophisticated programs are then used at each step to select a new geometry, aiming for rapid convergence to the geometry of the lowest energy. In the final, at minimum energy geometry, the force on each atom is zero. This procedure is not guaranteed to find the global minimum, i.e., the geometry with the lowest energy because the optimization procedure stops when it finds a stationary point, i.e., a point where forces on the atoms are zero. As one can imagine, existence of many local minima on a potential energy surface is possible. So, one has to consider and check all possible structures.

Various ab initio and density functional theories are used to find the structure, interaction energies, and other properties of the molecular complexes. In particular, though ab initio procedures provide highly reliable results, especially accurate interaction energies, they are time-consuming and also limited to small- to medium-sized molecular complexes. However, nowadays, the DFT plays a prominent role in studying the chemistry of molecules mainly because the DFT-based calculations are not time-consuming, unlike ab initio calculations, and they can be used to study very large molecules with even a thousand atoms. Also, they provide reliable results. Many studies have shown that various ion affinities obtained using hybrid density functionals are highly comparable in quality with the experimental values [76–82]. This is an interesting fact considering that very large systems, up to thousands of atoms, can be studied by using the DFT. Because of the availability of these accurate theories, studies of ion–molecule interactions involving very large biomolecules as well as materials are becoming possible.

Though there are many density functionals available to study the ion–molecule complexes using the DFT, as mentioned earlier, the Becke's three-parameter hybrid functional, B3LYP, which includes a mixture of HF exchange and DFT exchange correlation has been much popular and using which reliable to accurate results have been obtained on many systems. The Becke's three-parameter functional [83, 84] has the form $AE_x^{Slater} + (1-A)E_x^{HF} + BE_x^{Becke88} + E_C^{VWN} + C\Delta E_C^{non-local}$, where the nonlocal correlation is provided by the Lee–Yang–Parr expression [84]. The constants A, B, and C are those determined by Becke via fitting the results in the G1 molecular set (a large number of molecular set). Readers can refer to the parent references in computational chemistry books for details on other density functionals.

3.4.1 Structure, Bonding Energies, Rate Constants, and Trajectories

Normally, molecules, as well as ion–molecule complexes, are optimized first at a medium level of theory, say MP2 in ab initio and B3LYP functional in DFT, using a medium-size basis set, say 6–31G(*d*), followed by the frequency calculation to make sure that the optimized structures are minimum energy on the potential energy surface. These frequency calculations also yielded the zero-point energies (ZPE), which might be scaled by considering anharmonicity and the incompleteness of the basis set, according to the level of theory is used for the frequency calculations; thermal corrections (at 298.15 K) were needed for the calculation of enthalpies. Second, an appropriately higher level of theory, either B3LYP functional or any other suitable functional for the system of interest or a high-level ab initio theory such as coupled-cluster methods, with the possible large basis set, for example, 6-311+G(3*df*,2*p*) or cc-pVTZ, will be used to optimize the molecules as well as ion–molecule complexes. This should be done to obtain the reliable complex structures as well as energies. This will also help to derive the accurate binding energies. In rare cases, such as in large ion–molecule complexes, single-point energy calculations are performed instead of the full optimization (second step) at the medium-level optimized structures obtained in the first step.

By analogy with proton affinity (PA) and gas-phase basicity (GB) used for gas-phase BrØnsted basicity, alkali metal cation affinity (CA) and alkali metal cation basicity (CB) are, respectively, defined as the standard enthalpy ΔH and the standard Gibbs energy ΔG of dissociation of a bond formed between metal cation M^+ and a Lewis base (or ligand) L:

$$\left(\mathrm{L}-\mathrm{M}\right)^{+} \rightarrow \mathrm{L}+\mathrm{M}^{+}. \tag{18}$$

The binding energies (ΔE) are obtained from the difference between the total energy of the ion–molecule complex [$E(\mathrm{L{-}M})^+$] and the sum of the total energies of the ion [$E(\mathrm{M}^+)$] and molecule [$E(\mathrm{L})$], using the optimized or single-point energies:

$$\Delta E = \left[E\left(\mathrm{L}-\mathrm{M}\right)^{+}\right] - \left\{\left[E\left(\mathrm{L}\right)\right] + \left[E\left(\mathrm{M}^{+}\right)\right]\right\}. \tag{19}$$

Binding enthalpies (alkali metal cation affinities or simply, ion affinities of the molecules) are then calculated using the following relation:

$$\Delta H = \Delta E + \Delta E_{ZPE} + \Delta E_{thermal} + \Delta E\left(PV\right) \tag{20}$$

where $\Delta(PV) = nRT = -0.593$ kcal/mol at 298.15 K.

ZPE and thermal and entropic corrections at the appropriate experimental temperatures can be calculated using the frequencies in conjunction with the standard textbook formulas for the statistical thermodynamics of an ideal gas under the harmonic oscillator/rigid rotor approximation. Equations (4) and (5) relates the rate constant and equilibrium constant with the Gibbs free energy, which can be described in terms of the enthalpy (H) and the entropy (S) in the following equation:

$$\Delta G = \Delta H - T\Delta S. \tag{21}$$

These quantities, H and S, hence G, can be calculated using any computational chemistry methodology. From the calculated relevant G values, various rate constants can be derived.

Computational chemistry calculations can be used to determine the geometries, vibrational frequencies and infrared intensities for the normal modes, rotational constants of reactants, dissociating complexes, and the products. These calculations also provide the other molecular parameters necessary to evaluate the internal energy of the reactant complex and as input for the RRKM calculations used to estimate the rate constant and the lifetime for dissociation. So, the parameters necessary for kinetic modeling by using the above mentioned theories such as RRKM or VTST can be obtained by using computational chemistry procedures. A computer program such as POLYRATE [85] can be used to perform all kinetic calculations, especially VTST calculations. POLYRATE is designed to be used in conjunction with interfaces to electronic structure programs that makes it convenient to perform structure and dynamics studies together.

The PST calculations of the association rates for the ion–molecule reactions can be carried out using programs such as VariFlex [86]. Though the main aim of this program is the calculation of rates for barrierless reactions, it also allows convenient estimates of the rates using PST. The program employs Monte Carlo phase space integration to evaluate the statistical functions.

The QCT calculations need to be done in three stages: (1) structure and energetics using ab initio or DFT calculations, (2) potential energy surface (PES), from the structure and energy data obtained in the first stage, construction using some interpolation, such as modified Shepard interpolation, [87–89] method, and (3) finally trajectory calculations. For the final step, initial conditions are for the trajectories to simulate random collisions between an ion and a molecule and the criterion for the recombination to be set. Normally, the QCT calculations are performed using a computer program, such as VENUS96 [90], which is modified to incorporate PES. For further methodological details on QCT, readers can look up the references [91–93].

3.4.2 Nature of Bonding

There are a number of ways to analyze interactions in ion–molecule complexes. Among them, population analysis plays a vital role in understanding the bonding or nonbonding interactions. Population analysis is a mathematical way of partitioning a wave function or electron density into charges on the nuclei, bond orders, and other related information. These are the most used results that are not experimentally observable. Atomic charges cannot be observed experimentally because they do not correspond to any unique physical property. In reality, atoms have a positive nucleus and surrounded by negative electrons, and not partial charges, on each atom. However, condensing electron density and nuclear charges down to partial charges on the nucleus results in understanding the electron density distribution. These are not formal integer charges but rather fractions of an electron corresponding to the percentage of time an electron is near each nucleus. Though these values are artificial, they are very effective for predicting sites susceptible to nucleophilic or electrophilic attack and other aspects of molecular interaction. These partial charges correspond well to ionic or covalent bonds, polarity, and so on.

Molecular orbital pictures are informative; however, to quantify a specific interaction, deriving numerical values will be very useful. Calculating the atomic charges, quantifying the charge distribution, in the molecules and in the complexes will help to analyze the bonding nature of the ion–molecule interaction. There are several population schemes to do this job. Mulliken population analysis (MPA) [94, 95], which is based on the one-particle density matrix defined over standard nonorthogonal atomic basis sets, was the first scheme for predicting atomic charges and became very popular due to its simplicity. In this approach, atomic charge is defined as the difference between the total number of electrons in the ground state of the neutral atom and the gross atomic population on the atom within a molecule. The latter depends on the equally shared overlap population with the adjacent atoms that is best described as when the boundary between atoms is placed in the middle of the bond. Though MPA is the most common population analysis method, its results are largely basis set dependent and also it yields unnatural values in some cases. The natural population analysis (NPA) scheme [96, 97] was proposed to overcome the problems existing in the MPA scheme. In NPA, the nonorthogonal atomic orbitals are transformed into an orthonormal set by an occupancy-weighted symmetric orthogonalization procedure, where the orbitals having the highest occupancy are strongly preserved in form, whereas orbitals of negligible occupancy can distort freely to achieve orthogonality. This property renders the NPA inherent to the wave function rather than the quality of the basis set, and so NPA is insensitive to basis set. Because of this, the NPA scheme has been recognized as a reliable tool for calculating atomic charges and has been used for studying a number of chemical systems [98–100]. The other approach is the one-particle electron density approach which is based on fitting atomic point charges to reproduce the electrostatic potential (ESP) of a molecule [101–104]. Different algorithms are used for calculating ESP-driven charges, such as Chelp [105], ChelpG [106], Connolly [107–110], and Geodesic [111]. It should be noticed that in the GAUSSIAN software the Connolly algorithm is known as the MK scheme.

A very different approach, from those described above, is the Bader's atoms-in-molecules (AIM) method [112–116]. In the AIM scheme, electron densities are integrated over the volumes (basins) assigned to the individual atoms. The volumes are assigned in terms of zero flux surfaces defined by the gradients of the electronic density function. It was shown that within these basins the quantum (atomic) subsystems obeyed a local virial relation and therefore, atomic charges are well-defined within the quantum mechanical formulation. Bader's AIM theory has an additional advantage: It can be used to thoroughly analyze the nature of bonding by studying the electron density (ρ) and its Laplacian $\left(\nabla^2\rho\right)$. As has been shown by the Bader group [112–116], $\nabla^2\rho$ identifies regions of space wherein the electronic charge of a system is locally concentrated, $\nabla^2\rho < 0$ or depleted, $\nabla^2\rho > 0$. According to the AIM theory, negative values of $\nabla^2\rho$ at the bond critical point (BCP, where $\nabla^2\rho = 0$) are associated with shared interactions, typically covalent bonds, whereas positive $\nabla^2\rho$ values are associated with closed-shell interactions, corresponding to ionic bonds, electrostatic, hydrogen bonds, and van der Walls interactions. The concentration of the electronic charge at the BCP is relatively low for closed-shell systems and large for shared interactions. Therefore, by analyzing ρ and $\nabla^2\rho$ parameters at BCPs, one can identify the nature of ion–molecule bonding as well as bond activation effects in the molecules upon ion attachment.

3.5 Ion–Molecule Complexes

The aim of this chapter is not to make it a review-like material by collecting most of the ion–molecule interaction studies available in the literature, but to give an essence of the importance of studying the ion interactions with the molecules with a few examples.

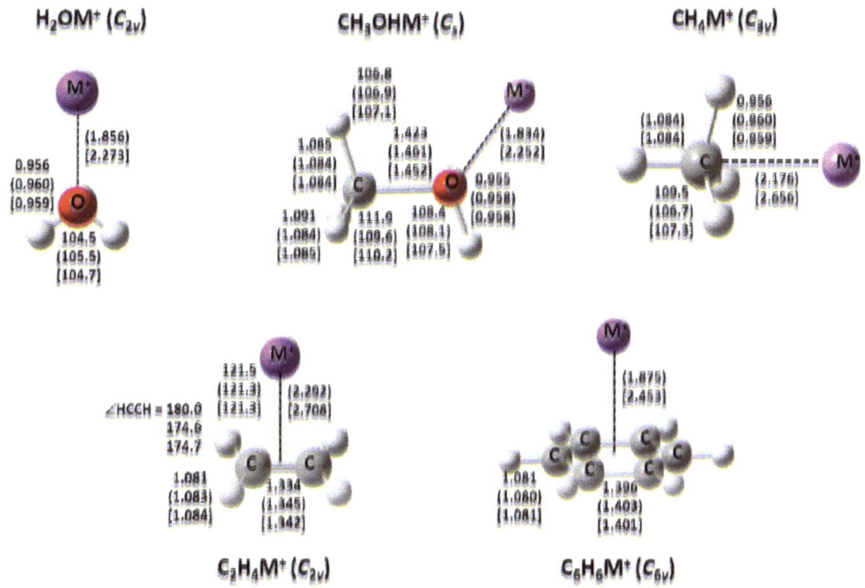

Fig. 3.2 CCSD/6-311 + G(2*d*,2*p*) optimized ion-molecule complex structures. The point groups given in the figure are for respective ion-molecule complexes. Structural parameters, without any brackets, represent the ligand monomers, whereas values in parenthesis represent complexes with Li^+ ion and the values in square brackets are for complexes with Na^+ ion

3.5.1 Interactions with Organic Molecules

For this work, a few simple organic molecules are considered for the complexation with two different alkali metal ions, Li^+ and Na^+ ions. The selected organics are H_2O, CH_4, C_2H_4, CH_3OH, and C_6H_6 molecules. Calculations have been done using two different density functionals, B3LYP and M06-2X, with 6-311 + G(3*df*,2*p*) basis set in DFT, and coupled-cluster methods with 6-311 + G(2*d*,2*p*) basis set in ab initio. In all the calculations, structures are optimized at their respective level of theory except at the CCSD(T) level, where CCSD optimized structures are used. The optimized geometries of the ion–molecule complex structures with the structural parameters, along with their respective ligands, are depicted in Fig. 3.2. Table 3.1 shows the Li^+ and Na^+ ion affinities for the selected organics calculated using various level of theories along with their corresponding experimental values available in the literature. As it is clear from the table, all the calculated values nicely agree with the experimental values. These calculations, like many other available studies, support the fact that not only ab initio but DFT too can be used to calculate the alkali metal ion affinities of various molecules reliably.

The structural parameters noted in Fig. 3.2 show minimum to moderate changes in the ligand geometries upon ion attachment. For example, C–O bond length in methanol is elongated by around 0.4 Å due to the complexation with the Li^+ ion and the same value is around 0.3 Å in the Na^+ addition. The C = C bond in ethylene

Table 3.1 Calculated ion affinities (ΔH in kcal/mol at 298 K) of various organics. Values in parenthesis are experimental uncertainties

Organics	B3LYP/6-311 + G(3df,2p)		M06-2X/6-311 + G(3df,2p)		CCSD/6-311 + G(2d,2p)		CCSD(T)/6-311 + G(2d,2p)		Expt.[117–123]	
	Li^+	Na^+	Li^+	Na^+	Li^+	Na^+	Li^+	Na^+	Li^+	Na^+
H_2O	−35.06	−24.48	−35.25	−24.99	−33.42	−22.91	−33.20	−22.76	−32.7 (3.3)	−23.4 (1.8)
CH_4	−13.76	−7.95	−14.13	−8.56	−11.78	−6.28	−11.87	−6.34	−11.0	
C_2H_4	−22.11	−15.31	−22.79	−15.98	−20.28	−12.92	−20.19	−12.87		−10.7 (1.1)
CH_3OH	−38.39	−26.40	−38.27	−26.72	−36.71	−24.69	−36.47	−24.52	−37.0 (2.1)	−22.3 (1.4)
C_6H_6	−38.45	−24.69	−41.22	−27.58	−36.93	−22.75	−36.57	−22.59	−39.3 (3.2)	−22.5 (1.5)

also has a significant elongation with interaction with the metal ions. CH_4 changes its symmetry from T_d to C_{3v} while forming an ion–molecule complex with an alkali metal ion. One other notable change is the movement of hydrogen atoms from the carbon plane in C_2H_4. Overall, the ion–molecule complex structures obtained using the DFT calculations agree well with the high-level ab initio calculations.

Simple population analysis, along with the structural parameters, can explain the bonding nature of the ion–ligand bond in these complex structures. While the alkali metal ion bonding in H_2OM^+ and CH_3OHM^+ (M^+ refers to metal ion) complexes are typical ion–dipole interactions (metal ion interacts with the dipole of these molecules by orienting itself in the same direction as the dipole vector of the ligand), the same bonding in the other molecules are different. The CH_4M^+ complex is a classic example. As a nonpolar molecule, CH_4 does not have permanent dipole moment; however, the calculated results show that CH_4Li^+ and CH_4Na^+ complexes possess dipole moment of 6.34 and 4.58 Debyes (at B3LYP/6-31G(*d*) level of theory), respectively. The analysis clearly reveals that the alkali metal ion bonding in CH_4M^+ is an ion–induced dipole interaction. The other two complexes, $C_2H_4M^+$ and $C_6H_6M^+$, are ideal examples for the complexes with cation–π interactions. MPA calculations reveal that charge-transfer acceptor–donor type interaction plays an important role in these complexes. Electronic charges, calculated at B3LYP/6-31G(*d*), of around 0.31 e and 0.23 e are transferred from ligand to metal ion in $C_2H_4Li^+$ and $C_2H_4Na^+$ complexes, respectively. The same values in $C_6H_6Li^+$ and $C_6H_6Na^+$ are 0.58 e and 0.35 e, respectively. These larger charge transfers in benzene-ion complexes reflect in their larger ion affinities (see Table 3.1). Though C_6H_6 has zero permanent electric dipole moment like C_2H_4, it has a larger quadruple moment. Apart from that, benzene has six π electrons, whereas ethylene has one double bond with two π electrons. Every π electron in both the molecules is involved in the interaction with the alkali metal ions. All these facts contribute to a stronger bonding with alkali metal ions and larger ion affinity of benzene. Bonding analyses conclude that the bonding mechanism in C_6H_6–M^+ complex, in a classical sense, is characterized by attractive forces due to ion–induced polarization and ion–quadruple interactions.

Knowing metal ion interactions with organics, especially their ion affinities, become an essential part of ion attachment spectroscopy. Fujii [124] has brilliantly utilized the metal ion attachment technique, ion attachment with the organic molecules/radicals, as a soft ionization method to overcome complications related to the fragmentation of the sample molecule in the normal electron ionization mass spectroscopy (EI-MS) due to the high energy of the electrons. Li^+ ion attachment mass spectrometry (IAMS) was developed two decades ago by Fujii as a novel soft ionization mass spectrometry technique [124–129]. Unlike the more commonly used EI-MS, the ion attachment process in IAMS is nondissociative and generates $[M+Li]^+$ ions that do not fragment. The fragment-free measurement of chemical samples allows the analysis of mixtures with spectra that are difficult to interpret using EI-MS. It was demonstrated that Li^+ MS produces only the molecular ions permitting the direct determination of unfamiliar and reactive species.

Since the ionization efficiency in $(M+Li)^+$ ion formation strongly relies on the Li^+ ion affinity of the molecule, to observe the Li^+ ion complex in IAMS the mol-

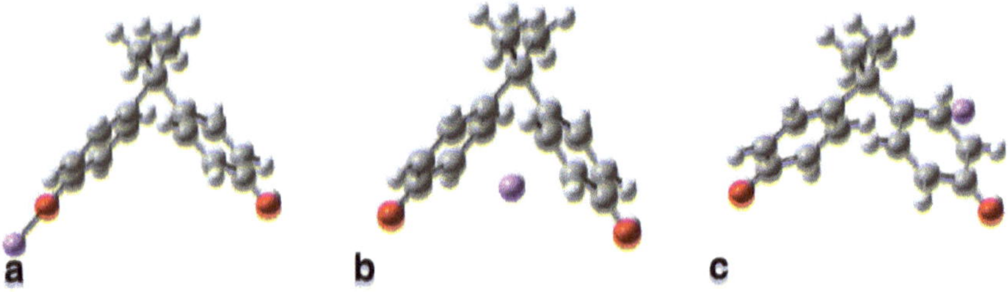

Fig. 3.3 Structures of the lithiated bisphenol A biradical. Carbon atoms are shown *brown*, oxygen *red*, lithium cation *violet*, and hydrogen *white*

ecule should have a large Li^+ ion affinity. So, to show the applicability of IAMS to detect the species of interest, one should know the Li^+ ion affinity of the species; hence, understanding the ion interaction energy becomes important. For this purpose, Li^+ ion affinities of various molecules were obtained using different theoretical methodologies to support the experimental studies. These studies helped in many ways in interpreting the experimental results. For example, an unknown species detected in the analysis of the products in a pyrolysis of polycarbonate in the IAMS was able to be assigned to a bisphenol A biradical [130]. The Li^+ ion affinities calculated using DFT at B3LYP/6-311+G(3*df*,2*p*) reveal that the bisphenol A diradical has large ion affinities ranging from 26.6 to 58.2 kcal/mol. The biradical–Li^+ complex structures are given in Fig. 3.3. Because of the biradical nature where unpaired electrons occupy the oxygen atoms, the complex structure with the Li^+ ion interacting with the oxygen atom (Fig. 3.3 (a)) has a larger interaction energy. The larger Li^+ ion affinities obtained for the BPA biradical derived, along with other theoretical analysis and experimental results, reveal that the Li^+ complexes could be detected in the Li^+ IAMS, and on the basis of these results the Li^+ adduct peak at m/z 233 detected in the pyrolysis if polycarbonate is assigned to the bisphenol A biradical. Similarly, with the help of theoretical calculations, many other rare organics detected in the experiments are also identified; for example, H_2O_2 species in the microwave (MW) discharge CH_4/O_2 plasma [131] and C_3N_4 in the MW discharge C_2H_2/N_2 plasma [132].

Also, in an attempt to understand the feasibility of Li^+ ion attachment mass spectrometry to estimate emissions from industry of the global-warming perfluorocarbons (PFCs), DFT studies using B3LYP functional incorporating various basis sets up to 6-311+G(3*df*) were performed to determine the Li^+ ion affinities of six PFCs [80]. These PFCs are CF_4, CHF_3, CH_2F_2, CH_3F, C_2F_6, and C_4F_8. The complex structures are given in Fig. 3.4.

Fluorine atom is more electronegative than a carbon atom and so negative charges reside in fluorine atoms in all of these PFCs. Because of that, alkali metal ion cation, here Li^+ ion, interacts with these PFCs through their fluorine atoms while becoming ion–molecule complexes. Some of these complex structures are particularly interesting: For example, CF_4 has a tetrahedral structure similar to CH_4; however, while

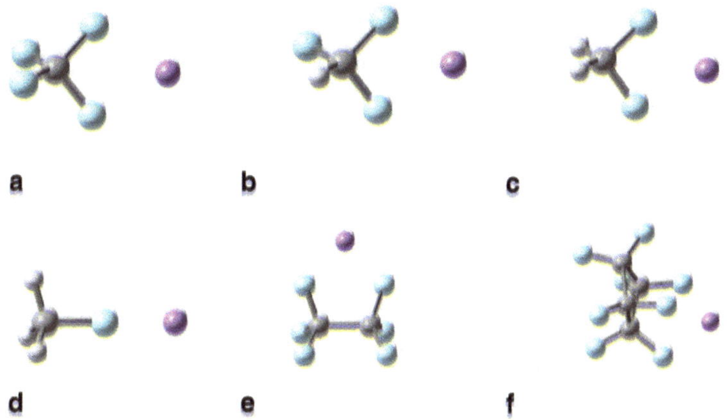

Fig. 3.4 B3LYP/6-31G(d) optimized complex structures of lithiated perfluorocarbons. **a** CF_4–Li^+, **b** CHF_3–Li^+, **c** CH_2F_2–Li^+, **d** CH_3F–Li^+, **e** C_2F_6–Li^+, **f** C_4F_8–Li^+. Carbon atoms are shown *brown*, fluorine *green*, lithium cation *violet*, and hydrogen *white*

the attachment with a Li^+ ion makes CH_4 to have a C_{3v} structure (Fig. 3.2), CF_4, on the other hand, becomes a C_{2v} symmetry structure upon complexation with a Li^+ ion. As is clear now, Li^+ ion interacts with the carbon atom when it forms a complex with CH_4 molecule; however, the same ion interacts with the fluorine atoms when it forms a complex with CF_4 molecule. Basically CF_4, like its counterpart CH_4, does not have a dipole or quadrupole moment, and its dipole polarizability is small. So, as in the case of CH_4–Li^+ complex, the ion–induced dipole interaction plays an important role in the interaction between CF_4 and the Li^+ ion. The ion induces a dipole in CF_4 and interacts with the induced dipole, hence it has weaker interaction. The calculated ion affinities are tabulated in Table 3.2.

Once again it is proved that the theoretically calculated ion affinities agree very well with the experimental values. The larger Li^+ ion affinities obtained for the PFCs studied here indicate that lithiated PFCs can be detected in the Li^+-IAMS. Later, through many experimental studies [135], it has been proved that the IAMS could be used to quantify the emissions of these PFCs from the semiconductor industries. Analyzing the calculated atomic charges and the bond analysis using Bader's AIM theory demonstrated that the ion–molecule interaction in most of these complexes is mainly due to dipole-induced electrostatic interactions.

Here, only a few representative studies on alkali–metal ion interactions with a few organic molecules are provided; however, as one can imagine, there have been hundreds of studies in the past. Also, only a few theoretical methods are used for the calculations and their results are given in this material though there are many other highly reliable methods, especially on the ab initio side. Composite methods [66–68], such as Gaussian-*n*, complete basis set, and Weizmann-*n* methods can be used to obtain very accurate results, especially alkali–metal ion affinities. Because these calculations are computationally expensive, even for a molecule with more

Table 3.2 Calculated Li^+ ion affinities of various perfluorocarbons using B3LYP functional

PFC	ΔH (in kcal/mol)		
	6-311+G(2*d*)	6-311+G(3*df*)	Expt. [133, 134]
CF_4	−12.1	−12.3	
CHF_3	−19.4	−19.6	
CH_2F_2	−26.6	−26.8	−26.5
CH_3F	−30.0	−30.1	−31.0
C_2F_6	−16.9	−17.2	
C_4F_8	−20.6	−21.1	

than ten atoms, normally the DFT approach is used as it has been shown that reliable results can be obtained using DFT. Most of the results provided here are calculated using DFT and these values agree with the experimental ones. Still, there is room to improve the results such as one can use better density functionals, instead of using the popular B3LYP functional, and also correlation-consistent basis sets, such as Dunning's basis sets. Experience shows that core correlation of Li^+–C is necessary to accurately measure the Li^+ ion affinities and so correlation-consistent basis set is a better choice. Also, the accuracy of the results can be improved further by considering scaling of zero-point energy corrections and basis-set super position errors. For further information on these general issues, mainly on obtaining the accurate alkali–metal ion affinities, readers can see the references [136] and [137].

3.5.2 Interactions with Biomolecules

The involvement of metal ions, especially alkali–metal ions, in biological processes has been identified early [138−145]. Metal ions play both direct, as in oxidation-reduction reactions, and indirect, by inducing conformational changes, roles [146]. Different metal ions exhibit different structural effects so that the nature of the metal contact in the nucleus can conceivably influence the course of genetic information transfer. Base-binding metal ions can cause more profound effects on the DNA conformation than metal ions that bind to the phosphate backbone [139]. Cations of sodium and potassium, and to a lesser extent of lithium, have important effects from the biological point of view [147, 148]. These cations involve several functions of living systems such as enzyme regulation, stabilization of structural elements, transmission of cellular signals (Na^+ or K^+ channels), and transport of glucides and amino acids to proteins through transmembrane channels [147, 149–157]. Metal ions also play key roles in enzymatic activity, cellular metabolism, and structural stabilization [158−161]. Thus, the understanding of the details of the local interactions of metal ions with amino acids and nucleic acids has therefore become fundamental in biology.

One of the main non-covalent interactions in biology is the cation–π interaction. Reviews on cation–π interactions by Dougherty coworkers [162, 163] and other researchers [164] provide a detailed overview of this interesting interaction, highlighting both its fundamental nature and its importance in the biological chemistry. The pioneering work by Sunner et al. [165] showed that the non-covalent interaction of K^+ with benzene was significant and stronger than the interaction of K^+ with a water molecule. It is well-known, now, that cation–π interactions play an important role in protein structural organization and the functioning of ionic channels in membranes [162, 163, 166−172]. Additionally, a number of studies have established that cation–aromatic interactions are very important in protein–ligand interactions, too [162]. It is also found that energetically significant cation–π interactions are common in proteins and that they probably contribute to protein stability. Because of these facts, cation–π interactions are now considered as important non-covalent interactions [173]. Among many metal ions, alkali metal ions are significantly involved in the cation–π interactions in the biological systems [174−176]. These metal ions can form cation–π interaction with nucleic acids and with proteins through aromatic amino acids, especially through tryptophan, phenylalanine, and tyrosine. Hence, understanding cation–π interactions involving alkali metal ions, both from a fundamental perspective as well as the detailed role that they play in biological systems, becomes important.

Because of this importance, studies on metal-binding affinities to biologically relevant molecules have attracted much attention at theoretical levels too [81, 82, 147, 149, 177−229]. These studies include alkali metal ion interactions with all the amino acids, nucleic acids, and peptides among the others. Theoretical investigations can overcome the problems faced by the experimental techniques allowing the determination of not only the minimum energy structures but also a means of following their formation mechanism through TS characterizations.

After recognizing the importance of metal ions, especially alkali metal ions, interactions with the DNA components, many researchers investigated various metal ion interactions with nucleic acids by using various theoretical methodologies. Let us consider one of these studies. Rodgers and Armentrout [225] studied the alkali–metal ion (Li^+, Na^+, and K^+) interactions of three nucleic acid bases, adenine, thymine, and uracil by using threshold collision-induced dissociation (TCID) and theoretical methods. Their calculations were done at the MP2(full)/6-311+G(2*d*,2*p*)//MP2(full)/6-31G(*d*) level of theory (single point calculations with MP2(full)/6-311+G(2*d*,2*p*) theory at the MP2(full)/6-31G(*d*) optimized geometries). The calculations provide insight into the structures and binding of the metal ions to the nucleic acid bases. Authors considered various possible binding sites on the bases. Complex structures are given in Fig. 3.5 (stable MP2(full)/6-31G(*d*) geometries of thymine–Na^+ complexes bound at the O2 and O4 sites, and adenine–Na^+ complexes bound at the N1, N3, and N7 sites). The figure shows that all the stable complex structures are almost planar. The calculated Na^+ ion interaction energies, in kcal/mol, for the most stable complex structures, with adenine, thymine, and uracil bases are 30.7, 32.3, and 32.1, respectively. They have also calculated the Li^+ ion and K^+ interaction energies with these three bases. These calculated energies compare favorably with the experimental values though the calculated values for Li^+ to all three bases and adenine with all three metal ions

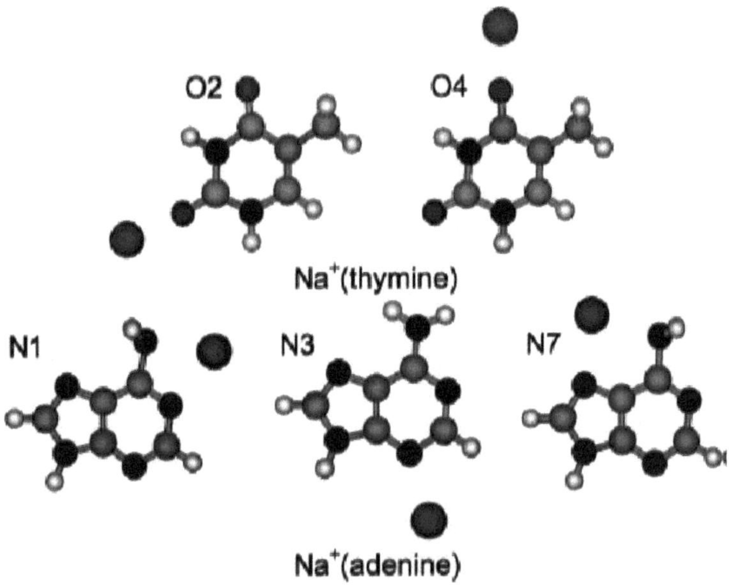

Fig. 3.5 Na^+ complex structures with thymine and adenine. (Taken from [225])

are systematically low by around 4 kcal/mol. Their calculations undoubtedly did not only complement their experimental results but also helped to identify the complex structures by providing the location of the interactions.

Kabelac and Hobza did a thorough study on the interactions of all 14 amino and imino tautomers of adenine with Na^+ ion using the MP2/TZVPP level of theory [223]. According to their study, cations interact mostly in a bidentate (interacting with two positions simultaneously) manner with all the nucleic acid bases. The metal ion—nucleic acid bonding energy in these cases is larger than those when ions bind in a unidentate manner. It can be easily understood by considering the fact that nucleic acids have oxygen or nitrogen lone pairs. For example, when an alkali metal ion interacts with the adenine through N7 (see N1 or N7 structure of Fig. 3.5), the amino group is distorted from the original position, allowing a direct interaction of the ion with the nitrogen lone pairs of the amino group. This makes the ion interacts with the nucleic acid in a bidentate manner. It is the same with the other nucleic acids too, though thymine and cytosine possess fewer binding sites than the other nucleic acids. The metal ion-base intermolecular distances in adenine and cytosine are shorter than those in guanine and thymine. An interesting consequence of such metal ion interaction with these DNA elements is the metal-induced stability changes in DNA duplexes. For example, as noticed earlier [217], the binding of alkali metal ions to adenine, which occurs preferentially at the $N7/NH_2$ chelation site (Fig. 3.5), would tend to disrupt the hydrogen bonding of A:T pairs. However, disruption of a single hydrogen bond in an AT base pair costs around 6 kcal/mol, more than the calculated difference in binding affinities of the N3 and N7 sites. There are at least two possibilities: one, nature tends to save the existence and in such a scenario, metal ions may preferentially bind at N3 site, such that hydrogen bonding between DNA elements is not disturbed. In fact, such a binding at the N3 site should make the amino hydrogen atoms to become more acidic that should result in an enhancement of the hydrogen bonding in duplex DNA. Otherwise, metal ion binding could cause the disruption of a hydrogen bond in a base pair, which could damage the DNA. There is a lot of scope of studying and understanding the metal–ion bonding with the DNA bases.

Understanding the metal–ion, especially alkali–metal ion, interactions with the amino acids is important because such interactions can influence, among others, the protein folding. An important as well as interesting fact is that the 20 amino acids vary considerably in their physicochemical properties such as polarity, acidity, basicity, aromaticity, bulk, and conformational flexibility. Due to these differences in nature (amino acids with nonpolar side chains—glycine, alanine, valine, leucine, isoleucine, methionine, proline, phenylalanine, tryptophan; amino acids with uncharged polar side chains—serine, threonine, asparagine, glutamine, tyrosine, cysteine; and amino acids with charged side chains—lysine, arginine, histidine, aspartic acid, and glutamic acid), the metal ion interactions vary among the amino acids. In addition to that phenylalanine, tryptophan, histidine, and tyrosine are the aromatics ones. These amino acids can also interact with the metal ions through cation–π interactions.

Many researchers studied the alkali–metal ion interactions with amino acids. Feng et al. [213] determined the lithium ion binding energies of 15 of the common amino acids using the kinetic method in a quadrupole ion trap mass spectrometer. The calculated binding energies scale from glycine to glutamic acid spans the range 41.6–52.9 kcal/mol. They concluded that alkali metal cations exalted binding energies for amino acids with side chains that include oxygen-bearing functional groups (i.e., alcohols and carboxyl acids). Armentrout and coworkers extensively studied the binding energies of alkali metal cations (Li^+, Na^+, K^+, Rb^+, and Cs^+) with all the amino acids using both experimental (both threshold collision-induced dissociation (TCID) and infrared multiple photon dissociation (IRMPD) methods) and theoretical methods (both ab initio and DFT) [187–207].

Let us briefly see alkali metal ion interactions with Histidine (His) amino acid [204, 205, 208, 212]. Histidine is the basic but an essential amino acid and is chemically one of the most flexible protein residues thanks to its imidazole side chain, which functions as both acid and base near neutral pH. His residue presents three potential coordination sites in aqueous solution. The imidazole nitrogen of histidine is a primary site for binding metal ions and the other two sites are carboxyl and amino nitrogen sites. Bojesen et al. [208] found that Na^+ affinities for His was the second highest, lying only below that of arginine in their relative measurements. This order was also confirmed by Kish et al. [212] in a more quantitative study, where the Na^+ ion binding affinity for His was found to be 52.3 kcal/mol. Recently, Armentrout and coworkers studied all alkali metal ion complex structures with His using IRMPD and theory [204]. The very significant fact in using theory to study such complex systems is that it not only gives the ion affinities but also it correctly

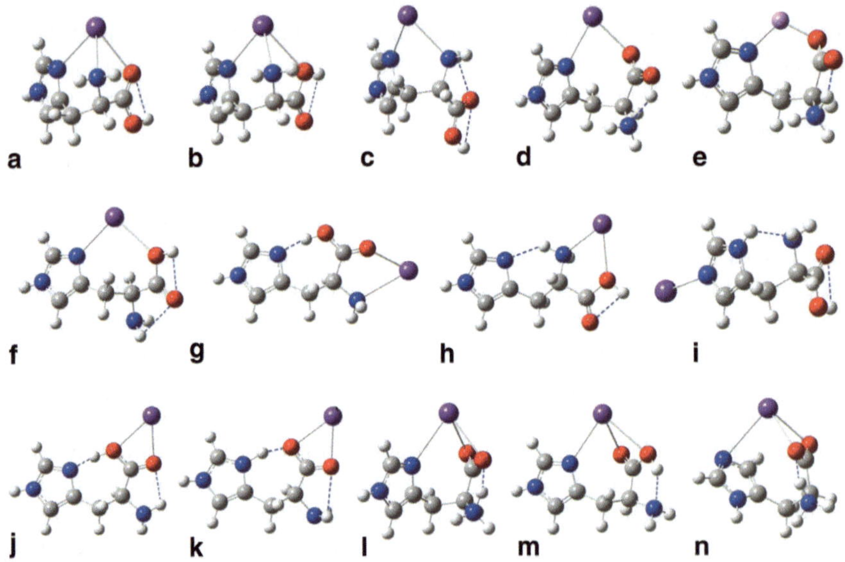

Fig. 3.6 M^+ (His) Complex structures. (Taken from reference [204])

identifies the complex structures that are present in the experiments. That is the strength of an accurate theory.

Armentrout and coworkers studied all possible alkali–metal ion complex structures with histidine. Two density functionals (B3LYP and B3P86) in DFT and MP2 in ab initio theory with different basis sets are used for their calculations [204]. Structures of the complexes calculated at the B3LYP/6-311+G(*d,p*) level of theory are given in Fig. 3.6.

There are many energy minima complex structures. It is common in many of the amino acids, for example, theoretical studies using B3LYP/6-311++G(*d,p*) level of theory have shown that eight minima are present on the potential energy surface for alkali–metal ion (Li^+, Na^+, K^+) cationized alanine [82]. At all levels of theory used, the authors [204] found that the ground state structure for Li^+ (His) and Na^+ (His) is the tridentate charge-solvated structure in which the metal ion binds to the backbone carbonyl oxygen, backbone amino nitrogen, and imidazole side chain nitrogen (structure **a** of Fig. 3.6). Again it is same with the K^+ (His) complex according to the MP2 level of theory though DFT supports a bidentate structure in which the metal ion binds to the backbone carbonyl oxygen and imidazole side chain nitrogen. For the larger alkali metal ion complexes with the histidine, the ground state structures are slightly different.

To identify the structures present in the experimental study, authors compared their IRMPD spectra with the single photon spectra calculated at the B3LYP/6-311+G(*d,p*) level. Figure 3.7 compares the experimental IRMPD action spectrum with the calculated IR spectra for three distinct conformers of Li^+ (His): the structure **a** (see Fig. 3.6), ground state at all levels of theory, and two other representative lower energy conformers. Though the calculated IR intensities may

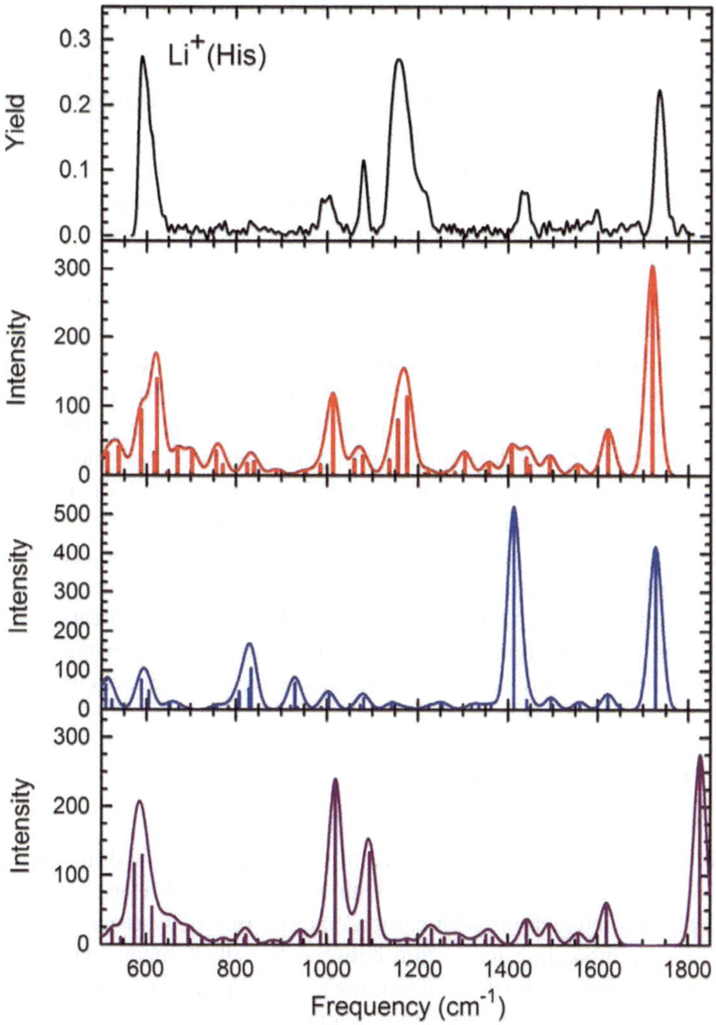

Fig. 3.7 Comparison of the experimental IRMPD action spectrum for Li^+ (His) with IR spectra for three low-lying conformation predicted at the B3LYP/6-311+G(*d,p*) level of theory. (Taken from Reference [204])

not be in direct accord with the IRMPD spectrum because of the latter being a multiple photon process, contrary to the theoretical IR spectra, the bands predicted for the structure **a** conformer correspond reasonably well with the observed spectrum (see Fig. 3.7) in terms of both band positions and relative intensities. The observed IR spectrum does not coincide well with none of the other spectra calculated for other conformers. On the basis of their calculations and comparing the spectra, the authors could conclude that the structure present in the experimental study is the structure **a**. Authors could also identify the other alkali metal ion complex structures with the histidine. Later, Armentrout and coworkers [205] studied the interactions of alkali metal ions with the amino acids. Their calculated bond energies are in reasonable agreement with their experimental values. For example, the experimental binding energy observed for the Na^+ (His) is 53.2 kcal/mol, whereas the corresponding calculated values are 55.0 (B3LYP), 52.6 (B3P86), and 52.7 (MP2) level of theory. Their calculated values include zero-point energy corrections (with scaling) and counterpoise corrections for basis set superposition errors. These studies conclude that the binding energies of M^+ (His) are larger than those of other M^+ (amino acid) complexes. This is attributed to contributions from the local dipole moment of the imidazole side chain functionality.

Though most of the most stable ion–molecule complexes involving alkali–metal ions and amino acids are σ-type complexes (planar or near-planar), in the actual cases, in proteins or in DNA, they could easily form π-type complexes in the presence of the neighboring counterparts. In fact, there were a few studies on cation–π complexes involving aromatic amino acids [211, 220, 230, 231]. Dunbar and coworkers conducted theoretical studies on complexation of Na^+ and K^+ with phenylalanine, tyrosine, and tryptophan using DFT calculations [230]. They conclude that the binding site formed by tridentate N/O/Ring (involving cation–π interaction) provides strong binding for both the alkali ions to all the three aromatic amino acids. An increment of around 7 kcal/mol was found in the alanine/phenylalanine comparison of bonding enthalpies, confirming the importance of cation–π binding enhancement in the phenylalanine case [231]. These studies show the vital role the cation–π interactions play in the ion–aromatic amino acid interactions.

3.5.3 Interactions with Materials

There is a lot of experimental evidence of the presence of alkali metal ion, especially, Li^+ and Na^+ ions, interactions with materials [164]. Here, we specifically briefly see such interactions with carbon material-related systems. Understanding the nature of nonbonding interaction between the alkali–metal ion with the graphene surface attracted much attention; mainly it helps to design new energy storage materials [232–238]. Sastry and coworkers mainly used DFT calculations to understand the binding of Li^+ with the graphene surface [233]. Their study concluded that Li^+ ion is oriented exactly above the center of the six-membered ring in the considered system. Marquez et al. [234] calculated the interaction between Li^+ and the hydrogen terminated cluster model ($C_{32}H_{18}$) using DFT and indicated that the Li^+ ion is preferentially bound outside of the cluster model. A study on Li^+ interaction on a C_{96} planar carbon cluster by Nakadaira et al. [235] using semiempirical MO theory suggested that the edge site is more stable than that of the bulk. Suzuki et al. [236] investigated the storage state of the Li^+ ion with a $C_{54}H_{18}$ cluster using the PM3 method. Since the elucidation of the diffusion processes of the Li^+ ion on the amorphous carbon is one of the important themes in the development of higher performance lithium secondary batteries, Tachikawa and Shimizu [237] have investigated the Li^+ ion interactions and its diffusion processes on a model surface of amorphous carbon ($C_{96}H_{24}$ system) using direct molecular orbital dynamics method

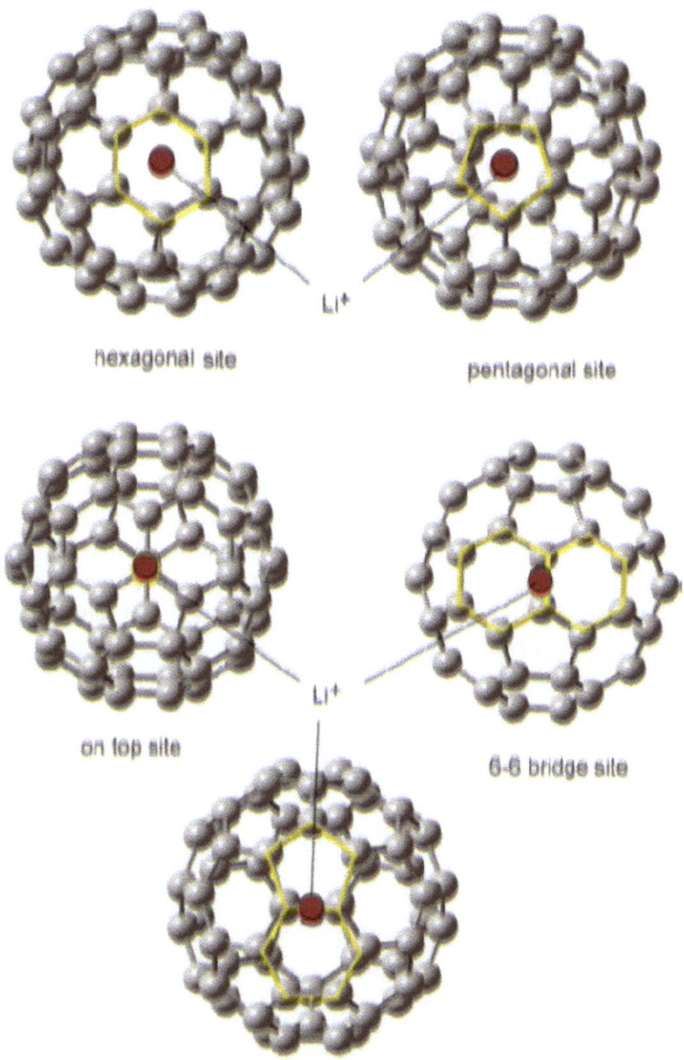

Fig. 3.8 Optimized structures of Li^+ C60 complex at the B3LYP/6-311G(*d*,*p*) level. (Taken from reference [242])

at the semiempirical AM1 level of theory. Their study showed the preference of binding of Li^+ at the edges of the cluster.

Alkali ion interaction with fullerenes have also attracted attention in recent years [238–243]. Moradi et al. [238] studied the binding energy between alkali metal ions (Li^+, Na^+, and K^+) with pristine C_{24} or doped fullerenes of BC_{23}. They found that the most suitable site of interaction is at the top of the center of a six-membered ring of the exterior surface of C_{24} molecule. Zhao et al. [239] obtained the electronic states of Li^+ C_{60} using DFT and they showed that the Li^+ ion can bind in hexagonal site of

C_{60} with a binding energy of 41.7 kcal/mol. Santos et al. [240] suggested that the Li^+ ion is bound at ~2.0 Å from the C_{60} surface. Tachikawa [241, 242] has studied the interaction of the Li^+ ion with fullerene C_{60} surface using DFT and direct molecular orbital-molecular dynamics methods.

A recent study by Tachikawa [242] concluded that the Li^+ ion can bind two stable binding sites, hexagonal and pentagonal sites where Li^+ ion binds to six- and five-membered rings, respectively (See Fig. 3.8). The Li^+ ion binding energy obtained for the most stable complex, pentagonal binding site, is 34.8 kcal/mol at the B3LYP/6-311G(*d,p*) level. The binding energy of hexagonal site is 34.5 kcal/mol. The (6-5) and (6-6) sites are 2.7 and 3.6 kcal/mol higher in energy than that of the pentagonal site, respectively; however, these two structures are found to be saddle points connecting the stable two structures. The relative energy of the on-top site is 3.9 kcal/mol higher than the pentagonal site. The distance between the Li^+ ion and the interacting surface in the most stable two structures is 1.920 Å (hexagonal) and 1.952 Å (pentagonal). The dynamics calculations showed that the Li^+ ion diffuses between stable points near the saddle points.

Studies on alkali metal ion interaction with carbon nanotubes also gets moment [243, 244]. Mpourmpakis et al. [243] used ab initio and molecular mechanics calculations to study the interaction between different types of carbon nanotubes with alkali metal cations (Li^+, Na^+, and K^+). The results showed that ions are located on top of a phenyl group of the nanotube, resulting in a strong cation–π interaction. Umadevi and Sastry studied the same alkali metal ions binding with carbon nanotubes and identified some interacting structures [244].

All these studies show that alkali metal ions interact with these carbon material systems in a cation–π fashion. With the emergence of reliable theoretical methodologies and because of the importance, coming years will see a lot of alkali metal ion bonding studies on a wide range of materials.

References

1. Eyring H. The activated complex in chemical reactions. J Chem Phys. 1935;3:107–14.
2. Evans MG, Polanyi M. Some applications of the transition state method to the calculation of reaction velocities, especially in solution. Trans Faraday Soc. 1935;31:875–94.
3. Wigner E. Calculation of the rate of elementary association reactions. J Chem Phys. 1937;5:720–5.
4. Horiuti J. On the statistical mechanical treatment of the absolute rate of chemical reaction. Bull Chem Soc Jpn. 1938;13:210–6.
5. Keck JC. Variational theory of reaction rates. Adv Chem Phys. 1967;13:85–121.
6. Truhlar DG, Garrett BC. Variational transition state theory. Annu Rev Phys Chem. 1984;35:159–89 and references therein.
7. Marcus RA, Rice OK. The kinetics of the recombination of methyl radical and iodine atoms. J Phys Colloid Chem. 1951;55:894–908.
8. Marcus RA. Unimolecular dissociation and free radical recombination reactions. J Chem Phys. 1952;20:359–63.
9. Robinson PJ, Holbrook KA. Unimolecular reactions. New York:Wiley; 1972.
10. Forest W. Theory of Unimolecular reactions. New York:Academic; 1973.

11. Rosentock HM, Wallentein MB, Wahrhaftig AL, Eyring H. Absolute rate theory for isolated systems and the mass spectra of polyatomic molecules. Proc Natl Acad Sci USA. 1952;38:667–78.
12. Pechukas P, Light JC. On detailed balancing and statistical theories of chemical kinetics. J Chem Phys. 1965;42:3281–90.
13. Keck JC. Statistical theory of chemical reaction rates. J Chem Phys. 1958;29:410–4.
14. Light JC. Statistical theory of bimolecular exchange reactions. Discuss Faraday Soc. 1967;44:14–29.
15. Nikitin EE. Teor. Eksp. Khim. Acak. Nauk. Ukr. SSR 1965;1:428.
16. Pechukas P, Light JC, Rankin R. Statistical theory of chemical kinetics: application to neutral atom-molecule reactions. J Chem Phys. 1966;44:794–805.
17. Chesnavich WJ, Bowers MT. Statistical phase space theory of polyatomic systems. Applications to the unimolecular reactions $C_6H_5CN^+ \rightarrow C_6H_4^+ + HCN$ and $C_4H_6^+ \rightarrow C_3H_3^+ + CH_3$. J Am Chem Soc. 1977;99:1705–11.
18. Chesnavich WJ, Bowers MT. Statistical phase space theory of polyatomic systems: rigorous energy and angular momentum conservation in reactions involving symmetric polyatomic species. J Chem Phys. 1977;66:2306–15.
19. Klots CE. Reformulation of the quasiequilibrium theory of ionic fragmentation. J Phys Chem. 1971;75:1526–32.
20. Bass LM, Kemper PR, Anicich VG, Bowers MT. Ion-molecule radiative association reactions. A statistical phase space theory model. J Am Chem Soc. 1981;103:5283–92.
21. Bass LM, Cates RD, Jarrold MF, Kirchner NJ, Bowers MT. Ion-molecule association reactions: reaction sequences initiated by protonated methanol ($CH_3OH_2^+$) in CH_3OH; experiment and theory. J Am Chem Soc. 1983;105:7024–33.
22. Porter RN, Raff, LM. Classical trajectory methods in molecular collisions. In: Miller WH, Editor. Dynamics of molecular collisions, Part B. Vol. 2. New York: Plenum Press; 1976. pp. 1–52.
23. Raff, LM, Thompson DL. The classical approach to reactive scattering. In: Baer M, Editor. Theory of chemical reaction dynamics, Vol. 3. Boca Raton: CRC Press; 1985. pp. 1–121.
24. Truhlar, DG, Muckerman JT. Reactive scattering cross section. III: Quasiclassical and semiclassical methods. In: Berntein RB, Editor. Atom-molecule collision theory. New York: Plenum Press; 1979, pp. 505–66.
25. Schatz, GC. Quasiclassical trajectory studies of state to state collisional energy transfer in polyatomic molecules. In: Bowman JM, Editor. Molecular collision dynamics. Topics in current physics. Heidelberg: Springer-Verlag; 1983. Vol. 33. pp. 25–60.
26. Hase WL, Feng D-F. Dynamics of ion solvation. $Li^+ + H_2O \rightarrow Li^+(H_2O)^*$. J Chem Phys. 1981;75:738–44.
27. Swamy KN, Hase WL. Dynamics of ion-molecule recombination. II. An alkali ion and a water molecule. J Chem Phys. 1982;77:3011–21.
28. Swamy KN, Hase WL. Dynamics of ion-molecule recombination. 3. Trends in the recombination efficiency. J Am Chem Soc. 1984;106:4071–7 and references therein.
29. Vande Linde SR, Hase WL. Dynamics of ion-molecule recombination IV. $Li^+ + (CH_3)_2O$ association. Comp Phys Commun. 1988;51:17–34.
30. Vande Linde SR, Hase WL. Non-RRKM kinetics in gas-phase S_N2 nucleophilic substitution. J Phys Chem. 1990;94:6148–50.
31. Vande Linde SR, Hase WL. Trajectory studies of S_N2 nucleophilic substitution. I. Dynamics of $Cl^- + CH_3Cl$ reactive collisions. J Chem Phys. 1990;93:7962–80.
32. Hase WL, Darling CL, Zhu I. Dynamics of ion-molecule recombination. V. A study of energy transfer pathways. J Chem Phys. 1992;96:8295–306.
33. Frisch MJ, Trucks GW, Schlegel HB, Scuseria GE, Robb MA, Cheeseman JR, Scalmani G, Barone V, Mennucci B, Petersson GA, Nakatsuji H, Caricato M, Li X, Hratchian HP, Izmaylov AF, Bloino J, Zheng G, Sonnenberg JL, Hada M, Ehara M, Toyota K, Fukuda R, Hasegawa J, Ishida M, Nakajima T, Honda Y, Kitao O, Nakai H, Vreven T, Montgomery Jr JA, Peralta JE, Ogliaro F, Bearpark M, Heyd JJ, Brothers E, Kudin KN, Staroverov VN, Keith

T, Kobayashi R, Normand J, Raghavachari K, Rendell A, Burant JC, Iyengar SS, Tomasi J, Cossi M, Rega N, Millam JM, Klene M, Knox JE, Cross JB, Bakken V, Adamo C, Jaramillo J, Gomperts R, Stratmann RE, Yazyev O, Austin AJ, Cammi R, Pomelli C, Ochterski JW, Martin RL, Morokuma K, Zakrzewski VG, Voth GA, Salvador P, Dannenberg JJ, Dapprich S, Daniels AD, Farkas O, Foresman JB, Ortiz JV, Cioslowski J, Fox DJ. Gaussian 09, Revision D. 01; Gaussian, Inc.: Wallingford, CT, 2013.
34. Schmidt MW, Baldridge KK, Boatz JA, Elbert ST, Gordon MS, Jensen JH, Koseki S, Matsunaga N, Nguyen KA, So SJ, Windus TL, Dupuis M, Montgomery JA. General atomic and molecular electronic structure system. J Comput Chem. 1993;14:1347–63.
35. Shao Y, Fusti-Molnar L, Jung Y, Kussmann J, Ochsenfeld C, Brown ST, Gilbert ATB, Slipchenko LV, Levchenko SV, O'Neill DP, Distasio Jr RA, Lochan RC, Wang T, Beran GJO, Besley NA, Herbert JM, Lin CY, Van Voorhis T, Chien SH, Sodt A, Steele RP, Rassolov VA, Maslen PE, Korambath PP, Adamson RD, Austin B, Baker J, Byrd EFC, Dachsel H, Doerksen RJ, Dreuw A, Dunietz BD, Dutoi AD, Furlani TR, Gwaltney SR, Heyden A, Hirata S, Hsu C-P, Kedziora G, Khalliulin RZ, Klunzinger P, Lee AM, Lee MS, Liang W, Lotan I, Nair N, Peters B, Proynov EI, Pieniazek PA, Rhee YM, Ritchie J, Rosta E, Sherrill CD, Simmonett AC, Subotnik JE, Woodcock III HL, Zhang W, Bell AT, Chakraborty AK, Chipman DM, Keil FJ, Warshel A, Hehre WJ, Schaefer III HF, Kong J, Krylov AI, Gill PMW, Head-Gordon M. Advances in methods and algorithms in a modern quantum chemistry program package. Phys Chem Chem Phys. 2006;8:3172–91.
36. Valiev M, Bylaska EJ, Govind N, Kowalski K, Straatsma TP, van Dam HJJ, Wang D, Nieplocha J, Apra E, Windus TL, de Jong WA. NWChem: a comprehensive and scalable open-source solution for large scale molecular simulations. Comput Phys Commun. 2010;181:1477–89.
37. Werner H-J, Knowles PJ, Knizia G, Manby FR, Schutz M, Celani P, Korona T, Lindh R, Mitrushenkov A, Rauhut G, Shamasundar KR, Adler TB, Amos RD, Bernhardsson A, Berning A, Cooper DL, Deegan MJO, Dobbyn AJ, Eckert F, Goll E, Hampel C, Hesselmann A, Hetzer G, Hrenar T, Jansen G, Koppl C, Liu Y, Lloyd AW, Mata RA, May AJ, McNicholas SJ, Meyer W, Mura ME, Nicklass A, O'Neill DP, Palmieri P, Peng D, Pfluger K, Pitzer R, Reiher M, Shiozaki T, Stoll H, Stone AJ, Tarroni R, Thorsteinsson T, Wang M. MOLPRO version 2012.1.
38. Aquilante F, De Vico L, Ferré N, Ghigo G, Malmqvist P-A, Neogrády P, Pedersen TB, Pitonak M, Reiher M, Roos BO, Serrano-Andrés L, Urban M, Veryazov V, Lindh R. Software news and update MOLCAS 7: the next generation. J Comput Chem. 2010;31:224–47.
39. Veryazov V, Widmark P-O., Serrano-Andres L, Lindh R, Roos BO. MOLCAS as a development platform for quantum chemistry software. Inter J Quan Chem. 2004;100:626–35.
40. Stanton JF, Gauss J, Perera SA, Watts JD, Yau AD, Nooijen M, Oliphant N, Szalay PG, Lauderdale WJ, Gwaltney SR, Beck S, Balkov'a A, Bernholdt DE, Baeck KK, Rozyczko P, Sekino H, Huber C, Pittner J, Cencek W, Taylor D, Bartlett RJ. Integral packages included are VMOL (Almlof J, Taylor PR.), VPROPS (Taylor P), ABACUS (Helgaker T, Aa. Jensen HJ, Jørgensen P, Olsen J, Taylor PR), HONDO/GAMESS (Schmidt MW, Baldridge KK, Boatz JA, Elbert ST, Gordon MS, Jensen JJ, Koseki S, Matsunaga N, Nguyen KA, Su S, Windus TL, Dupuis M, Montgomery JA.).
41. TURBOMOLE V6.3 2011, a development of University of Karlsruhe and Forschungszentrum Karlsruhe GmbH, 1989–2007, TURBOMOLE GmbH, since 2007; http://www.turbomole.com.
42. HyperChem ™ Professional 7.51, Hybercube, Inc., 1115 NW 4th Street, Gainesville, Florida 32601, USA.
43. Hehre WJ, Radom L, Schleyer PvR., Pople J. *Ab initio* molecular orbital theory. New York: Wiley; 1986.
44. Parr RG, Yang W. Density functional theory of electronic structure of molecules. Annu Rev Chem. 1995;46:701–28, and references therein.
45. Parr RG, Yang W Density-functional theory of atoms and molecules. New York: Oxford University Press; 1994.

46. Bingham RC, Dewar MJS, Lo DH. Ground states of molecules. 25. MINDO-3: Improved version of MINDO semiempirical SCF-MO method. J Am Chem Soc. 1975;97:1285–93.
47. Dewar MJS, Zoebisch EG, Healy EF. Development and use of quantum mechanical molecular models. 76. AM1: a new general purpose quantum mechanical molecular model. J Am Chem Soc. 1985;107:3902–9.
48. Stewart JJP. Optimization of parameters for semiempirical methods. I. Method. J Comput Chem. 1989;10:209–20.
49. Case DA, Babin V, Berryman JT, Betz RM, Cai Q, Cerutti DS, Cheatham III TE, Darden TA, Duke RE, Gohlke H, Goetz AW, Gusarov S, Homeyer N, Janowski P, Kaus J, Kolossváry I, Kovalenko A, Lee TS, LeGrand S, Luchko T, Luo R, Madej B, Merz KM, Paesani F, Roe DR, Roitberg A, Sagui C, Salomon-Ferrer R, Seabra G, Simmerling CL, Smith W, Swails J, Walker RC, Wang J, Wolf RM, Wu X, Kollman PA. AMBER 14. San Francisco: University of California; 2014.
50. Brooks BR, Brooks III CL, Mackerell AD, Nilsson L, Petrella RJ, Roux B, Won Y, Archontis G, Bartels C, Boresch S, Caflisch A, Caves L, Cui Q, Dinner AR, Feig M, Fischer S, Gao J, Hodoscek M, Im w, Kuczera K, Lazaridis T, Ma J, Ovchinnikov V, Paci E, Pastor RW, Post CB, Pu JZ, Schaefer M, Tidor B, Venable RM, Woodcock HL, Wu X, Yang W, York DM, Karplus M. CHARMM: The biomolecular simulation program. J Comput Chem. 2009;30:1545–614.
51. Rappe AK, Casewit CJ, Colwell KS, Goddard III WA, Skiff WM. UFF, A full periodic-table force-field for molecular mechanics and molecular-dynamics simulations. J Am Chem Soc. 1992;114:10024–35.
52. Allinger NL, Chen K, Lii J-H. An improved force field (MM4) for saturated hydrocarbons. J Comput Chem. 1996;17:642–68.
53. Halgren TA. Merck molecular force field. I. Basis, form, scope, parameterization, and performance of MMFF94. J Comput Chem. 1996;17:490–519.
54. Schrodinger E. Quantisierung als Eigenwertproblem. Ann Phys. 1926;79:361–76.
55. Born M, Oppenheimer R. On the quantum theory of molecules. Ann Phys. 1927;389:457–84.
56. Slater JC. The theory of complex spectra. Phys Rev. 1929;34:1293–323.
57. Ditchfield R, Hehre WJ, Pople JA. Self-consistent molecular-orbital methods. IX. An extended Gaussian-type basis for molecular-orbital studies of organic molecules. J Chem Phys. 1971;54:724–8.
58. Dunning TH. Gaussian basis sets for use in correlated molecular calculations. I. The atoms boron through neon and hydrogen. J Chem Phys. 1989;90:1007–23.
59. Hartree DR. The wave mechanics of an atom with a non-coulomb central field. Part 1—theory and methods. Proc Cam Phil Soc. 1928;24:89–110.
60. Fock V. Näherungsmethode zur Lösung des quantenmechanischen Mehrkörperproblems. Z Physik. 1930;61:126–48.
61. Moller C, Plesset MS. Note on an approximation treatment for many-electron systems. Phys Rev. 1934;46:0618–22.
62. Hegarty D, Robb MA. Application of unitary group methods to configuration-interaction calculations. Mol Phys. 1979;38:1795–812.
63. Pople JA, Seeger G, Krishnan R. Variational configuration interaction methods and comparison with perturbation theory. Int J Quantum Chem. 1977;Suppl. Y-11:149–63.
64. Bartlett R, Purvis III GD. Many-body perturbation theory, coupled-pair many-electron theory, and the importance of quadruple excitations for the correlation problem. Int J Quantum Chem. 1978;14:561–81.
65. Pople JA, Krishnan R, Schlegel HB, Binkley JS. Electron correlation theories and their application to study of simple reaction potential surfaces. Int J Quantum Chem. 1978;14:545–60.
66. Pople JA, Head-Gordon M, Fox DJ, Raghavachari K, Curtiss LA. Gaussian-1 theory. A general procedure for prediction of molecular energies. J Chem Phys. 1989;90:5622–9.
67. Nyden MR, Petersson GA. Complete basis set correlation energies. 1. The asymptotic convergence of pair natural orbital expansions. J Chem Phys. 1981;75:1843–62.

68. Martin JML, de Oliveira G. Towards standard methods for benchmark quality ab initio thermochemistry. J Chem Phys. 1999;111:1843–56.
69. Hosenberg P, Kohn W. Inhomogeneous electron gas. Phys Rev. 1964;136:B864–71.
70. Thomas LH. The calculation of atomic fields. Proc Camb Phil Soc. 1927;23:542–8.
71. Fermi E. Un metodo statistico per la determinazione di alcune priopreta dell atomo. Rend Accad Lincei. 1927;6:602–7.
72. Kohn W, Sham LJ. Self-Consistent equations including exchange and correlation effects. Phys Rev. 1965;140:A1133–A38.
73. Cramer CJ. Essentials of computational chemistry. England: John Wiley & Sons, Ltd; 2002.
74. Levine IN. Quantum chemistry. Englewood Cliffs: Prentice-Hall; 1991.
75. Jensen F. Introduction of computational chemistry. England: John Wiley & Sons Ltd.; 1999.
76. Stockigt D. Application of density functional theory Hartree-Fock hybrid methods. Geometries and bond dissociation energies of Al+ complexes. Chem Phys Lett. 1996;250:387–92.
77. Remko M. Structure and gas phase stability of complexes L-M, where M = Li^+, Na^+, $Mg2^+$ and L is formaldehyde, formic acid, formate anion, formamide and their sila derivatives. Mol Phys. 1997;91:929–36.
78. Miklis PC., Ditchfield R., Spencer TA. Carbocation-π interaction: computational study of complexation of methyl cation with benzene and comparisons with related systems. J Am Chem Soc. 1998;120:10482–89.
79. Arulmozhiraja S, Fujii T, Tokiwa H. In+ cation interactions with some organics: Ab initio molecular orbital and density functional theory. Chem Phys. 1999;250:237–42.
80. Arulmozhiraja S, Fujii T. Li+ ion affinities of global-warming perfluorocarbons. J Phys Chem A. 2000;104:9613–8.
81. Srinivas A, Sastry GN. Cation [M = H^+, Li^+, Na^+, K^+, Ca^{2+}, NH_{4+}, and NMe_{4+}] interactions with the aromatic motifs of naturally occurring amino acids: a theoretical study. J Phys Chem A. 2005;109:8893–903.
82. Marino T, Russo N, Toscano M. Potential energy surface for the gas-phase interaction between α-alanine and alkali metal ions (Li^+, Na^+, K^+). A density functional study. Inorg Chem. 2001;40:6439–43.
83. Becke A. Density-functional thermochemistry. III. The role of exact exchange. J Chem Phys. 1993;98:5648–52.
84. Lee C, Yang W, Parr RG. Development of the Colle-Salvetti correlation-energy formula into a functional of the electron density. Phys Rev B. 1988;37:785–9.
85. Zheng J, Zhang S, Lynch BJ, Corchado JC, Chuang Y-Y, Fast PL, Hu W-P, Liu Y-P, Lynch GC, Nguyen KA, Jackels CF, Ramos AF, Ellingson BA, Melissas VS, Villa J, Rossi I, Coitino EL, Pu J, Albu V, Steckler R, Garrett BC, Isaacson AD, Truhlar DG. POLYRATE, Version 2010-A.
86. Klippenstein SJ, Wagner AF, Robertson SH, Dunbar R, Wardlaw DM. Variflex, Version 1.0, http://chemistry.anl.gov/variflex.
87. Thomson KC, Jordon MJT, Collins MA. Polyatomic molecular potential energy surfaces by interpolation in local internal coordinates. J. Chem Phys. 1998;108:8302–16.
88. Collins MA. Molecular potential-energy surfaces for chemical reaction dynamics. Theor Chem Acc. 2002;108:313–24.
89. Bettens RPA, Collins MA. Learning to interpolate molecular potential energy surfaces with confidence. A Bayesian approach. J Chem Phys. 1999;111:816–26.
90. Hase WL. MERCURY: a general Monte-Carlo classical trajectory computer program, QCPE 3, 453, 1983. An updated version of this code is VENUS96. Hase WL, Duchovic RJ, Hu X, Komornik A, Lim KF, Lu D-H, Peslherbe GH, Swamy KN, van de Linde SR, Varandas AJC, Wang H, Wolf RJ. QCPE Bull. 1996;16:43.
91. Sewell TD, Thompson DL. Classical trajectory methods for polyatomic molecules. Int J Mod Phys B. 1997;11:1067–112.
92. Aoiz FJ, Banares L, Herrero VJ. Recent results from quasiclassical trajectory computations of elementary chemical reactions. J. Chem Soc Faraday Trans. 1998;94:2483–500.

93. Peslherbe GH, Wang H, Hase WL. Monte Carlo sampling for classical trajectory simulations. Adv Chem Phys. 1999;105:171–201.
94. Mulliken R. Electronic population analysis on LCAO-MO molecular wave functions. II. Overlap populations, bond orders, and covalent bond energies. J Chem Phys. 1955;23:1841–6.
95. Mulliken R. Criteria for the construction of good self-consistent field molecular orbital wave functions, and the significance of LCAO-MO population analysis. J Chem Phys. 1962;36:3428–40.
96. Reed AE, Weinstock RB, Weinhold F. National population analysis. J Chem Phys. 1985;83:735–46.
97. Reed AE, Curtiss LA, Weinhold F. Intermolecular interactions from a natural bond orbital, donor-acceptor viewpoint. Chem Rev. 1988;88:899–926.
98. Martin F, Zipse H. Charge distribution in the water molecule—A comparison of methods. J Comput Chem. 2005;26:97–105.
99. Clark AE, Sonnenberg JL, Hay PJ, Martin RL. Density and wave function analysis of actinide complexes: what can fuzzy atom, atoms-in-molecules, Mulliken, Lowdin, and natural population analysis tell us? J Chem Phys. 2004;121:2563–70.
100. De Proft F, Alsenoy CV, Peeters A, Langenaeker W, Geerlings P. Atomic charges, dipole moments, and Fukui functions using Hirshfeld partitioning of the electron density. J Comput Chem. 2002;23:1198–209.
101. Williams DE. Net atomic charge and multipole models for the *ab initio* molecular electric potential. In: Lipkowitz KB, Boyd DB, Editors. Reviews of computational chemistry, Vol. II. New York: VCH Publishers; 1991. pp. 219–71.
102. Wang B, Ford GP. Atomic charges derived from a fast and accurate method for electrostatic potentials based on modified AM1 calculations. J Comput Chem. 1994;15:200–7.
103. Bayly CI, Cieplak P, Cornell WD, Kollman PA. A well-behaved electrostatic potential based method using charge restraints for deriving atomic charges—The RESP model. J Phys Chem. 1993;97:10269–80.
104. Cornell WD, Cieplak P, Bayly CI, Kollman PA. Application of RESP charges to calculate conformational energies, hydrogen-bond energies, and free energies of solvation. J Am Chem Soc. 1993;115:9620–31.
105. Chirlian LE., Francl MM. Atomic charges derived from electrostatic potentials—A detailed study. J Comput Chem. 1987;8:894–905.
106. Breneman CM., Wiberg KB. Determining atom-centered monopoles from molecular electrostatic potentials. The need for high sampling density in formamide conformational analysis. J Comput Chem. 1990;11:361–73.
107. Besler BH, Merz KM, Kollman PA. Atomic charges derived from semi-empirical methods. J Comput Chem. 1990;11:431–9.
108. Merz KM. Analysis of a large data base of electrostatic potential derived atomic charges. J Comput Chem. 1992;13:749–67.
109. Reynolds CA, Essex JW, Richards WG. Atomic charges for variable molecular conformations. J Am Chem Soc. 1992;114:9075–9.
110. Singh UC, Kollman PA. An approach to computing electrostatic charges for molecules. J Comput Chem. 1984;5:129–45.
111. Spackman MA. Potential derived charges using a geodesic point selection scheme. J Comput Chem. 1996;17:1–18.
112. Bader RFW, Essen H. The characterization of atomic interactions. J Chem Phys. 1984;80:1943–60.
113. Bader RFW, MacDougall PJ, Lau CDH. Bonded and nonbonded charge concentrations and their relation to molecular geometry and reactivity. J Am Chem Soc. 1984;106:1594–605.
114. Bader RFW. Atoms in molecules. Acc Chem Res. 1985;18:9–15.
115. Wiberg KB, Bader RFW, Lau CDH. Theoretical analysis of hydrocarbon properties. 1. Bonds, structures, charge concentrations, and charge relaxations. J Am Chem Soc. 1987;109:985–1001.

116. Bader RFW. Atoms in molecules: a quantum theory. Oxford: Oxford University Press; 1990.
117. Keesee RG, Castleman Jr AW. Thermochemical data on gas-phase ion-molecule association and clustering reactions. J Phys Chem Ref Data. 1986;15:1011–071.
118. Rodgers MT, Armentrout PB. Collision-induced dissociation measurements on $Li^+ (H_2O)_n$, $n = 1–6$: the first direct measurements of the $Li^+–OH_2$ bond energy. J Phys Chem A. 1997;101:1238–49.
119. Rodgers MT, Armentrout PB. Absolute binding energies of lithium ions to short chain alcohols, $C_nH_{2n+2}O$, $n = 1–4$, determined by threshold collision-induced dissociation. J Phys Chem A. 1997;101:2614–25.
120. Rodgers MT, Armentrout PB. Absolute binding energies of sodium ions to short chain alcohols, $C_nH_{2n+2}O$, $n = 1–4$, determined by threshold collision-induced dissociation experiments and *ab initio* theory. J Phys Chem A. 1999;103:4955–63.
121. Burk P, Koppel IA, Koppel I, Kurg R, Gal J-F, Maria P-C, Herreros M, Notario R, Abboud J-LM, Anvia F, Taft RW. Revised and expanded scale of gas-phase lithium cation basicities. An experimental and theoretical study. J Phys Chem A. 2000;104:2824–33.
122. Armentrout PB, Rodgers MT. Ab absolute sodium cation affinity scale: threshold collision-induced dissociation experiments and *ab initio* theory. J Phys Chem A. 2000;104:2238–47.
123. Amicangelo JC, Armentrout PB. Absolute binding energies of alkali-metal cation complexes with benzene determined by threshold collision-induced dissociation experiments and *ab initio* theory. J Phys Chem A. 2000;104:11420–32.
124. Fujii T, Ogura M, Jimba H. Chemical ionization mass-spectrometry with lithium ion attachment to the molecule. Anal Chem. 1989;61:1026–9.
125. Fujii T. A novel method for detection of radical species in the gas phase: usage of Li^+ ion attachment to chemical species. Chem Phys Lett. 1992;191:162–8.
126. Fujii T. Diagnostics of microwave plasmas of C_2H_2: mass spectrometric investigations of ionic and neutral species. Phys Rev. 1998;58:6495–502.
127. Fujii T. Alkali-metal ion/molecule association reactions and their applications to mass spectrometry. Mass Spectrom Rev. 2000;19:111–38.
128. Sablier M, Fujii T. Mass spectrometry of free radicals. Chem. Rev. 2002;102:2855–924.
129. Fujii T. Ion attachment mass spectrometry. In: Gross LM, Caprioli, RM. Editors. Encyclopaedia of mass spectrometry. American Society for Mass Spectroscopy. Vol. 6. New York: Elsevier; 2007. pp. 327–34.
130. Arulmozhiraja S, Coote ML, Kitahara Y, Juhasz M, Fujii T. Is the bisphenol A biradical formed in the pyrolysis of polycarbonate? J Phys Chem A. 2011;115:4874–81.
131. Fujii T, Iijima S, Iwase K, Arulmozhiraja S. The production of H_2O_2 in the microwave discharge plasma of CH_4/O_2. J Phys Chem A. 2001;105:10089–92.
132. Fujii T, Muraki J, Arulmozhiraja S, Kareev M. Possible production of C_3N_4 in the microwave-discharge plasma of C_2H_2/N_2. J Appl Phys. 2000;88:5592–6.
133. Dzidic I, Kebarle P. Hydration of the alkali ions in the gas phase. Enthalpies and entropies of reactions $M^+(H_2O)_{n-1} + H_2O = M^+(H_2O)_n$. J Phys Chem. 1970;74:1466–74.
134. Staley RH, Beauchamp JL. Intrinsic acid-base properties of molecules. Binding energies of Li^+ to π- and n-donor bases. J Am Chem Soc. 1975;97:5920–1.
135. Fujii T, Arulmozhiraja S, Nakamura M, Shiokawa Y. Mass spectrometry for on-line monitoring of perfluoro compounds using Li^+ ion attachment techniques. Anal Chem. 2001;73:2937–40.
136. Rodger MT, Armentrout PB. A critical evaluation of the experimental and theoretical determination of lithium cation affinities. Int J Mass Spectrom. 2007;267:167–82, and references therein.
137. Mayeux C, Burk P. Evaluation of alkali metal cation affinities and basicities using extrapolation to the complete basis set limit. J Phys Chem A. 2014;118:1906–17, and references therein.
138. Eichhorn GL. In: Eichhorn GL, Marzilli LG, Editors. Advances in inorganic biochemistry, Vol. 3, New York: Elsevier; 1981. pp. 2–46.

139. Shin YA, Eichhorn GL. Reversible change in ψ structure of DNA-poly(Lys) complexes induced by metal binding. Biopolymers 1977;16:225–30.
140. Shack J, Jenkins RJ, Thompsett JM. The interaction of ions and desoxypentose nucleic acid of calf thymus. J Biol Chem. 1953;203:373–87.
141. Dove WF, Davidson N. Cation effects on the denaturation of DNA. J Mol Biol. 1962;5:467–78.
142. Eichhorn GL. Metal ions as stabilizers or destabilizers of the deoxyribonucleic acid structure. Nature. 1962;194:474–5.
143. Eichhorn GL., Shin YA. Interaction of metal ions with polynucleotides and related compounds. XII. The relative effect of various metal ions on DNA helicity. J Am Chem Soc. 1968;90:7323–8.
144. Krakauer H. The binding of Mg^{++} ions to polyadenylate, polyuridylate, and their complexes. Biopolymers. 1971;10:2459–90.
145. Blagoi YP, Sorokin VA, Valeyev VA, Khomenko SA, Gladchenko GO. Magnesium-ion effects on helix-coil transition of DNA. Biopolymers. 1978;17:1103–8.
146. Ivanov VI, Minchenkova LE, Minyat EE, Frank-Kamenetskii MD, Schyolkina AK. BBAR to ABAR transition of DNA in solution. J. Mol Biol. 1974;87:817–33.
147. Hoyau S, Norrman K, McMahon TB, Ohanessian G. A quantitative basis for a scale of Na^+ affinities of organic and small biological molecules in the gas phase. J Am Chem Soc. 1999;121:8864–75.
148. Lippard SJ, Berg JM. Principles of bioinorganic chemistry. Mill Valley: University of Science Books; 1994.
149. Hoyau S, Ohanessian G. Interaction of alkali metal cations (Li^+, Cs^+) with glycine in the gas phase: a theoretical study. Chem Eur J. 1998;4:1561–9.
150. Stryer L. Biochemistry. New York: W. H. Freeman and Co.; 1988.
151. Barbry P, Lazdunski M. Structure and regulation of the amiloride-sensitive epithelial sodium channel. In: Narahashi T, Editor. Ion channels. New York: Plenum Press; 1996. vol. 4, pp. 115–67.
152. Lingueglia E, Deval E, Lazdunski M. FMRFamide-gated sodium channel and ASIC channels: a new class of ionotropic receptors for FMRFamide and related peptides. Peptides. 2006;27:1138–52.
153. Diochot S, Salinas M, Baron A, Escoubas P, Lazdunski M. Peptides inhibitors of acid-sensing ion channels. Toxicon. 2007;49:271–84.
154. Chemin J, Patel A, Duprat F, Sachs F, Lazdunski M, Honore E. Up- and down-regulation on the mechano-gated K-2P channel TREK-1 by PIP2 and other membrane phospholipids. Pflugers Arch—Eur J Physio. 2007;455:97–103.
155. Silberberg SD, Swartz K. Structural biology trimeric ion-channel design. Nature. 2009;460:580–1.
156. Lippard SJ, Berg JM, Editors. Principles of bioinorganic chemistry. Mill Valley: University Science Books; 1994.
157. Hsiao C, Tannenbaum E, VanDaeusen H, Hershkovitz E, Perng G, Tannenbaum AR, Williams LD. Complexes of nucleic acids with group I and II cations. In: Hud NV, Editor. Nucleic acid-metal ion interactions. Cambridge: Royal Society of Chemistry; 2009. pp 1–38.
158. Permyakov EA. Metalloproteins. New Jersey: Wiley; 1009.
159. Holm RH, Kennepohl P, Solomon EI. Structural and functional aspects of metal sites in biology. Chem Rev. 1996;96:2239–314.
160. Page MJ, Di Cera E. Role of Na^+ and K^+ in enzyme function. Physiol Rev. 2006;86:1049–92.
161. Reyes-Caballero H, Campanello GC, Giedroc DP. Metalloregulatory proteins: metal selectivity and allosteric switching. Biophy Chem. 2011;156:103–14.
162. Dougherty DA. Cation-π interactions in chemistry and biology: a new view of benzene, Phe, Tyr, and Trp. Science. 1996;271:163–8.
163. Ma JC, Dougherty DA. The cation-π interaction. Chem Rev. 1997;97:1303–24.
164. Mahadevi AS, Sastry GN. Cation–π interaction: its role and relevance in chemistry, biology, and materials science. Chem Rev. 2013;113:2100–38, and see references therein.

165. Sunner J, Nishizawa K, Kebarle P. Ion-solvent molecule interaction in the gas-phase: the potassium-ion and benzene. J Phys Chem. 1981;85:1814–20.
166. DeVos AM, Ultsch M, Kossiakoff AA. Human growth hormone and extracellular domain of its receptor: crystal structure of the complex. Science. 1991;255:306–12.
167. Karlin A. Structure of nicotinic acetylcholine receptors. Curr Opin Neurobiol. 1993;3:299–309.
168. Raves ML, Harel M, Pang YP, Silman I, Kozikowski A, Sussman JL. Structure of acetylcholinesterase complexed with the nootripic alkaloid, (–)-huperzine A. Nat Struct Biol. 1997;4:57–63.
169. Stauffer DA, Karlin A. Electrostatic potential of the acetylcholine binding sites in the nicotinic receptor probed by reactions of binding-site cysteines with charged methanethiosulfonates. Biochemistry. 1994;33:6840–9.
170. Mitchell JB, Nandi CL, McDonald IK, Thornton JM, Price SL. Amino/aromatic interactions in proteins—Is the evidence stacked against hydrogen-bonding? J Mol Biol. 1994;239:315–31.
171. Zhong W, Gallivan JP, Zhang Y, Li L, Lester HA, Dougherty DA. From *ab initio* quantum mechanics to molecular neurobiology: a cation-π binding site in the nicotinic receptor. Proc Natl Acad Sci USA. 1998;95:12088–93.
172. Donini O, Weaver DF. Development of modified force field for cation-amino acid interactions: *ab initio*-derived empirical correction terms with comments on cation-π interactions. J Comput Chem. 1998;19:1515–25.
173. Gallivan JP, Dougherty DA. Cation-π interactions in structural biology. Proc Natl Acad Sci USA. 1999;96:9459–64.
174. Gokel GW, De Wall SL, Meadows ES. Experimental evidence for alkali metal cation-π interactions. Eur J Org Chem. 2000;2000:2967–78.
175. Lippard SJ, Berg JM. Principles of bioinorganic chemistry. Mill Valley California: University Science Books; 1994.
176. Caldwell JW, Kollman PA. Cation-π interactions—Nonadditive effects are critical in their accurate representation. J Am Chem Soc. 1995;117:4177–8.
177. Tomlinson MJ, Scott JR, Wilkins CL, Wright JB, White WE. Fragmentation of an alkali metal-attached peptide probed by collision-induced dissociation Fourier transform mass spectrometry and computational methodology. J Mass Spectrom. 1999;34:958–68.
178. Bouchonnet S, Flament JP, Hoppilliard Y. Desorption ionization and in-flight fragmentation of monoalkali-glycine adducts—plasma-desorption mass-spectrometry and *ab initio* calculations. Rapid Commun Mass Spectrom. 1993;7:470–6.
179. Cerda BA, Hoyau S, Ohanessian G, Wesdemiotis C. Na^+ binding to cyclic and linear dipeptides. Bond energies, entropies of Na^+ complexation, and attachment sites from the dissociation of Na^+-bond heterodimers and *ab initio* calculations. J Am Chem Soc. 1998;120:2437–48.
180. Marino T, Russo N, Toscano M. Gas-phase metal ion (Li^+, Na^+, Cu^+) affinities of glycine and alanine. J Inorg Biochem. 2000;79:179–85.
181. Bouchonnet S, Hoppilliard Y. Proton and sodium-ion affinities of glycine and its sodium-salt in the gas-phase—*ab initio* calculations. Org Mass Spectrom. 1992;27:71–6.
182. Bertran J, Rodriguez-Santiago L, Sodupe M. The different nature of bonding in Cu^+-glycine and Cu^{2+}-glycine. J Phys Chem B. 1999;103:2310–7.
183. Hoyau S, Ohanessian G. Absolute affinities of alpha-amino acids for Cu^+ in the gas phase. A theoretical study. J Am Chem Soc. 1997;119:2016–24.
184. Jensen F. Structures and stability of complexes of glycine and glycine methyl analogs with H^+, Li^+, and Na^+. J Am Chem Soc. 1992;114:9533–7.
185. Wyttenbach T, Witt M, Bowers MT. On the stability of amino acid zwitterions in the gas phase: The influence of derivatization proton affinity, and alkali ion addition. J Am Chem Soc. 2000;122:3458–64.
186. Klassen JS, Anderson SG, Blades AT, Kebarle P. Reaction enthalpies for $M^+L = M^+ + L$, where $M^+ = Na^+$ and K^+ and L = acetamide, N-methylacetamide, N, N-dimethylacetamide,

glycine, and glycylglycine, from determinations of the collision-induced dissociation thresholds. J Phys Chem. 1996;100:14218–27.
187. Moision RM, Armentrout PB. Experimental and theoretical dissection of sodium cation/glycine interactions. J Phys Chem A. 2002;106:10350–62.
188. Moision RM, Armentrout PB. An experimental and theoretical dissection of potassium cation/glycine interactions. Phys Chem Chem Phys. 2004;6:2588–99.
189. Ruan C, Rodgers MT. Cation-π interactions: structures and energetics of complexation of Na^+ and K^+ with the aromatic amino acids, phenylalanine, tyrosine, and tryptophan. J Am Chem Soc. 2004;126:14600–10.
190. Moision RM, Armentrout PB. The special five-membered ring of proline: an experimental and theoretical investigation of alkali metal cation interactions with proline and its four- and six-membered ring analogues. J Phys Chem A. 2006;110:3933–46.
191. Ye SJ, Clark AA, Armentrout PB. Experimental and theoretical investigation of alkali metal cation interactions with hydroxyl side-chain amino acids. J Phys Chem B. 2008;112:10291–302.
192. Heaton AL, Moision RM, Armentrout PB. Experimental and theoretical studies of sodium cation interactions with the acidic amino acids and their amide derivatives. J Phys Chem A. 2008;112:3319–27.
193. Heaton AL, Armentrout PB. Experimental and theoretical studies of potassium cation interactions with the acidic amino acids and their amide derivatives. J Phys Chem B. 2008;112:12056–65.
194. Armentrout PB, Gabriel A, Moision RM. An experimental and theoretical study of alkali metal cation/methionine interactions. Int J Mass Spectrom. 2009;283:56–68.
195. Armentrout PB, Armentrout EI, Clark AA, Cooper TE, Stennett EMS, Carl DR. An experimental and theoretical study on alkali metal cation interaction with cysteine. J Phys Chem B. 2010;114:3927–37.
196. Bowman VN, Heaton AL, Armentrout PB. Metal cation dependence of interactions with amino acids. Bond energies of Rb^+ by Gly, Ser, Thr, and Pro. J Phys Chem B. 2010;114:4107–14.
197. Armentrout PB, Chen Y, Rodgers MT. Metal cation dependence of interactions with amino acids: bond energies of Cs^+ to Gly, Pro, Ser, Thr, and Cys. J Phys Chem A. 2012;116:3989–99.
198. Armentrout PB, Rodgers MT, Oomens J, Steill JD. Infrared multiphoton dissociation spectroscopy of cationized serine: effects of alkali-metal cation on gas-phase conformation. J Phys Chem A. 2008;112:2248–57.
199. Rodgers MT, Armentrout PB, Oomens J, Steill JD. Infrared multiphoton dissociation spectroscopy of cationized threonine: effects of alkali-metal cation size on gas-phase conformation. J Phys Chem A. 2008;112:2258–67.
200. O'Brien JT, Prell JS, Steill JD, Oomens J, Williams ER. Interactions of mono- and divalent metal ions with aspartic and glutamic acid investigated with IR photodissociation spectroscopy and theory. J Phys Chem A. 2008;112:10823–30.
201. Heaton AL, Bowman VN, Oomens J, Steill JD, Armentrout PB. Infrared multiple photon dissociation spectroscopy of cationized asparagine: effects of metal cation size on gas-phase conformation. J Phys Chem A. 2009;113:5519–30.
202. Carl DR, Cooper TE, Oomens J, Steill JD, Armentrout PB. Infrared multiple photon dissociation spectroscopy of cationized methionine: effects of alkali-metal cation size on gas-phase conformation. Phys Chem Chem Phys. 2010;12:3384–98.
203. Citir M, Stennett EMS, Oomens J, Steill JD, Rodgers MT, Armentrout PB. Infrared multiple photon dissociation spectroscopy of cationized cysteine: effects of metal cation size on gas-phase conformation. Int J Mass Spectrom. 2010;297:9–17.
204. Citir M, Hinton CS, Oomens J, Steill JD, Armentrout PB. Infrared multiple photon dissociation spectroscopy of cationized histidine: effects of metal cation size on gas-phase conformation. J Phys Chem A. 2012;116:1532–41.

205. Armentrout PB, Citir M, Chen Y, Rodgers MT. Thermochemistry of alkali metal cation interaction with histidine: influence of the side chain. J Phys Chem A. 2012;116:11823–32.
206. Armentrout PB, Yang Bo, Rodgers MT. Metal cation dependence of interaction with amino acids: bond energies of Rb^+ and Cs^+ to Met, Phe, Tyr, and Trp. J Phys Chem B. 2013;117:3771–81.
207. Heaton AL, Ye SJ, Armentrout PB. Experimental and theoretical studies of sodium cation complexes of the deamidation and dehydration products of asparagine, glutamine, aspartic acid, and glutamic acid. J Phys Chem A. 2008;112:3328–38.
208. Bojesen G, Breindahl T, Andersen UN. On the sodium and lithium ion affinities of some alpha amino acids. Org Mass Spectrom 1993;28:1448–52.
209. Andersen UN, Bojesen G. The order of lithium ion affinities for the 20 common alpha-amino acids. Calculation of energy-well depth of ion-bound dimmers. J Chem Soc Perkins Trans. 1997;2:323–7.
210. Cerda BA, Wesdemiotis C. Zwitterionic vs. charge-solvated structures in the binding arginine to alkali metal ions in the gas phase. Analyst. 2000;125:657–60.
211. Ryzhov V, Dunbar RC, Cerda BA, Wesdemiotis C. Cation-π effects in the complexation of Na^+ and K^+ with Phe, Tyr, and Trp in the gas phase. J Am Soc Mass Spectrom. 2000;11:1037–46.
212. Kish MM, Ohanessian G, Wesdemiotis C. The Na^+ affinities of alpha-amino acids: side-chain substituent effects. Inter J Mass Spectrom. 2003;227:509–24.
213. Feng WY, Gronert S, Lebrilla CB. The lithium cation binding energies of gaseous amino acid. J Phys Chem A. 2003;107:405–10.
214. Ye SJ, Armentrout PB. Absolute thermodynamic measurements of alkali metal cation interactions with a simple dipeptide and tripeptide. J Phys Chem A. 2008;112:3587–96.
215. Roux B., Karplus M. Potential-energy function for cation-peptide interactions—An *ab initio* study. J Comput Chem. 1995;16:690–704.
216. Marino T, Russo N, Toscano M. Interaction of Li^+, Na^+, and K^+ with the proline amino acid. Complexation modes, potential energy profiles, and metal ion affinities. J Phys Chem B. 2003;107:2588–94.
217. Russo N, Toscano M, Grand AE. Lithium affinity for DNA and RNA nucleobases. The role of theoretical information in the elucidation of the mass spectrometry data. J Phys Chem B. 2001;105:4735–41.
218. Russo N, Toscano M, Grand A. Bond energies and attachments sites of sodium and potassium cations to DNA and RNA nucleic acid bases in the gas phase. J Am Chem Soc. 2001;123:10272–9.
219. Russi N, Toscana M, Sicilia E, Grand A. On the interaction between manganese cation (Mn^{2+}) and nucleic acid bases (T, U, C, A, G) in the gas phase. Int J Quantum Chem. 2002;90:903–9.
220. Polfer NC, Oomens J, Dunbar RC. IRMPD spectroscopy of metal-ion/tryptophan complexes. Phys Chem Chem Phys. 2006;8:2744–51.
221. Forbes MW, Bush MF, Polfer NC, Oomens J, Dunbar RC, Williams ER, Jochusch RA. Infrared spectroscopy of arginine cation complexes: direct observation of gas-phase zwitterions. J Phys Chem A. 2007;111:11759–770.
222. Rodgers MT, Armentrout PB. Influence of d orbital occupation on the binding of metal ions to adenine. J Am Chem Soc. 2002;124:2678–91.
223. Kabelac M, Hobza P. Na^+, Mg^+, and Zn^{2+} binding of all tautomers of adenine, cytosine, and Thymine and the eight most stable keto/enol tautomers of guanine: a correlated *ab initio* quantum chemical study. J Phys Chem B. 2006;110:14515–23.
224. Rajabi K, Gillis EAL, Fridgen TD. Structures of alkali metal ion-adenine complexes and hydrated complexes by IRMPD spectroscopy and electronic structure calculations. J Phys Chem A. 2012;114:3449–56.
225. Rodgers MT, Armentrout PB. Noncovalent interaction of nucleic acid bases (Uracil, thymine, and adenine) with alkali metal ions. Threshold collision-induced dissociation and theoretical studies. J Am Chem Soc. 2000;122:8548–58.

226. Rodgers MT, Armentrout PB. Absolute alkali metal ion binding affinities of several azoles determined by threshold collision-induced dissociation. Int J Mass Spectrom. 1999;185–187:359–80.
227. Cerda BA, Wesdemiotis C. Li^+, Na^+, and K^+ binding to the DNA and RNA nucleobases. Bond energies and attachment sites from the dissociation of metal ion-bound heterodimers. J Am Chem. Soc. 1996;118:11884–92.
228. Dilger JM, Valentine SJ, Glover MS, Ewing MA, Clemmer DE. A database of alkali metal-containing peptide cross section: influence of metals on size parameters for specific amino acids. Int J Mass Spectrom. 2012;330–332:35–45.
229. Dunbar RC, Polfer NC, Berden G, Oomens. Metal ion binding to peptides: oxygen or nitrogen sites? Int J Mass Spectrom. 2012;330–332:71–7.
230. Dunbar RC. Complexation of Na^+ and K^+ to aromatic amino acids: a density functional computational study of cation-π interactions. J Phys Chem A. 2000;104:8067–74.
231. Gapeev A, Dunbar RC. Cation-π interactions and the gas-phase thermochemistry of the Na^+/phenylalanine complex. J Am Chem Soc. 2001;123:8360–5.
232. Reddy AS, Sastry GM, Sastry GN. Cation-aromatic database. Proteins: Struct Funct Bioinf. 2007;67:1179–84.
233. Umadevi D, Sastry GN. Molecular and ionic interaction with graphene nanoflakes: a computational investigation of CO_2, H_2O, Li, Mg, Li^+, and Mg^{2+} interaction with polycyclic aromatic hydrocartbons. J Phys Chem C. 2011;115:9656–67.
234. Marquez A., Vargas A, Balbuena PB. Computational studies of lithium interaction in model graphite in the presence of tetrahydrofuran. J Electrochem Soc. 1998;145:3328–34.
235. Nakadaira M, Saito R, Kimura T, Dresselhaus G, Dresselhaus MS. Excess Li ions in a small graphite cluster. J Mater Res. 1997;12:1367–75.
236. Suzuki T, Hasegawa T, Mukai SR, Ta'om H. A theoretical study on storage states of Li ions in carbon anodes of Li ion batteries using molecular orbital calculations. Carbon. 2003;41:1933–9.
237. Tachikawa H, Shimizu A. Diffusion dynamics of the Li^+ ion on a model surface of amorphous carbon: a direct molecular orbital dynamics study. J Phys Chem B. 2005;109:13255–62.
238. Moradi M, Peyghan AA, Bagheri Z. Kamfiroozi M, Cation-π interaction of alkali metal ions with C_{24} fullerene: a DFT study. J Mol Model. 2012;18:3535–40.
239. Zhao YL, Pan XM, Zhou DF, Su ZM, Wang RS. Theoretical study on C_{60}-doped polyacenic semiconductor (PAS) interacting with lithium. Synth Met. 2003;135:227–8.
240. Santos JD, Bullhoes LOS, Longo E, Varela JA. Interaction between Li^+ and C_{60} molecules. Theochem. 1995;335:149–52.
241. Tachikawa H. Diffusion dynamics of the Li ion on C_{60}: a direct molecular orbital-molecular dynamics study. J Phys Chem C. 2007;111:13087–91.
242. Tachikawa H. Diffusion of the Li^+ ion on C_{60}: a DFT and molecular dynamics study. J Phys Chem C. 2011;115:20406–11.
243. Mpourmpakis G, Tylianakis E, Papanikolaou D, Froudakis G. Theoretical study of alkaline metal cations in carbon nanotubes. Rev Adv Mater Sci. 2006;11:92–7.
244. Umadevi D, Sastry GN. Metal ion binding with carbon nanotubes and graphene: effects of chirality and curvature. Chem Phys Lett. 2012;549:39–43.

Chapter 4
Experimental Methods and Instrumentation

Wilfried M.A. Niessen

4.1 Introduction

This chapter discusses a variety of experimental methods and introduces instrumentation for the study of gas-phase ions and ion–molecule chemistry in general and of ion attachment processes in particular, with an emphasis on mass spectrometry (MS) methods. The section starts with a discussion on common methods to generate gas-phase alkali metal ions, to be applied in various ion attachment experiments. Subsequently, (tandem) MS methods are discussed that enable the investigation of the structure of ions and the results of ion–molecule reactions. Separate sections are presented for beam instruments and ion-trapping instruments. In some cases, these instruments can be used to perform ion–molecule reactions as part of the measurement protocol. Emphasis is put on commercially available MS and MS–MS instruments. In subsequent sections, various other tools are discussed that may be combined with MS and allow the study of gas-phase reactions of ions. These tools comprise (a) flowing afterglow methods (FA-MS), including derived methods like selected ion flow tubes (SIFT-MS) and proton-transfer reaction devices (PTR-MS), (b) drift tubes, ion-mobility spectrometry (IMS) and IMS–MS, and (c) high-pressure MS instruments. The text provides ample references for further reading.

Currently, MS is primarily used in analytical applications. In fact, rather than the study of ion structures, gas-phase ion–molecule reactions, or the structure elucidation of unknown compounds, the routine quantitative analysis of target analytes in complex (biological) matrices using combined gas chromatography (GC–MS) or liquid chromatography (LC–MS) is by far the most important application area of MS. Nevertheless, mass spectrometers have proven to be powerful tools for studying the kinetics, mechanisms, and product distributions of gas-phase bi- and termolecular organic reactions. A wide variety of ion–molecule reactions may be studied [1].

W. M. A. Niessen (✉)
hyphen MassSpec, de Wetstraat 8, 2332 XT Leiden, The Netherlands
e-mail: mail@hyphenms.nl

T. Fujii (ed.), *Ion/Molecule Attachment Reactions: Mass Spectrometry*,
DOI 10.1007/978-1-4899-7588-1_4

These studies provide us with fundamental understanding of the organic reactions. Gas-phase chemistry can reveal details of reaction mechanisms that are obscured by solvation and ion pairing, when studied in the condensed phase.

4.2 Production of Gas-phase Alkali Metal ions

4.2.1 Thermoionic Emission

Thermionic emission involves the heat-induced emission of charge-carrying particles from a surface. The process occurs when the thermal energy given to the carrier overcomes the work function of the metal. The charge-carrying particles may be electrons or ions. The emission of electrons, known as the Edison effect, can be achieved from a heated filament in vacuum, like with the hot (tungsten or rhenium) filament used as primary source of ionization in electron ionization. In the current context, the emission of (alkali) metal ions is of more interest. This can be achieved by heating aluminum silicates doped with (alkali) metal oxides [2–5]. A schematic setup of a thermoionic emitter of K^+ mounted onto a solid probe is shown in Fig. 4.1 [5]. In this way, chemical ionization with alkali metal ions as primary ionization source was performed, e.g., studying gas-phase reactions of Li^+ with fluoroethane and hydrocarbons [6] or of K^+ with a wide variety of analytes, including ketones, ethers, esters, crown ethers, and small peptides [5]. Li^+-emitters are also applied in ion attachment MS experiments [3].

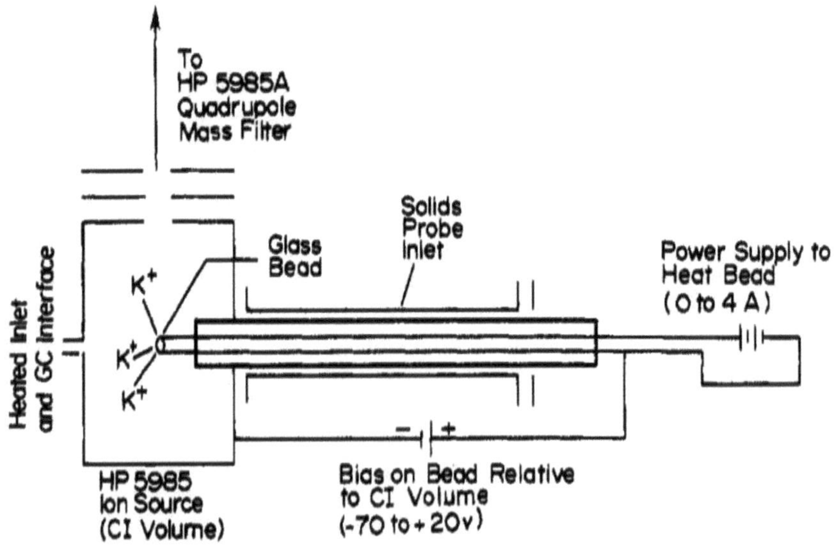

Fig. 4.1 Ion source for thermoionic emission of alkali metal ions. (Reprinted with permission from ref. 5. ©1984, American Chemical Society)

4.2.2 *Laser Ionization*

Laser ablation is the process of removing material from a solid surface by irradiating it with a (pulsed) laser beam. At low laser flux, the material evaporates or sublimates due to heating by the absorbed laser energy. At high laser flux, the material is converted into plasma. If ions are produced in the plasma, the process is usually called laser ionization. In this way, gas-phase metal ions can be produced, which can subsequently be applied in gas-phase ion–molecule studies. Initially, Cu^+ and Ag^+ were generated from their respective metals, and Cr^+, Fe^+, and Ni^+ from a stainless steel sample [7]. The method seems to be especially useful to generate gas-phase ions and ion clusters of transition metals [8, 9]. As such, it was, for instance, applied to characterize pollutants on dust particles [10].

4.2.3 *Desorption Ionization: Doping with Alkali Metal Salts*

In desorption ionization methods, like field desorption [11], fast-atom bombardment [12], and matrix-assisted laser desorption (MALDI) [13], adding small amounts of alkali metal salts to the sample on the emission wire (in field desorption) or target (in fast-atom bombardment and MALDI) may result in the observation of adduct ions $[M + Alkali]^+$ (see also Sect. 7.2). This is generally termed "doping with alkali metal salts." In principle, when a platinum or tungsten wire, to be used in thermoionic emission, is doped with alkali metal salts, gas-phase alkali metal ions may be produced as well. It may be questioned whether this is due to a thermoionic or desorption/ionization effect.

A typical example of this type of doping is a fragmentation study of the flavonoid glycoside rutin, adducted with different alkali metal ions (Li^+, Na^+, or K^+), by post-source decay MALDI time-of-flight (TOF) MS [14]. Differences in fragment ions and especially the relative abundance of fragment ions were observed (see Sect. 7.5.2). Numerous other examples of this type of studies are available.

4.2.4 *Solvent Additives in Electrospray Ionization*

Adduct formation of analytes to alkali metal ions is also frequently observed in electrospray ionization (ESI) MS [15]. For compounds with high affinities to alkali metal ions, the residual concentrations of 10^{-5}–10^{-4} M of alkali metal ions, commonly present in solvents used in LC-MS, is sufficient for this type of adduct formation. Alternatively, low concentrations of alkali metal ion salts (up to 1 mM) may be added to the mobile phase; higher concentrations (> 1 mM) result in significant ionization suppression. Because in many applications of ESI-MS compounds are analyzed from biological matrices; these matrices also act as a source of alkali metal ions, especially if no rigorous desalting protocol is adapted.

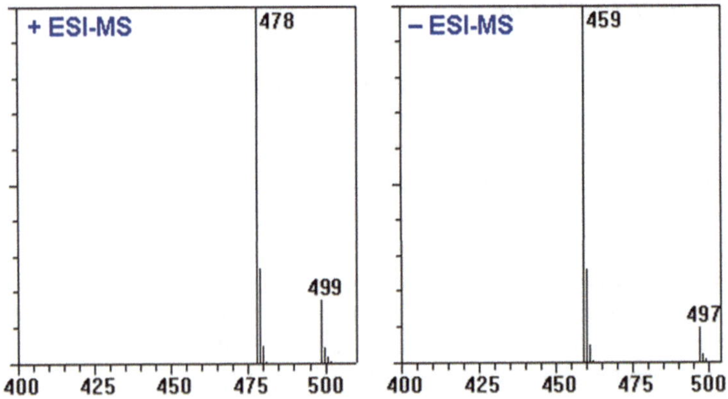

Fig. 4.2 Electrospray mass spectra of a compound with two carboxylic acid functions analyzed as dipotassium salt in a mobile phase containing ammonium acetate. (©2006, hyphen MassSpec)

It appears that adduct formation especially takes place for compounds with a number of oxygen (or sulfur) atoms in such an orientation that chelation or complex formation with the alkali metal ions is possible (see Chap. 7 for a more detailed discussion). This is, for instance, the case with oligosaccharides. However, next to "real" adduct formation, that is formation of $[M+Na]^+$, apparent adduct formation in ESI-MS may be due to liquid-phase H^+/Na^+-exchange reactions in analytes with acidic functions, resulting in salts which are subsequently transferred to the gas phase as protonated molecules [15]. Thus, the analyte RCOOH is converted into RCOONa, protonated, and transferred to the gas-phase for mass analysis as $[(RCOOH–H+Na)+H]^+$ rather than as $[M+Na]^+$ [15, 16]. This behavior is observed with any compound with acidic functions, e.g., peptides and (oligo) nucleotides. The possibility to generate such H^+/Na^+-exchange reaction products by a gas-phase reaction in the vacuum interface of an ESI-MS system has also been investigated [17]. As expected, methyl ester formation in peptides reduces the adduct formation, because the H^+/Na^+-exchange is taken away [18].

An example of this behavior is shown in Fig. 4.2, where the positive-ion and negative-ion ESI-MS spectra of a compound are shown with two carboxylic acid functions, analyzed as di-potassium salt in a mobile phase containing ammonium acetate. In positive-ion mode, both the ammoniated molecule $[M+NH_4]^+$ and the potassiated molecule $[M+K]^+$ (or the H^+/K^+-exchange product $[(M–H+K)+H]^+$) are observed, whereas in negative-ion mode, the H^+/K^+-exchange product $[(M–H+K)–H]^-$ is observed next to the deprotonated molecule $[M–H]^-$. Thus, apparent alkali metal ion adducts may be observed in negative-ion ESI-MS as well.

4.3 Tandem Mass Spectrometry

4.3.1 Introduction

From instrumental point of view, tandem mass spectrometry (MS–MS) comprises a combination of two mass analyzers in series (in time or space, see below) with a reaction chamber in between. In MS–MS, the *m/z* values of ions are measured before and after a chemical reaction within the mass spectrometer. In most cases, a change in mass and thus in *m/z* is involved, although a change in charge, e.g., charge stripping of multiple-charge ions, is also possible. For positive ions, the precursor or parent ion m_p^+ is converted into the product or daughter ion m_d^+ via the loss of a neutral fragment m_n. Whereas the neutral fragment m_n is generally not detected in the mass spectrometer, its mass can be inferred from the difference in *m/z* of m_p^+ and m_d^+. In the product-ion analysis mode, which is the most basic MS–MS experiment, the precursor ion m_p^+ is selected in the first stage of mass analysis within the instrument (MS1), while the product ions m_d^+ are mass analyzed and detected in the second stage of mass analysis (MS2). Thus, MS–MS involves the detection of ions that, after their initial formation in the ion source, have undergone a change in *m/z* and/or charge during the course of their analysis with a mass spectrometer [19].

The observation and explanation in the 1940s of the occurrence of metastable ions in a mass spectrum acquired using a magnetic-sector instrument can be considered as the starting point of the history of MS–MS [20]. In the 1960s, it was discovered that the abundance of the metastable ions can be increased by the introduction of a collision gas in a collision cell. From the mid-1970s onwards, instruments were especially designed for MS–MS experiments. Important landmarks in the development of MS–MS are: (a) the introduction in 1978 of the triple quadrupole (TQ) instrument by Yost and Enke [21]; (b) the demonstration in 1987 of (multiple stages of) MS–MS in an ion trap instrument by Louris et al. [22]; (c) the introduction in the early 1990s of various MS–MS technologies in Fourier transform ion cyclotron resonance (FT-ICR) instrument [23]; (d) the introduction in 1996 of a hybrid quadrupole-time-of-flight instrument (Q-TOF) by Morris et al. [24]; (e) the introduction in 2002 of the hybrid quadrupole-linear ion trap instrument (Q-LIT) by Hager [25]; (f) the introduction in 2002 of tandem TOF–TOF instrument [26], and (g) the introduction in 2005 of the hybrid linear ion trap–Orbitrap instrument by the group of Makarov [27, 28].

Convenient symbolism and terminology for the wide variety of MS–MS and MS^n experiments that can be performed in various data acquisition modes have been proposed by Schwartz et al. [29] and are frequently used ever since.

Nowadays, partly initiated by the advent of ESI as a powerful soft ionization technique for highly polar biomolecules and as a convenient method to couple LC and MS, MS–MS is frequently applied. For routine quantitative bioanalysis of target compounds, the selected reaction monitoring (SRM) mode is extensively applied. In the SRM mode, both stages of mass analysis perform the selection of ions with a particular *m/z* value, i.e., in MS1 a precursor ion, mostly the protonated or deprotonated molecule of the target analyte, is selected, subjected to dissociation in the collision cell, while in MS2 a preferably structure-specific product ion of the selected precursor is selected and detected. Due to the high selectivity involved in SRM, excellent sensitivity may be achieved in target quantitative analysis. The SRM mode is the method of choice in quantitative bioanalysis using LC-MS [30], e.g., in (pre)clinical studies for drug development within the pharmaceutical industry, and has been implemented in quantitative analytical strategies using GC-MS as well [31].

4.3.2 Ion Dissociation Techniques

Either metastable ions or activated ions may be involved in MS–MS experiments. Metastable ions are ions with sufficient internal energy that survive long enough to be extracted from the ion source before they fragment, but may then fragment in the mass analyzer region prior to detection. The charged fragments of metastable ions that dissociate in the reaction region of the instrument may be detected. Alternatively, ions may be activated after they have left the ion source. Collision activation is the most widely applied method to increase the internal energy of ions. Upon acceleration and collision of an ion with a target gas (He, N_2, or Ar) in a collision cell, part of the ion translational energy is converted into internal energy. If subsequent dissociation of the ion occurs in the collision cell, the process is called collision-induced dissociation (CID). Next to collision activation with a target gas, there is a wide variety of other activation methods [32, 33], including surface-induced dissociation (SID), laser photodissociation (LPD), infrared multiphoton photodissociation (IRMPD), sustained off-resonance irradiation (SORI), black-body infrared radiative dissociation (BIRD), electron-capture dissociation (ECD), and electron-transfer dissociation (ETD) [34]. Most of these alternative techniques are primarily applied to induce fragmentation in FT-ICR-MS instruments, although ETD can also be implemented on ion trap (and other) instruments [34].

The CID, being the most widely applied method to induce fragmentation in MS–MS, is a two-step process, where in the first step, ion translational energy is converted into ion internal energy due to the collision event, while in the second step unimolecular decomposition of the excited ions may yield various product ions by competing reaction pathways. The first step is much faster than the second one. In between the two steps of the process, energy redistribution within the ion may take place. CID can be performed in two different energy regimes [33]. With most instruments, low-energy CID is performed involving multiple collisions with a target gas such as He, N_2, or Ar ($\sim 10^{-3}$ mbar) with a laboratory collision energy generally not exceeding 60 eV. In sector and TOF–TOF instruments, high-energy CID can be performed, which involves single keV collisions with He as a target gas. High-energy collisions open a wider range of fragmentation reactions, thus resulting in more informative and more complex MS–MS spectra. In the low-energy CID regime, one may further discriminate between collision cell CID and ion trap CID. In collision cell CID, that is, in TQ and Q–TOF instruments, after acceleration of the precursor ions with 10–60 V, collisions are performed with N_2 or Ar.

4.3.3 General Aspects of MS–MS Instrumentation

The MS–MS instrument comprises a combination of two mass analyzers. The first and second stage of mass analysis may be performed by the same type of mass analyzer, like in a triple quadrupole or an ion trap instrument. In TQ instruments, the three steps of the MS–MS process (precursor ion selection, CID, and mass analysis of product ions) are performed in spatially separated devices ("tandem-in-space"), whereas in an ion trap instrument, the three steps are performed one-after-another in the same device ("tandem-in-time") [35]. In hybrid instruments, the first and second stage of mass analysis is performed in two different types of mass analyzers, e.g., in a first-stage quadrupole and second-stage TOF in a Q–TOF instrument, or a first-stage quadrupole and second-stage linear ion trap in a Q–LIT instrument.

An MS–MS instrument may be used to study the fragment ions of selected precursor ions and is therefore an indispensable tool in fundamental studies on ion generation, ion–molecule reactions, unimolecular fragmentation reactions, and identity of ions. It also plays an important role in analytical applications of MS, both in qualitative and in quantitative analysis, e.g., in applications involving the online coupling of MS as a detector to GC–MS and LC–MS.

4.3.4 MS–MS in Sector Instruments

As indicated before, the basis of MS–MS lies in the observation of metastable ions or, perhaps more accurately, of the fragment ions of metastable ions. In order to detect these ions, linked scan procedures are required. In a magnetic sector instrument, featuring kV-acceleration of ions from the ion source and thus providing significant kinetic energy to the ions, an in-source-generated fragment ion with a particular *m/z* has a higher kinetic energy than a post-source-generated fragment ion, the so-called metastable ion. Adjustment of the acceleration voltage (V), or the electric (E) and magnetic (B) fields in a double-focusing sector instrument is required to observe the metastable ions or their fragments. In a linked scan, the two fields are automatically adjusted at the same time. In most cases, linked scan procedures are applied in the first field-free region, e.g., E^2/V or B/E to observe the fragment ions and B^2/E to observe the precursor ions [19, 36].

As a result of the introduction of alternative MS–MS instruments, which are more cost-effective and easier to operate, the double-focusing sector instruments are hardly used in MS–MS. The same holds for hybrid MS–MS systems comprising sector instruments combined with quadrupole or ion trap building blocks.

4.3.5 MS–MS in Tandem Quadrupole Instruments

Probably, the most widely used MS–MS configuration is the TQ instrument, where mass analysis is performed in the first and third quadrupoles, while the second

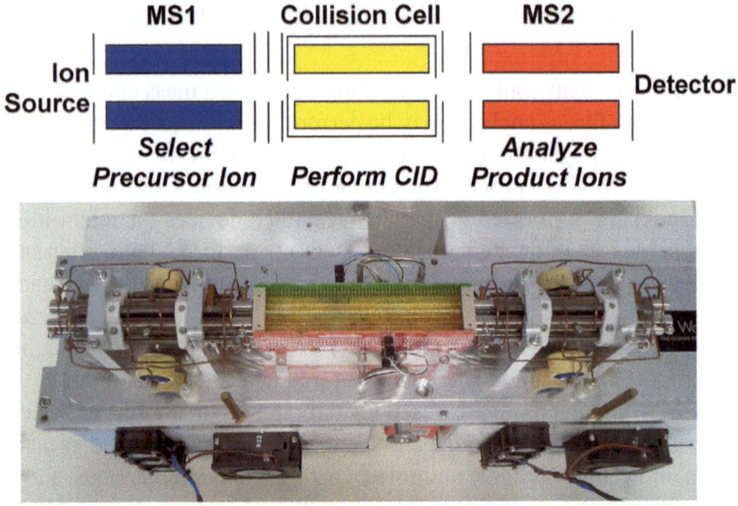

Fig. 4.3 Schematic diagram and photo of tandem quadrupole mass spectrometer. The photo is taken from a Waters tandem-quadrupole instrument, featuring a travelling wave stacked ring RF-only collision cell. (©2013, hyphen MassSpec)

quadrupole is used as collision cell in the radiofrequency (RF)-only mode, i.e., in a Q–q_{coll}–Q configuration (Fig. 4.3). The TQ instrument was initially developed by Yost and Enke [21] in the late 1970s. Upon its introduction, the TQ instrument yielded significantly better product-ion resolution than the sector instruments, used for MS–MS at that time. The acquisition of a product-ion spectrum was greatly facilitated.

In addition, various structure-specific screening procedure of triple quadrupole instruments were introduced, e.g., the precursor ion and neutral loss analysis mode [37, 38]. In the precursor ion analysis mode, MS1 is operated in scanning or full-spectrum mode, whereas in MS2 a structure-specific product ion is monitored. In the neutral loss analysis mode, both quadrupole mass analyzers are operated in scanning mode, but with a fixed *m/z* offset corresponding to a structure-specific neutral loss in the fragmentation reaction. These scan modes have been successfully applied for structure-specific screening in order to search for specific compound classes in complex matrices, as demonstrated by the screening for two classes designer drugs in urine [39], or for glutathione and cyanide-trapped reactive drug metabolites [40], to quote two recent examples. Apart from these analytical applications, these analysis modes can be very useful in the study of fragmentation reactions. The precursor ion analysis mode allows to determine which is (or are) the precursor ion(s) of a particular product ion, thus answering the question whether a particular product ion is formed from a particular precursor ion in a one-step dissociation reaction or an intermediate step has been involved. This enables more detailed studies on fragmentation pathways, as for instance demonstrated for naloxonazine and naloxone, zwitterionic morphine opiate antagonists [41], and for morphine and related opiates [42].

Although usually called a triple quadrupole instrument, because of the initial lineup of two analyzing quadrupoles and a quadrupole collision cell, the term tandem quadrupole (TQ) would nowadays be more appropriate to describe the commercially available instruments. The gas-filled collision cells operated in RF-only mode and enabling refocusing of ions scattered by collisions result in significant transmission losses. In attempts to reduce these losses, alternative collision cells have been implemented by various instrument manufacturers, e.g., RF-only hexapoles or octapoles, a linear acceleration high-pressure collision cell (LINAC) [43], or a stacked ring collision cell (see Fig. 4.3), featuring an axial travelling wave or transient DC voltage to propel the ions and to reduce the transit times [44].

4.3.6 MS–MS in Q-TOF Instruments

The Q–TOF instrument can be considered as a modified TQ instrument, where the third quadrupole has been replaced by an orthogonal acceleration reflectron-TOF mass analyzer. The first commercially available Q–TOF instrument was produced in 1996, especially aiming at peptide sequencing analysis [24], but the instrument has found much wider application, especially in the structure elucidation studies. Q–TOF instruments are now available from various instrument manufacturers. Where fragmentation characteristics in the collision cell are the same for TQ and Q–TOF, a significant advantage of Q–TOF over TQ in structure elucidation is the ability to determine the *m/z* values of both precursor and product ions with higher mass accuracy (<5 ppm), which is due to the high resolving power of the reflectron-TOF analyzer. Principles and applications of Q–TOF hybrid instruments have been reviewed by Chernushevich et al. [45].

4.3.7 MS–MS in Q–LIT instruments

The hybrid quadrupole-linear ion trap (Q–LIT) instrument, introduced in 2002, also has the general layout of a TQ instrument, but in the Q–LIT, the third quadrupole can be operated (under software control) as either a normal linear quadrupole or a linear ion trap [25]. When used as a linear ion trap, it provides accumulation of ions prior to detection, thus enabling ion trap full-spectrum sensitivity, while still acquiring collision cell CID spectra. Hardware, electronics, and software control of the Q–LIT instrument have been optimized to allow very rapid switching between various MS and MS–MS experiments. The potential of Q–LIT instruments in structure elucidation can be readily demonstrated, e.g., by comparing the information content between collision cell and ion trap product ion mass spectra [46]. Commercial Q–LIT systems provide a wide variety of typical operating modes [47] (see also: Sect. 4.4.5).

4.3.8 Reactive Gases in Collision Cells

Instead of using inert collision gases like He, N_2, or Ar in CID, one might also consider using reactive gases. The use of reactive collision gases like H_2, CH_4, and NH_3 in RF-only quadrupoles, hexapoles, or octapoles has become a common practice in inductively coupled plasma (ICP) MS in order to reduce contributions of interfering ions in trace level determination of elements by ICP-MS [48, 49]. Different setups have been described by different instrument manufacturers.

Reactive collision gases have also been used in organic MS. Pioneering experiments involved the study of the formation of an adduct ion between protonated esters and ammonia in the collision cell of a TQ instrument [50]. Since then, numerous examples have been reported such as the study of gas-phase reactions of dilactones, psorospermin, and quabalactone diterpenes [51], different phospholipid classes [52], ethyl vinyl ether, the transacetalization with gaseous carboxonium and carbosulfonium ions by collision cell reactions with cyclic acetals and ketals [53], and ketalization of phosphonium ions using 1,4-dioxane [54].

This type of gas-phase ion–molecule reactions have especially been used to discriminate between isomeric species. Collision cell reactions of isomeric tetrachlorodibenzo-*p*-dioxins (TCDDs) molecular anions, generated by electron capture negative ionization, and O_2 were applied to discriminate isomeric forms by determining the number of chlorine atoms on each ring [55]. Other examples comprise reactive collisions between isomeric $C_2H_3O^+$ and $C_2H_6O^+$ ions and 1,3-butadine or benzene [56], and differentiation of isomeric 1,2-cyclopentadiols via reactive collision with NH_3 [57].

Data acquisition in the neutral gain analysis mode, where both quadrupole mass analyzers are operated in scanning mode but with a fixed *m/z* offset depending on the reaction performed, has been applied to efficiently monitor these type of gas-phase reactions in complex mixtures of, e.g., phospholipids [52] and phosphonium ions in relation to chemical warfare agents [54]. The ethyl vinyl ether reaction with phospholipids yields a neutral gain of 26 Da for phosphatidylglycerols and of 57 Da for phosphatidylinositols [52], whereas the reaction of the phophonium ion with 1,4-dioxane yields a neutral gain of 44 Da [54].

A pentaquadrupole (PQ-MS) instrument, thus featuring Q–q_{coll}–Q–q_{coll}–Q, has been developed and extensively used for the study of gas-phase ion–molecule reactions [58, 59]. The PQ-MS instruments obviously allow sequential product-ion analysis, i.e., MS^3 experiments, but other types of experiments are possible as well. The PQ-MS instrument can be used to perform reactive collision of selected ions either in the first or the second collision cell, as illustrated with a wide variety of examples [59].

4.4 Ion-Trapping Devices: Quadrupole Ion Traps and FT-ICR

4.4.1 Introduction

In most of its applications, the mass spectrometer can be considered as a detector, providing a tool to perform mass analysis, that is to separate ions according to their *m/z* value either in time, as is mostly done, or in space. Some instruments, especially those providing the trapping of ions over a considerable period of time, allow additional experiments as during trapping some parameters may be changed, reactive species may be introduced, and the results of such actions may be monitored. The instruments showing best perspectives in this respect are FT-ICR and ion trap instruments. This section provides an introduction to MS using ion traps and FT-ICR instruments and highlights some examples in fundamental ion chemistry studies.

4.4.2 Ion Trap Mass Analysis

A typical ion trap, also called quadrupole ion trap or three-dimensional ion trap, consists of a cylindrical ring electrode to which a quadrupole RF field is applied, and two end-cap electrodes [60, 61] (Fig. 4.4). The end-cap electrodes contain holes for the introduction of ions from an external ion source and for the ejection of ions out of the trap towards the external electron multiplier detector. The ion trajectories in the trap are stabilized by a He bath gas (~1 mbar). With respect to ion trap actions, the individual steps in the mass analysis process are performed

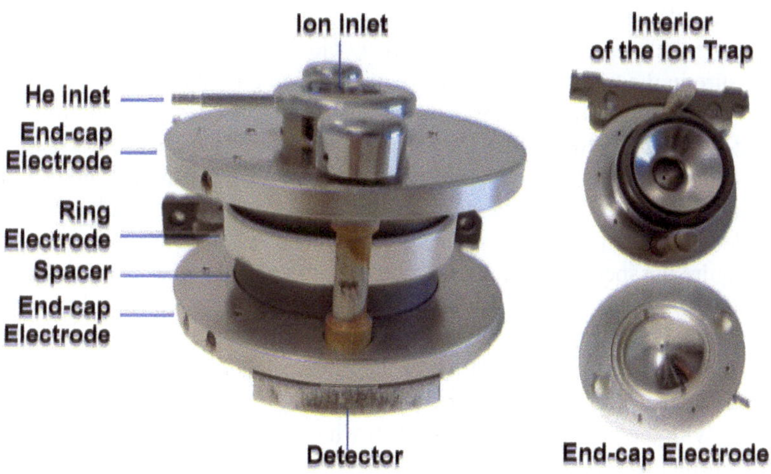

Fig. 4.4 Explanatory photo of a three-dimensional ion trap featuring essential elements of the device. (©2013, hyphen MassSpec)

consecutively in time. Thus, the acquisition of a mass spectrum requires two steps: (a) injection of ions by means of an ion injection pulse of variable duration and storage of the ions in the trap by application of an appropriate low RF voltage to the ring electrode, (b) ramping the RF voltage at the ring electrode (and eventually the application of additional waveforms to the end-cap electrodes) to consecutively eject ions with different *m/z*-values (resonant ion ejection) from the trap towards the external detector [60].

As the number of ions that can be stored in the trap is limited by space charge effects, software procedures have been developed to control the number of ions in the ion trap by varying the duration of the ion injection pulse from the external ion source with the ion current at the time [61]. Too high numbers of ions in the ion trap will adversely influence mass resolution and accuracy. Ion ejection and subsequent detection can be achieved with unit-mass resolution, or at enhanced resolution by slowing down the scan rate.

More recently, instruments with a linear two-dimensional ion trap (LIT), i.e., a linear quadrupole as ion trap, have become commercial available [24, 62, 63]. As a LIT is less prone to space charging effects, a higher number of ions can be accumulated, and enhanced sensitivity can be achieved. Initially, LITs were applied in hybrid Q–LIT [24] a LIT and hybrid LIT–FT-ICR-MS [64] instruments, but later on stand-alone versions of an LIT were introduced too, thus competing the three-dimensional ion traps. A dual-pressure two-stage LIT has been reported as well: the first high-pressure ion trap serves to capture, select, and fragment ions, whereas the second low-pressure ion trap is used to perform fast scanning of product ions, eventually at enhanced resolution [65].

4.4.3 MS^n in Ion Trap Instruments

Multistage MS–MS (MS^n) in ion traps is based on three features of the ion trap technology: the possibility to *m/z*-selectively eject ions from the trap, the constant He pressure in the trap which may serve as collision gas, and the possibility to apply an *m/z*-selective RF waveform to the end-cap electrodes to excite ions of the selected *m/z*. Thus, after filling the ion trap with ions in the usual way, MS–MS can be achieved by (a) selective ejection of all ions except the precursor ion, (b) excitation of the selected precursor ion while a low RF voltage is applied to the ring electrode to trap the product ions generated, (c) scan out the ions towards the detector to acquire the spectrum. Excitation of the ions means that the selected ions move with wider amplitude, and thus at greater speed through the ion trap. The resulting more energetic collisions with the He atoms of the bath gas lead to a gain in internal energy and subsequent fragmentation of the precursor ion.

However, the process can also be considered in more general terms (Fig. 4.5): one starts with the population of ions, from which the precursor ion is selected, excited, and fragmented, resulting in a new population of (product or daughter) ions. The latter population can either be scanned out to be detected, or can serve in a new

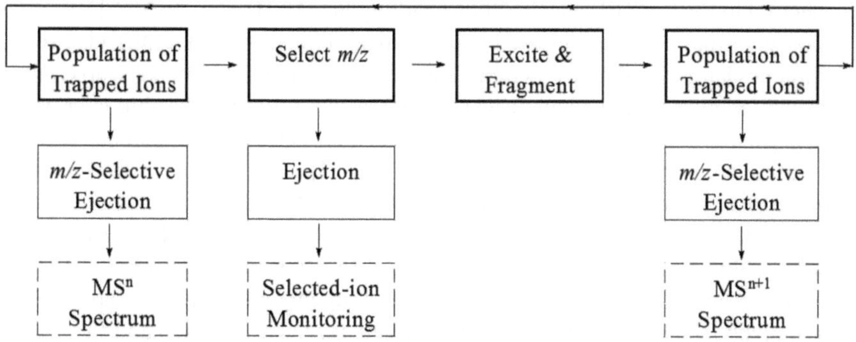

Fig. 4.5 Diagram of the steps involved in multistage MS–MS (MS^n) in an ion trap mass spectrometer. *Top row* represents the processes in the ion trap. By means of the processes in the second row, output may be generated, either the MS^n full spectrum or selected ion (MS^1), selected reaction (MS^2) or consecutive-reaction monitoring ($MS^{>2}$). (©2014, hyphen MassSpec)

series of subsequent steps of the process: selection of a product ion as precursor ion in a new MS–MS experiment, to be excited and fragmented, and leading to a new population of (granddaughter) ions. In this way, multiple stages of MS–MS or MS^n (up to 10 stages in most instruments) can be achieved consecutively ("tandem-in-time" [21]) with the ion trap instrument.

The ion trap MS^n thus enables step-wise fragmentation and the acquisition of fragmentation trees [66]. Such a fragmentation tree is generated by further fragmenting selected fragment ions of a particular stage of MS^n into a next stage, i.e., MS^{n+1}. This provides a wealth of information in structure elucidation and identification of unknowns, as is nicely exemplified for polyphenols [67].

4.4.4 Applications of Ion Trap MS^n

Currently, ion trap MS instruments are primarily used for analytical applications, i.e., in combination with GC or LC in, among other, environmental, food-safety, clinical, pharmaceutical, or biochemical application areas. In many of these applications, the potential of multistage-MS^n plays an important role. The acquisition of fragmentation trees enables detailed study on the fragmentation behavior of target compounds and/or compound classes, e.g., [67].

However, ion trap MS can also be applied in more fundamental ion chemistry studies [68], especially because the reaction time can be varied over several orders of magnitude. Due to the He bath gas in the ion trap, the pressure in the ion trap is several orders of magnitude higher than in an FT-ICR cell, but much lower than in high-pressure MS and FA-MS.

Giving the residence time of ions in the trap, one may anticipate that ion–molecule reactions would readily take place in the ion trap. In fact, self-protonation of several compounds, present at higher concentrations in the ion trap, was observed in electron ionization (EI) spectra, thus resulting in $[M+H]^+$ next to $M^{+\bullet}$ [69]. From this observation, it was anticipated that chemical ionization (CI), either via proton-transfer or charge-exchange reactions, should be possible. Initially, ion trap CI reactions with methane and isobutane as reagent gas were studied [70], but later on the headspace vapor of solvents like water, methanol, acetonitrile, acetone, or furan were used in CI on ion trap MS instruments [71–74]. Given the ease at which such experiments can be done and the interaction time can be varied, gas-phase ion–molecule reactions with unusual reagent gases can be readily performed in an ion trap. Similarly, this type of procedures can be used to determine the proton affinity or gas-phase basicity of compounds via the bracketing method using ion trap MS [75]. Alternatively, the kinetic method can be applied. By mass-selection and subsequent dissociation of proton-bound dimers (A–H^+–B), the proton affinity (PA) can be determined, provided the PA of A is known and several proton-bound dimers with similar functionalities are available [76].

It allows reaction products to be observed from reactions under either kinetic or thermodynamic control. Numerous reports are available, where ion trap MS is applied in ion chemistry studies [68], e.g., involving reactions between 1,4-benzodiazepines and dimethyl ether ions [77], dissociation of $[\text{Alanine} + \text{Alkali cation}]^+$-ions to study the role of the metal cation [78], or regioselective ion–molecule reactions to enable MS differentiation of protonated isomeric aromatic diamines [79]. The three-dimensional ion trap mass spectrometer has even been described as a complete chemical laboratory for fundamental gas-phase studies of metal-mediated chemistry [80].

Ion trap instruments also provide the possibility to study ion–ion reactions, e.g., reactions of multiple-charge peptide or protein ions with ions of opposite polarity. Mostly, proton-transfer reaction are performed, but electron transfer, fluoride transfer reactions, and even attachment reactions may occur as well [81]. Ion traps, either three-dimensional or linear ones, with multiple ion sources are applied in such experiments [82].

As indicated before, the ion trap CID process differs from collision cell CID in a number of ways, including collisions with a smaller target (He instead of Ar), ion excitation by an RF waveform pulse rather than by acceleration of ions in an electric field, and the interaction time, which is milliseconds in ion trap CID rather than microseconds in collision CID. As a result, different fragmentation pathways may be accessible in ion trap CID compared to collision cell CID. Ion trap CID is generally considered to be softer, thus both requiring and enabling multistage MS^n experiments to generate a wealth of structural information. The lower energy involved in ion trap CID also is at the basis of the generation of fragmentation trees and stepwise fragmentation strategies [66]. Another feature is that $[M+Na]^+$-ions may be fragmented in ion trap MS^n, whereas they are generally not in TQ and Q–TOF instruments. This is extensively applied in the structure characterization of oligosaccharides [83]. Interestingly, $[M+Li]^+$-ions, e.g., of (phospho)lipids, may be fragmented by both TQ and ion trap instruments [84, 85], suggesting that it is the short ion residence time that prevents $[M+Na]^+$ to be fragmented in TQ instruments. Other interesting features of the fragmentation of lithiated (phospho)lipids is the loss of lithium salts of fatty acids, next to the loss of fatty acids [84, 85], and the ability to apply charge remote fragmentation of lithiated lithium salts of unsaturated fatty acids to determine double bond positions [86].

4.4.5 *Hybrid Systems Involving Ion Traps*

A number of hybrid tandem mass spectrometer systems, in which ion traps are involved, are commercially available. Ion traps can be applied either in the first stage (MS1) or in the second stage (MS2) of mass analysis.

In the Q–LIT hybrid [25] instruments, an ion trap is implemented as the second stage of mass analysis, either for accumulation of ions to achieve improved sensitivity after collision cell CID [25, 46], and/or to perform MS^n [25, 46, 47]. The Q–LIT instrument can either be operated as a conventional TQ instrument or as the hybrid instrument. In TQ mode, the instrument is capable of all acquisition modes of a TQ, including SRM. In the hybrid mode, full-spectrum data can be acquired in the enhanced product ion (EPI) mode with up to 60-fold enhanced sensitivity compared to TQ instruments. Next to the enhanced multiple-charge scan and the time-delayed fragmentation scan, the system allows the acquisition of MS^3 spectra, with the second dissociation step to be performed in the LIT [46, 47].

If the ion trap is implemented as the first stage of mass analysis (MS1), it generally serves several functions. Next to acting as a "filter" with respect to the number of ions that are transferred to the second stage, based on the duration of the ion accumulation in the ion trap, MS^n experiments may be performed prior to transferring a package of ions to the second stage. Thus, in such hybrid systems, no collision cell has to be present between the first and second stage of mass analysis.

This is true for ion trap hybrids with FT-ICR-MS (see Sect. 4.4.7) and Orbitrap (see Sect. 4.4.9) instruments, but also for the ion trap–time-of-flight hybrid system. The latter system has been pioneered by the group of Lubman [87, 88]. It has become commercial available for both MALDI and LC–MS applications [89]. All these hybrid MS system are frequently applied in combination with LC and electrospray ionization in, for instance, drug metabolite identification studies and in various proteomics-related studies.

4.4.6 *Mass Analysis in Fourier Transform Ion Cyclotron Resonance Instruments*

A Fourier transform ion cyclotron resonance mass spectrometer (FT-ICR-MS) can be considered as an ion trap system, where the ions are trapped in the magnetic field rather than in a quadrupole electric field. The ICR cell is a cubic or cylindrical cell positioned in a strong magnetic field B (up to 15 T). The cell consists of two opposite trapping plates, two opposite excitation plates, and two opposite receiver plates (Fig. 4.6). Extreme high vacuum should be achieved in the cell, e.g., 10^{-9} mbar.

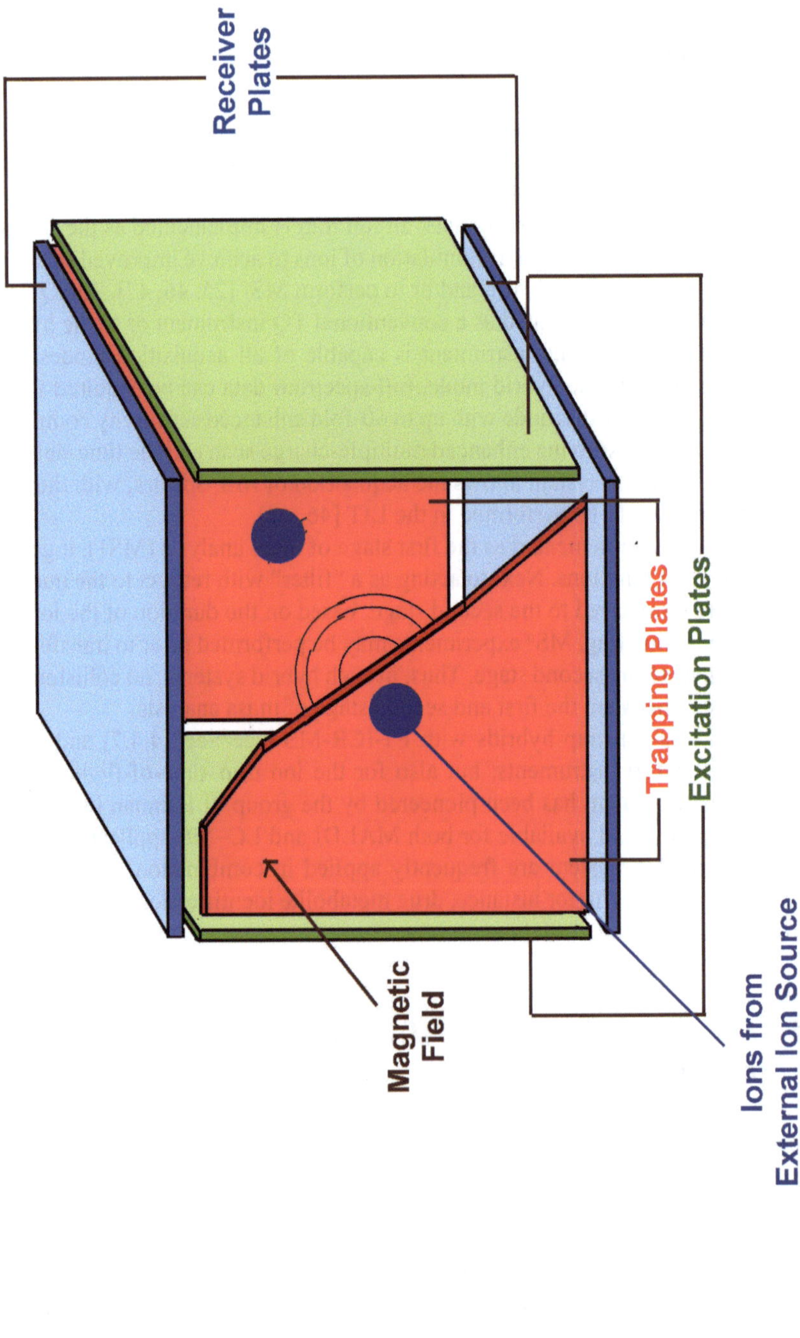

Fig. 4.6 Schematic diagram of a cubic FT-ICR cell. (©2014, hyphen MassSpec)

An ion of mass *m*, with velocity *v*, and *z* elementary charges describes a circle of radius *r*, perpendicular to and around the magnetic field lines. The resulting cyclotron frequency ω_c can be written as:

$$\omega_c = 2\pi f = v / r = Bez / m$$

where *f* is the frequency in Hz. The cyclotron frequency is thus inversely proportional to the *m/z* value. When ions, trapped in their cyclotron motion in the cell, are excited by means of an RF pulse at the excitation plates, the radius of the cyclotron motion increases and ions with the same *m/z* values start moving in phase. This coherent movement of the ions generates an image current at the receiver plates. As the coherency of the ion movement is disturbed in time, the image current signal decays in time as well. The time domain signal from the receiver plates contains all frequency information of the moving ions present in the cell. By applying Fourier transformation, the time domain signal can be transformed into a frequency domain signal, which can subsequently be transformed in a regular mass spectrum by application of the equation above [23].

Characteristic features of an FT-ICR-MS instrument are an extremely high (mass-dependent) resolution, i.e., in excess of 10^5 (FWHM), and a dynamic range of five orders of magnitude. For many years, FT-ICR-MS has primarily been used in fundamental studies of gas-phase ion–molecule reactions only. Due to its high-resolution and MS–MS capabilities, the application of FT-ICR-MS in combination with electrospray ionization for the characterization of large biomolecules has been investigated more recently [90]. At present, FT-ICR-MS plays an important role in top-down strategies to characterize proteins [91, 92]. For this purpose, more user friendly instruments have been introduced by the instrument manufacturers, featuring hybrid systems with either front end quadrupole [93] or ion trap systems [64, 94].

4.4.7 MS–MS in FT-ICR-MS

As targeted ions can be selectively trapped in the ICR cell, while unwanted ions can be eliminated by the application of RF pulses, the MS^n procedures in an FT-ICR-MS instrument greatly resemble those in an ion trap instrument. However, successful operation of an FT-ICR-MS instrument requires extreme low pressures in the cell. Thus, the vacuum constraints hamper the possibilities of performing CID in the FT-ICR cell [23]. This problem can be elegantly solved by the use of hybrid systems where fragmentation is performed prior to transfer of ions to the ICR cell [64, 93, 94]. Alternatively, alternative ion activation methods can be applied to induce fragmentation in the FT-ICR, such as infrared multiphoton photodissociation (IRMPD) and sustained off-resonance irradiation (SORI) [32, 33]. More recently, electron-capture (ECD) and electron-transfer dissociation (ETD) have been introduced as powerful ion dissociation tools, especially for peptides and proteins [34].

4.4.8 Application of FT-ICR-MS in Fundamental Studies

Numerous examples are available in the scientific literature, where the application of FT-ICR-MS in fundamental studies concerning ion–molecule reactions is reported. FT-ICR-MS instruments are highly versatile tools in the study of ion–molecule reactions, as they enable control of reaction time and energy, selection of reactant and product ions, as well as structure characterization of reactants and products using MS^n procedures [95]. Gas-phase ion–molecule reactions of organic anions have been discussed by Nibbering [96]. Gas-phase reaction molecules of transition metal ions and biomolecules have been discussed by Freiser [9].

Combined with electrospray ionization or MALDI, FT-ICR-MS is a very attractive tool for gas-phase studies of biomolecules such as peptides and proteins, oligonucleotides, and oligosaccharides. The possibility to trap ions for prolonged periods of time, even up to thousands of seconds, can be applied in the study of gas-phase ion–molecule reactions. Application of proton-transfer reactions in ICR cells in the study of biomolecules has been reviewed [97]. Detailed structural as well as conformational studies on biomolecules rely on H/D-exchange experiments, for which FT-ICR is an excellent tool, e.g., [98, 99].

4.4.9 Orbitrap Mass Spectrometry

Although both FT-ICR-MS and Orbitrap-MS are based on the acquisition of mass spectra by the Fourier transformation of image currents of trapped ions, the FT-ICR can be considered as a high-vacuum gas-phase reaction cell, whereas the Orbitrap is just applied to perform high-resolution measurement of populations of ions. However, the ion optical tools required to deliver the package of analyte ions in the Orbitrap provides ample possibilities for advanced ion chemistry experiments and or ion dissociation steps. The initial instrumental setup of the Orbitrap consisted of a LIT–Orbitrap hybrid configuration, featuring a LIT to control the number of ions transferred to the Orbitrap and to perform MS^n, when necessary, a so-called C-trap which essentially is a curved high-pressure quadrupole to direct the ion package into the Orbitrap, and the Orbitrap itself [28]. As the ion trap system in this commercial LIT–Orbitrap instrument is equipped with separate off-axis detectors, simultaneous acquisition of high-resolution precursor ion and unit-mass resolution product-ion spectra can be achieved. Subsequently, it was demonstrated that CID could be achieved in the C-trap, which turned to be more like collision cell CID than like ion trap CID [100]. Separate higher energy collision RF-only octapoles (higher energy CID, HCD) were mounted on LIT–Orbitrap systems to make optimum use of this feature. Such a system can be considered as a gas-phase chemistry laboratory by its own, featuring different ways to perform fragmentation, i.e., ion trap CID, HCD, and eventually ETD, as well as different ways to measure the *m/z* values of the resulting ion (unit-mass resolution with the ion trap, up to ultra-high resolution at the Orbitrap). The HCD-cells also lead to stand-alone Orbitrap (Exactive ™) and Q–Orbitrap hybrid systems (Q-Exactive ™) [101].

4.4.10 Comparison of MS–MS Strategies

It is obviously difficult to compare the different instruments and ion dissociation techniques. Some comparison between high-energy CID and low-energy CID are given in Sect. 7.5. However, in this respect, a review paper of Wuhrer et al. [102] on glycopeptide characterization by MS–MS is of interest. Wuhrer et al. [102] compared a range of fragmentation techniques with respect to the information content upon fragmentation of a tryptic glycopeptide (Ser^{295}–Arg^{313} from horseradish peroxidase). For this glycopeptide, CID in ion trap instruments primarily provided information on the glycan sequence, whereas in collision cell CID, e.g., in a Q–TOF instrument, cleavage of glycosidic bonds are induced at low-collision energy and peptide backbone cleavages at higher collision energies. Electron-transfer dissociation (ETD) in an ion trap instrument or electron-capture dissociation (ECD) in FT-ICR-MS leaves the glycan unaffected and provides peptide backbone cleavages. Infrared multiphoton dissociation (IRMPD) in FR-ICR-MS again provides information on the glycan structure. In high-energy CID in a TOF–TOF instrument after MALDI, both peptide sequence ions and fragmentation of glycosidic bonds is obtained [102].

4.5 Flowing Afterglow Mass Spectrometry

4.5.1 Introduction

The flowing afterglow (FA) is a flow reactor tube. Ions are produced by an ion source at the upstream end of the tube. These ions are carried by a buffer gas (He or Ar) and thermalized to room temperature down the flow tube. On their way down, they react with neutral molecules added downstream in the tube. The (ionic) reaction products can be monitored in a number of ways, including optical spectroscopy and MS. In the latter case, the resulting swarm of ions is sampled through an orifice into a high-vacuum chamber where they are mass analyzed and detected.

The FA-MS technique was originally developed in the early 1960s by the group of Ferguson [103] at the Environmental Science Services Administration (ESSA) Aeronomy Laboratory (now the National Oceanic and Atmospheric Administration, NOAA) in Boulder, CO. Their FA device, coupled to a quadrupole mass spectrometer, was initially used to study reactions of He^+-ions with atmospheric components, such as O_2, N_2, CO, NO, and NO_2. It provided new insights on ion–molecule dissociative charge-transfer reactions [103].

Over the years, the FA technique has undergone continuous refinement and development and found a wide variety of applications, e.g., in fundamentals of ion–molecule reactions and in atmospheric and interstellar chemistry [104]. The FA technique enables the generation of high-density, steady state populations of ions and reactive neutral species with well-defined thermal energy distributions. The reaction conditions can be carefully controlled among others due to the temporal and spatial separation of the source and reaction regions. The inherent simplicity and flexibility of the FA technique allows for straightforward adaptation to other experiments.

An important adaptation of the FA technique comprises the implementation of ion separation methods, which allows for more advanced flow drift tubes and selected ion flow tubes (SIFT) [105, 106] (see below). More recently, flow drift tubes (see Sect. 4.6) and flowing atmospheric pressure afterglow (FAPA) devices [107] have been developed.

4.5.2 *Instrumentation*

A typical FA-MS instrument consists of a 1-m long, 7–10-cm inner diameter (ID) pyrex, quartz, or stainless steel tube. A typical setup is shown in Fig. 4.7. By means of a high-speed, high-capacity mechanical pump, a pressure between 0.2 and 2.0 mbar is maintained, while a continuous flow of buffer gas (He or Ar, 100–200 STP ml/s) is introduced. At the upstream end, ions are generated, which are thermalized by collisions with the buffer gas. Initially, ions were generated by a weak microwave discharge in the He or Ar carrier gas, resulting in $He^{+\bullet}$ or $Ar^{+\bullet}$ ions. As the linear velocity of the buffer gas is in the range of 50–100 m/s, the ion residence time in the flow reactor is a few ms. The primary ions may then react in ion–molecule reactions with neutrals, introduced via an inlet. As a result, the afterglow may be observed, resulting from visible photoemission from excited ionic and neutral

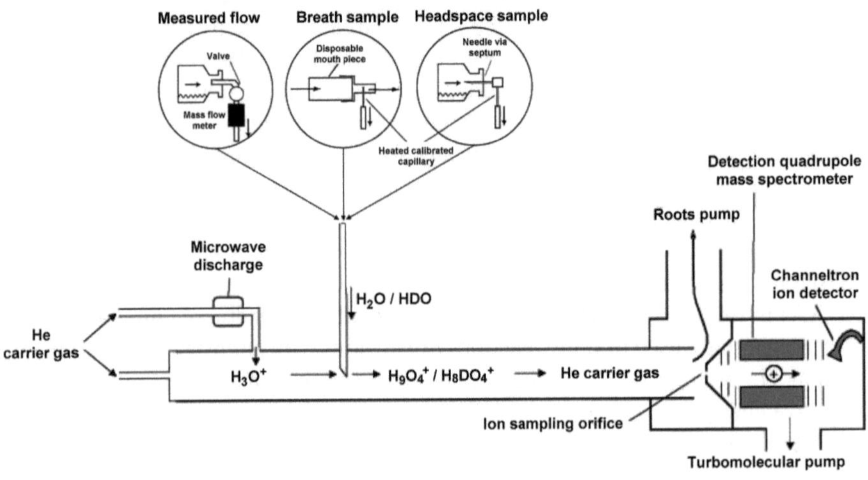

Fig. 4.7 Schematic diagram of a flowing afterglow mass spectrometry setup. The system is setup to perform reactions with H_3O^+. Other primary ions can be generated by other types of ion sources. (Adapted from ref. [121] and Smith, D.; Španěl, P.: Selected ion flow tube mass spectrometry (SIFT-MS) for online trace gas analysis. Mass Spectrom Rev. 24, 2005, 661–700] with permission of Wiley. ©2001 and 2005, Wiley, Ltd.)

species. The reaction mixture is sampled by an ion sampling orifice (0.2–2 mm ID) into a differentially pumped housing of the mass analyzer. While the pressure is reduced to high vacuum, the ions are focused and mass analyzed by a quadrupole mass analyzer and detected by an ion multiplier. Next to establishing the identity of the reaction products based on the *m/z* of the ions detected, the setup allows determining reaction rate coefficients by monitoring the decay of reactant ion intensity as a function of the flow of neutral reactants at a constant reaction time.

From the basic setup, just described, one may conclude that an FA instrument is a modular instrument, consisting of an ion source, a flow reactor tube, and a (mass spectrometric) detector as the building blocks. Each of the building blocks may be modified or adapted to the specific needs of the experiment to be performed.

Different types of ion sources may be introduced, e.g., instead of the microwave discharge ion source, ions may be generated by electron ionization, using electron emission from a heated filament. Ion generation in a high-pressure ion source allows the generation of cluster ions by termolecular association of ions with one or more neutrals, which may then be studied in FA-MS. Thermoionic emission filaments doped with alkali metal salts allows the generation of alkali metal ions [108] or even Ag^+ or Cu^+ ions [109]. Even, ion generation by electrospray ionization in combination with FA-MS has been reported [110]. In SIFT [105, 106], the ion source is combined with a quadrupole mass filter in order to select a particular ion from the various ions generated in the source to be introduced into the flow reactor. This allows an even wider range of potential ion sources to be used.

The inlet for the neutrals may have a fixed or a variable position at the flow reactor. The sample inlet may be a controlled flow of a trace gas, a breath sample, or a headspace sample, introduced via a heated sampling line (Fig. 4.7). Variation of temperature, by external heating of the flow tube, and ion kinetic energy may be applied as well.

Instead of the quadrupole mass analyzer, other types of detectors may be used, e.g., other types of mass analyzers like TOF-MS, tandem mass spectrometers (TQ instruments), but also devices for optical and photoelectron spectroscopy or neutral product analysis, ion photodissociation, or photoemission [104].

4.5.3 Application of FA-MS

The FA-MS has been applied in a wide variety of areas, including atmospheric and interstellar chemistry [111, 112], the study of a wide variety of ion–molecule reactions in order to determine parameters like rate constants and temperature dependence [113], the study of ion chemistry of, for instance, triazoles [114], the study of electron attachment to a wide variety of compounds including halogenated alkenes and alkanes [115, 116], transitions metal carbonyls [117], and sulfuroxyhalides [118], the study of gas-phase proton-transfer and hydride-transfer reactions [119]. Analytical applications of FA-MS involve, among others, online breath analysis [120–122] and screening for pesticides in food [123].

The online breath analysis by FA-MS is based on measuring the H/D ratio before and after introduction of a known amounts of D_2O of HDO into the system. Initially, a swarm of primary H_3O^+ ions is generated in the FA-MS ion source. These primary ions are allowed to react with the water in the exhaled breath analyzed, that is, with the H_2O, HDO, and D_2O molecules in the water vapor (Fig. 4.7). As a result, hydrated ions, $H_3O^+.(H_2O)_3$ or $H_9O_4^+$ with *m/z* 73 are formed, together with their isotopic variants $H_8DO_4^+$ with *m/z* 74 and eventually $H_7D_2O_4^+$ with *m/z* 75 (correction is required for the known amounts of $H_2{}^{17}O$ and $H_2{}^{18}O$ in natural water supplies). By comparing ion intensities for *m/z* 73, *m/z* 74 and eventually *m/z* 75 before and after introduction of D_2O or HDO, a noninvasive determination of total body water can be performed [121, 122]. More recently, monitoring of H/D abundance in breath has found clinical applications, e.g., in determining the extent of abnormal accumulation of fluid in the extravascular space in the alveoli of patients with pulmonary edema [124, 125].

Reactions between Li^+, Na^+, and K^+-ion clusters with O_3, N_2O_5, and SO_2 with $NO^\bullet$ and CO were studied. In these reactions, the alkali metal ions did not play a chemical role, i.e., in forming bonds for instance. Reaction products of the reactions with $NO^\bullet$ were NO_2, O_2, and the alkali metal ion. In the reaction with CO, the products were CO_2, O_2, and the alkali metal ions. Interestingly, the measured reaction rate constants are much larger than the rate constants for the analogous gas-phase reaction in the absence of the alkali metal ion [108].

4.5.4 Selected Ion Flow Tube Mass Spectrometry

As indicated before, SIFT-MS can be considered as an adaptation of FA-MS [105, 106]. The major adaptation concerns the selection of the precursor ion by means of a quadrupole mass spectrometer prior to its injection in the fast-flowing He carrier gas of the FA-MS. In most cases, H_3O^+, NO^+, or $O_2^{+\bullet}$ are generated in a microwave discharge and used as selected precursor ion. The precursor ion is used to react to and ionize trace gases in air or breath samples, introduced downstream in the flow tube. The characteristic product ions identify the trace gas constituents; their count rates allow quantification. Computer-searchable libraries have been compiled to facilitate the identification of the products from reactions of the selected precursor ions with various classes of compounds, including hydrocarbons, alcohols, aldehydes, and ketones, which may be present in breath samples. For identification of reaction products, the downstream mass spectrometer is operated in full-spectrum mode. More accurate quantitation of target compounds can be achieved by operating the downstream mass spectrometer in selected ion monitoring (SIM) mode [105, 106]. Developments in SIFT-MS instrumentation have been reviewed in detail [106]. Portable SIFT-MS instruments have been developed for diagnostic breath analysis in a clinical setting. Other applications concern headspace analysis of volatiles in the skin or in urine, and the emission of toxic compounds and explosives. The SIFT-MS technique and its applications are discussed in more detail in Sect. 8.3. SIFT-MS systems are commercially available.

4.5.5 *Proton-Transfer Reaction Mass Spectrometry*

Proton-transfer reaction mass spectrometry (PTR-MS) can be considered as a derivative of the SIFT technique [126, 127]. Two changes are made, relative to SIFT. In PTR-MS, the front-end mass analyzer to select the reactant ion has been replaced by a hollow-cathode discharge source, capable of a highly efficient generation of primary H_3O^+ ions, taking away the need for the mass filter. The long FA or SIFT flow tube has been replaced by a short drift tube, where the air from the sample is applied as carrier gas. As a result, PTR-MS has become a very easy-to-use and sensitive tool for the detection of volatile organic compounds (VOCs) in air. As such, it is frequently applied for environmental studies related to industrial and anthropogenic emission of VOCs, in plant studies, in relation to flavors in food, food quality, as well as in (clinical) breath analysis [127]. A recent adaptation of PTR-MS is the so-called selective reagent ionization approach (SRI-MS), where other primary ions such as O_2^{+}, NO^+, Kr^+, or Xe^+ can be chosen as reagent ions [128]. PTR-MS systems are commercially available.

4.6 Drift Tubes and Ion Mobility

4.6.1 *Introduction*

Whereas FA-MS is applied to study reactive collisions between ions and neutrals, static drift tubes are used for the study of the nonreactive ion-neutral interactions to determine diffusion and mobility characteristics [104, 129, 130]. A flow drift tube consists of (a) an ion-generation region, (b) a charge-separation region, (c) an ion shutter, (d) the actual drift-reaction region, and (e) an analyzer, which in the present discussion is a mass spectrometer. Ions can, in principle, be generated by any means. A radioactive ^{63}Ni-foil is the most common ionization source in drift-tube experiments, although techniques like photoionization, thermionic emission, and corona discharges have been used as well. The charge-separation region is applied to assure only either positive or negative ions to enter the drift region via the ion shutter. The ion shutter, consisting of two fine-mesh grids connected to a modulating power supply, enables pulse-wise introduction of ions into the drift tube, thus allowing for drift time measurements. In the drift reaction region, a uniform axial electric field gradient is maintained with a series of guard rings, separated by electrically insulating spacers and connected with appropriate precision resistors. Once in the drift tube, the ions are subjected to this homogeneous electric field (typically 1–1000 V/cm). This electric field drives the ions through the drift tube, where they interact with the (countercurrent) buffer gas (He, N_2, Ar). In absence of the drift field, the device functions as a flowing-afterglow device. However, with the drift field on, the ion velocity is the sum of flow velocity and drift velocity, the latter being determined by the mobility of the ion. The drift tube experiments may be considered as gas-phase electrophoresis.

Drift-tube experiments have been applied to study the energy dependences of a wide range of reactions and to correlate these to temperature dependences. Reactions like proton transfer and charge exchange, which proceed at or near the limiting collision rate, generally show little variation with ion kinetic energy. Slow reactions that show significant temperature dependence, often show substantial kinetic energy dependence. However, data must always be interpreted with care [104].

4.6.2 *Ion Mobility*

In its simplest form, the drift-tube ion-mobility system measures how fast a given ion moves in a uniform electric field through a given atmosphere. Thus, an ion-mobility system separates ions by shape and charge. The flow drift technique can be applied to determine quantities like ion mobility and diffusion coefficient, as these are functions of the nonreactive attractive and repulsive ion-neutral interactions [129, 130]. Ion mobilities have been measured for a wide range of ions in several buffer gases (He, N_2, Ar). With Ar as buffer gas, the mobility can be predicted with reasonable accuracy, whereas the measured mobility in He shows poor agreement with theoretical predictions [131].

Ion mobility is a measure of how fast an ion moves through the buffer gas under the influence of an electric field. As larger ions undergo more collisions with the buffer gas, they will have longer drift times than smaller ions. Higher charge states of an ion experience a greater effective drift force, and thus show higher mobility than the lower charge states. In practice, expressing ion mobility as reduced mobility is more useful, as it allows comparison between different experimental conditions. From the reduced mobility, the experimental collision cross-section can be determined. There is a good correlation between experimentally determined collision cross-sections and theoretically predicted ones [132, 133].

Based on these concepts, ion-mobility spectrometry initial called plasma chromatography or gaseous electrophoresis [134], or gas-phase ion chromatography [135], have been developed in the 1970s [129, 130]. Both atmospheric pressure and low-pressure ion-mobility spectrometry (IMS) systems have been used to study gas-phase chemistry, including differentiation between isomeric species [129, 130]. As chemical species can be separated based on their ion mobility, the drift time can be used to generate a response characteristic for the chemical composition of the measured sample. Given the speed, at which the separation and detection occurs (ms range), its ease of use, relatively high sensitivity, and the highly compact design, commercial ion-mobility spectrometers are used as a routine tool for the field detection of explosives [136], drugs of abuse, and chemical weapons [137], for instance at airports. Handheld IMS-based systems have been developed for this purpose. IMS systems have also been used as detectors for chromatography [130], including GC, LC, and supercritical fluid chromatography (SFC).

4.6.3 Ion-Mobility Spectrometry–Mass Spectrometry

The online combination of ion-mobility spectrometry devices with mass spectrometry obviously results in a very attractive tool to combine analysis of conformation and shape, as performed in IMS, with the analysis of *m/z* and structural features, as performed in MS. Ion-mobility spectrometry–mass spectrometry (IMS–MS) has been pioneered by the groups of Bowers [138] and Clemmer [139, 140]. In most cases, ion-mobility devices have been interfaced to quadrupole or TOF instruments. IMS–MS provides a rapid gas-phase separation step prior to MS analysis, enabling the identification of ions with different drift time, thus with different collisional cross-sections, as well as the measurement of collisional cross-sections and correlations to size, shape, and conformation. It may also help understanding ionization characteristics and fragmentation pathways, as a better understanding of gas-phase ion structures is obtained [141]. Instrumental developments in IMS–MS have been reviewed by the group of Hill [142]. The technology is intensively used to study protein conformations in the gas phase, which, for instance, provides interesting insights in the development and possible causes of several neurodegenerative and neuropathic diseases like Parkinson's and Alzheimer's disease [143]. However, IMS–MS is also very useful in the study of small molecules [141] (Fig. 4.8).

Ion-mobility may be implemented in IMS–MS systems in the form of (a) drift tubes, as already discussed [138, 139, 144], (b) differential ion-mobility or high-field asymmetric waveform ion mobility spectrometry (FAIMS) devices [145–147], and (c) traveling-wave ion guide devices [44, 148]. Tandem IMS–IMS–MS systems

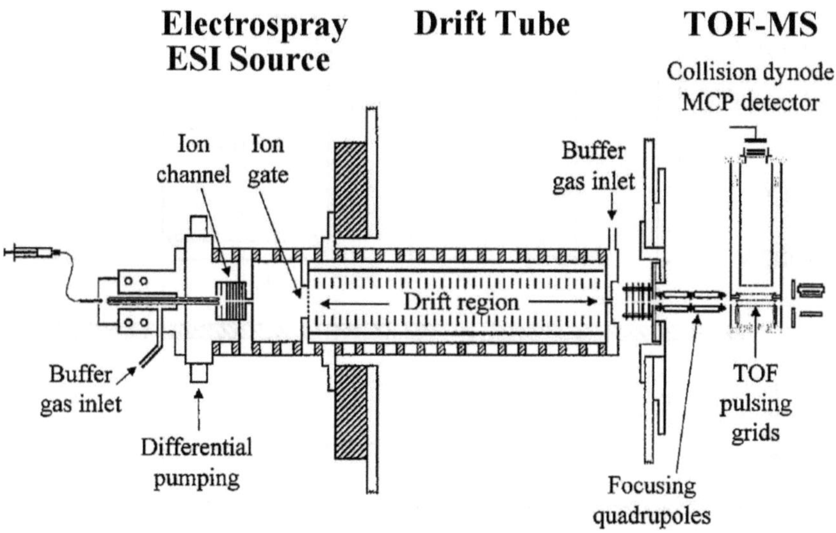

Fig. 4.8 Schematic diagram of a setup for ion-mobility spectrometry–mass spectrometry. (Reprinted with permission from ref. [140]. Copyright 1999, American Chemical Society)

have also been described [149]. This instrument performs an initial ion-mobility separation of individual components in a mixture of ions, collisional activation of selected ions inside the drift tube, transfer of product ions and subsequent ion-mobility separation to a second drift tube, and mass analysis and detection of the ions by TOF-MS, eventually after another CID step [149]. Whereas drift-tube devices have been the initial devices to measure and apply ion mobility in combination with MS, drift-tube IMS–MS system has only recently become commercially available. Nevertheless, successful application of and fundamental studies using IMS–MS have been developed in various laboratories [141–144]. In its various forms, IMS–MS plays an important role in many forefront application areas of biological MS, including structural proteomics [150, 151] characterization of protein assemblies [152], and chiral and structural analysis of biomolecules [153].

4.6.4 High-Field Asymmetric Waveform Ion-Mobility Spectrometry (FAIMS)

In FAIMS, the gas-phase mobility separation of ions in an electric field is achieved at atmospheric pressure [145–147]. In its simplest design, the FAIMS device comprises two parallel rectangular electrodes at a uniform distance (Fig. 4.9). One of the electrodes is grounded, while at the other an asymmetrical waveform is applied. The asymmetric waveform is characterized by a significant difference in voltage in the positive and negative polarities of the waveform. FAIMS uti-

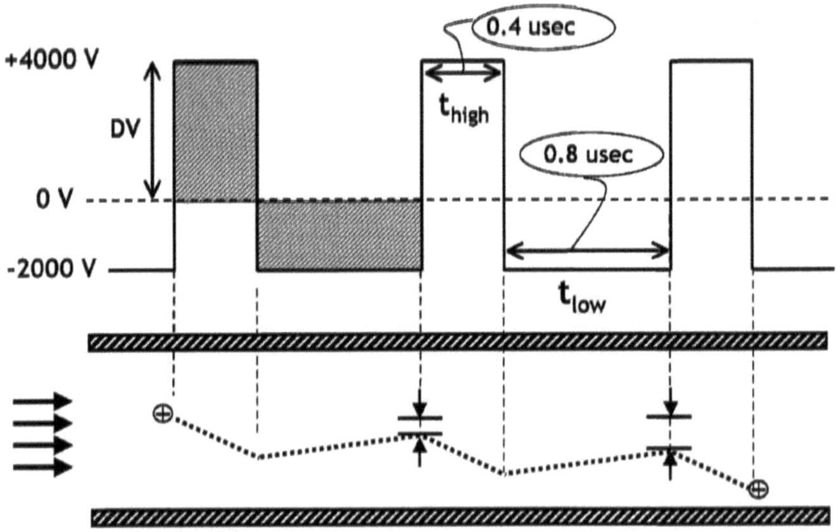

Fig. 4.9 Operation of an FAIMS device. (Reprinted with permission from [145]. Copyright 2004, Elsevier Science Publishers)

lizes much higher electrical fields (typically 10,000 V/cm) than conventional IMS (typically 200 V/cm). Ions drift through the gas between the electrodes and are separated depending on their mobility. At low field, the ion mobility is independent of the electric field: the drift velocity is proportional to the field strength, whereas at high field, the ion mobility becomes dependent on the applied electric field. This electric-field dependence of the ion mobility is the basis of FAIMS. As a result of the applied asymmetric waveform to one of the electrodes, the ions would show a net displacement towards the grounded plate. This net displacement can be corrected or compensated for by a DC voltage (compensation voltage), and assures that the ions remain between the plates. Scanning of the compensation voltage allows ions with different mobilities to be monitored (Fig. 4.9). In this way, for instance, the mobility separation of phthalic acid isomers has been demonstrated [145]. Next to ion mobility separation of ions, focusing of the beam is achieved, thus improving the sensitivity in an FAIMS–MS system. In current FAIMS devices, various electrode configurations, including various geometries of cylindrical electrode devices, are applied [145–147]. Currently, FAIMS devices are commercially available and primarily applied in combination with electrospray ionization and MS to improve sensitivity and to reduce matrix effects in quantitative analysis [147, 154].

4.6.5 IMS–MS Using Traveling Wave Ion Guide Devices

Traveling-wave stacked ring ion guides were initially developed to replace RF-only hexapole ion guides, present in commercial MS systems equipped with atmospheric pressure ion sources (electrospray or atmospheric pressure chemical ionization), to achieve high-transmission ion transport in the vacuum interface. Similarly, they were used to replace RF-only hexapole collision cells in tandem quadrupole systems, both to improve ion transmission and to reduce crosstalk by reducing the ion-residence time in the collision cell [44]. The very nature of the device, which greatly resembles a drift tube, suggests its application for ion-mobility experiments. To this end, a Q–TOF instrument featuring travelling wave ion guides in both the vacuum-interface region and the collision cell region was constructed. Initially, ion-mobility spectra were acquired by storing ions in the source ion guide and gating them periodically to the collision cell ion guide. The mobility-separated ions were subsequently analyzed using the TOF-MS system [44]. For ion-mobility measurement, the collision cell was operated at 0.2 mbar pressure of Ar.

The initial setup was developed into a hybrid quadrupole–ion-mobility–TOF instrument [148]. The collision cell region of this instrument features three traveling-wave stacked ring ion guides, of which the middle one is used as ion-mobility drift tube and the other two may be used as collision cell, when applicable. The 185 mm long IMS part is operated at pressures up to 1 mbar with up to 200 ml/min Ar gas, whereas the collision cells are 100 mm long and operated at 10^{-2} mbar with up to 10 ml/min gas [148]. The system can be used for a wide variety of applications, involving ion-mobility studies, such as protein conformation studies and differentiation of heterogeneities in glycoproteins [155]. By correlating theoretically predicted collision cross-sections to measured drift times of a parent drug and its fragments, Shimizu et al. [156] generated a calibration plot, which could subsequently be used to differentiate between chromatographically separated isomeric glucuronic acid conjugates, generated in biotransformation of the parent drug. For isomeric hydroxylated metabolites, the differences in drift time are too small, compared to the resolution of the IMS separation [157, 158]. Therefore, Shimizu and Chiba [158] introduced selective derivatization of aromatic hydroxyl groups of drug metabolites to increase the differences in collision cross-sections, and thus in drift times. This allowed differentiation between isomeric forms in the same way as demonstrated for the glucuronic acid conjugates [158].

4.7 High-Pressure Mass Spectrometry

The term "high-pressure mass spectrometry" (HP-MS) has been used in various contexts over the years. In many studies, the term refers to performing MS experiments with a high-pressure or atmospheric pressure ion source in combination with a high-vacuum mass analyzer (in contrast to the high-vacuum ion source), whereas in some studies the actual mass analysis is also performed at higher pressures as well, e.g., at ~ 1 mbar.

4.7.1 HP-MS: Mass Analysis at Higher Pressure

Mass analysis at higher pressures than the conventional high-vacuum conditions is especially of interest in the construction of portable mass spectrometers [159, 160]. The most widely used approach in portable mass spectrometers seems to be the use of (miniaturized) ion-trap instruments, as these are generally operated at relatively higher pressure (~ 1 mbar) due to the He bath gas present [159, 161, 162]. These types of instruments are primarily used for environmental monitoring and other field studies (explosives, chemical weapons, etc.). As such, they are frequently equipped with atmospheric pressure ion sources. Nevertheless, portable mass spectrometers based on double-focusing sector, quadrupole, and TOF technology have been described as well [159, 163],

4.7.2 HP-MS: High-Pressure Ion Sources

Historically, the introduction of chemical ionization (CI) by Munson and Field [164, 165] may be considered as the starting point of MS using high-pressure ion sources. They studied gas-phase ion–molecule reactions between reagent gas ions, generated in a ~1-mbar ion source, with analyte molecules introduced into the source. The general purpose of subsequent HP-MS experiments is to perform fundamental studies of gas-phase reactivity in ion–molecule reactions, preferably under equilibrium thermochemical conditions. Several HP-MS instruments have been built and applied by various groups.

The group of Kebarle [166, 167] initially applied an HP-MS instrument, consisting of a pulsed electron beam, a high-pressure ion source, and a sector or quadrupole mass analyzer. The high-pressure ion source comprises a temperature-controlled (−190–650 °C) reaction chamber in which a suitable reaction mixture at pressures of 0.5–10 mbar is ionized by short electron pulses. The ions created react and reach equilibrium while diffusing towards the walls of the ion source vessel. The ion source is sampled by means of a 3 mm × 10 μm slit towards a high-vacuum chamber containing a magnetic sector or quadrupole mass analyzer. A similar system was developed to study gas-phase ion equilibria of alkali metal ions like K^+ with various molecules. The alkali metal ions were generated by thermoionic emission from platinum filaments doped with appropriate alkali metal salts [168]. Later on, similar studies were performed to get a better understanding of electrospray ionization [169]. Also, alkali metal ions generated by electrospray ionization at atmospheric pressure were pulsed into a ~10-mbar reaction chamber to study reactions with amines and small peptides, introduced in the reaction chamber [170].

A very similar experimental setup was applied by the group of Castleman [171, 172] in their gas-phase ion-equilibrium studies involving gas-phase hydration of metal cations like Pb^+ and Bi^+. A thermocouple in the reaction chamber enabled continuous monitoring of the temperature. A gating grid between ion source and reaction chamber allowed more accurate control of the energy of the ions entering the reaction cell.

Instrumentation for pulsed ionization HP-MS has been developed by the group of McMahon [173] and applied in thermochemical studies of relatively nonvolatile biomolecules. The instrument consists of a laboratory-built high-pressure ion source, focusing and alignment optics, and a double-focusing reversed geometry (BE) sector instrument. The ion source is kept in a separate vacuum chamber. The setup can be used with either continuous ionization to acquire mass spectra or in pulsed mode (5–40 Hz), where the mass spectrometer is operated in the selected ion monitoring mode. The temperature-controlled high-pressure ion source is operated at 1–10-mbar pressure of a gas mixture prepared in an external reservoir. Ionization is achieved by means of a 2-kV electron gun. The high pressure is sampled through a 200-μm ID ion-exit aperture. When operated in the pulsed ion mode, the time evolution of the ion populations can be monitored to determine reaction rates of the ion–molecule reactions. Thermochemical properties and structures of, for instance, protonated dimers and trimers of glycine are studied in this way [173]. Later on, other gas-phase reactions of, among others, other amino acids are studied [174]. Instrumentation and experimental results with HP-MS have been reviewed and compared with FT-ICR-MS results [95].

4.8 Conclusion and Perspectives

This chapter deals with instrumentation and experimental methods to study gas-phase ion–molecule reactions, especially in relation to ion attachment with alkali metal ions. The obvious tools for this are mass spectrometers and tandem mass spectrometers. These are discussed in Sects. 4.3 and 4.4. Alternatively, gas-phase reactions may be studied in a variety of reaction chambers, either in static mode or in flow systems. Examples of the static reaction chambers are high-pressure MS devices (see Sect. 4.7). Examples of flow devices are flowing afterglow and drift-tube systems (see Sects. 4.5 and 4.6, respectively). However, next to these tools, there are a number of other tools that may be helpful.

In a crossed beam instrument [175], a beam of mass analyzed low-energy ions is intersected with a supersonic molecular beam of neutrals of a known chemical species. Products of the resulting ion–molecule reaction are energy and mass analyzed by a second mass spectrometer, which can be partly, typically for an octant, moved around the collision point. In this way, information is gathered on the chemical composition and energy of the fragments of the ion–molecule reaction involved.

In neutralization-reionization mass spectrometry (NRMS) [176], the mass spectrometer is used to study neutrals rather than ions. The neutrals are generated from mass-selected ions by intermolecular charge exchange or intramolecular dissociation reactions. Reionization of the neutrals is subsequently achieved by collision with gas-phase target atoms (although using photons or electrons has been described as well). NRMS can provide valuable information on the structures of the neutrals, especially if they are prepared by intramolecular dissociation reactions, as well as of their ionic precursors. To exemplify the wealth of information that can be obtained in NRMS, NR-spectra can be obtained for all four precursor ion–product ion combinations, i.e., positive-ion precursor and product ions, positive-ion precursor ion and negative-ion product ion, negative-ion precursor and product ions, or negative-ion precursor ion and positive-ion product ion.

Next to using a mass spectrometer to monitor and detect the product ions generated in ion–molecule reactions, these ions can also be studied using (laser) spectroscopy. One of the challenges in such studies is the low concentration of ions present in the source or reaction chamber. Informative reviews on the spectroscopy of molecular ions have been published [177–179].

References

1. Gronert S. Mass spectrometric studies of organic ion/molecule reactions. Chem Rev. 2001;101:329–60.
2. Blewett J, Jones EJ. Filament sources of positive ions. Phys Rev. 1936;50:464–8.
3. Selvin PC, Fujii T. Lithium ion attachment mass spectrometry: instrumentation and features. Rev Sci Instrum. 2001;72:2248–52.
4. Campbell EEB, Levine RD. Delayed ionization and fragmentation en route to thermoionic emission: statistics and dynamics. Annu Rev Phys Chem. 2000;51:65–98.

5. Bombick D, Pinkston JD, Allison J. Potassium ion chemical ionization and other uses of an alkali thermoionic emitter in mass spectrometry. Anal Chem. 1984;56:396–402.
6. Hodges RV, Beauchamp JL. Application of alkali ions in chemical ionization mass spectrometry. Anal Chem. 1976;48:825–9.
7. Cody RB, Burnier RC, Reents WD Jr, Carlin TJ, McCrery DA, Lengel RK, Freiser BS. Laser ionization source for ion cyclotron resonance spectroscopy. Application to atomic metal ion chemistry. Int J Mass Spectrom Ion Phys. 1980;33:37–43.
8. Aubriet F, Muller JF. Laser ablation mass spectrometry of inorganic transition metal compounds. Additional knowledge for the understanding of ion formation. J Am Soc Mass Spectrom. 2008;19:488–501.
9. Freiser BS. Gas-phase metal ion chemistry, J Mass Spectrom. 1996;31:703–15.
10. Carré V, Aubriet F, Scheepers PT, Krier G, Muller JF. Potential of laser ablation and laser desorption mass spectrometry to characterize organic and inorganic environmental pollutants on dust particles. Rapid Commun Mass Spectrom. 2005;19:871–80.
11. Lattimer RP, Schulten HR. Field ionization and field desorption mass spectrometry: past, present, and future. Anal Chem. 1989;61:1201A–15A.
12. Fenselau C, Cotter RJ. Chemical aspects of fast atom bombardment. Chem Rev. 1987;87:501–12.
13. Zenobi R, Knochenmuss R. Ion formation in MALDI mass spectrometry. Mass Spectrom Rev. 1998;17:337–66.
14. Kéki S, Deák G, Zsuga M. Fragmentation study of rutin, a naturally occurring flavone glycoside cationized with different alkali metal ions, using post-source decay matrix-assisted laser desorption/ionization mass spectrometry. J Mass Spectrom. 2001;36:1312–6.
15. Niessen WMA. Information from atmospheric-pressure ionization mass spectra. In: Niessen WMA, editor. Encyclopedia of mass spectrometry, Vol. 8. Oxford: Elsevier; 2006. p. 202–32.
16. Jang S, Song MJ, Kim H, Choi SS. Formation of metal complex ions from amino acid in the presence of Li^+, Na^+ and K^+ by electrospray ionization: metal replacement of hydrogen in the ligands. J Mass Spectrom. 2011;46:496–501.
17. Rodriquez CF, Guo X, Shoeib T, Hopkinson AC, Siu KW. Formation of $[M{-}nH + mNa]^{(m-n)+}$ and $[M{-}nH + mK]^{(m-n)+}$ ions in electrospray mass spectrometry of peptides and proteins. J Am Soc Mass Spectrom. 2000;11:967–75.
18. Poon C, Kaplan H, Mayer PM. Methylating peptides to prevent adduct ion formation also directs cleavage in collision-induced dissociation mass spectrometry. Eur J Mass Spectrom. 2004;10:39–46.
19. Busch KL, Glish GL, McLuckey SA. Mass spectrometry/mass spectrometry: Techniques and applications of tandem mass spectrometry. New York: VCH; 1988.
20. Hipple JA, Fox RE, Condon EU. Metastable ions formed by electron impact in hydrocarbon gases. Phys Rev. 1946;69:347–356.
21. Yost RA, Enke CG. Selected ion fragmentation with a tandem quadrupole mass spectrometer J Am Chem Soc. 1978;100:2274–5.
22. Louris JN, Cooks RG, Syka JEP, Kelley PE, Stafford GC Jr, Todd JFJ. Instrumentation, applications, and energy deposition in quadrupole ion-trap tandem mass spectrometry. Anal Chem. 1987;59:1677–85.
23. Marshall A, Hendrickson CL, Jackson GS. Fourier transform ion cyclotron resonance mass spectrometry: a primer. Mass Spectrom Rev. 1998;17:1–35.
24. Morris HR, Paxton T, Dell A, Langhorne J, Berg M, Bordoli RS, Hoyes J, Bateman RH. High-sensitivity collisionally-activated decomposition tandem mass spectrometry on a novel quadrupole–orthogonal acceleration time-of-flight mass spectrometer. Rapid Commun Mass Spectrom. 1996;10:889–96.
25. Hager JW. A new linear ion trap mass spectrometer, Rapid Commun Mass Spectrom. 2002;16:512–26.
26. Bienvenut WV, Déon C, Pasquarello C, Campbell JM, Sanchez JC, Vestal ML, Hochstrasser DF. Matrix-assisted laser desorption/ionization-tandem mass spectrometry with high resolution and sensitivity for identification and characterization of proteins. Proteomics. 2002;2:868–76.

27. Hu Q, Noll RJ, Li H, Makarov A, Hardman M, Cooks GR. The Orbitrap: a new mass spectrometer. J Mass Spectrom. 2005;40:430–43.
28. Makarov A, Denisov E, Kholomeev A, Balschun W, Lange O, Strupat K, Horning S. Performance evaluation of a hybrid linear ion trap/orbitrap mass spectrometer. Anal Chem. 2006;78:2113–20.
29. Schwartz JC, Wade AP, Enke CG, Cooks RG. Systematic delineation of scan modes in multidimensional mass spectrometry. Anal Chem. 1990;62:1809–18.
30. van Dongen WD, Niessen WMA. LC-MS systems for quantitative bioanalysis. Bioanalysis. 2012;4:2391–9.
31. Cherta L, Portolés T, Beltran J, Pitarch E, Mol JG, Hernández F. Application of gas chromatography-(triple quadrupole) mass spectrometry with atmospheric pressure chemical ionization for the determination of multiclass pesticides in fruits and vegetables. J Chromatogr A. 2013;1314:224–40.
32. Sleno L, Volmer DA. Ion activation methods for tandem mass spectrometry. J Mass Spectrom. 2004;39:1091–12.
33. Laskin J, Futrell JH. Activation of large ions in FT-ICR mass spectrometry. Mass Spectrom Rev. 2005;24:135–67.
34. Zhurov KO, Fornelli L, Wodrich MD, Laskay ÜA, Tsybin YO. Principles of electron capture and transfer dissociation mass spectrometry applied to peptide and protein structure analysis. Chem Soc Rev. 2013;42:5014–30.
35. Johnson JV, Yost RA, Kelley PE, Bradford DC. Tandem-in-space and tandem-in-time mass spectrometry: triple quadrupoles and quadrupole ion traps. Anal Chem. 1990;62:2162–72.
36. De Hoffmann E, Stroobant V. Mass spectrometry. Principles and applications. 3. ed. Chichester: Wiley; 2007.
37. Johnson JV, Yost RA. Tandem mass spectrometry for trace analysis Anal Chem. 1985;57:758A–68A.
38. Hunt DF, Shabanowitz J, Harvey TM, Coates ML. Analysis of organics in the environment by functional group using a triple quadrupole mass spectrometer. J Chromatogr. 1983;271:93–105.
39. Montesano C, Sergi M, Moro M, Napoletano S, Romolo FS, Del Carlo M, Compagnone D, Curini R. Screening of methylenedioxyamphetamine- and piperazine-derived designer drugs in urine by LC-MS/MS using neutral loss and precursor ion scan. J Mass Spectrom. 2013;48:49–59.
40. Jian W, Liu HF, Zhao W, Jones E, Zhu M. Simultaneous screening of glutathione and cyanide adducts using precursor ion and neutral loss scans-dependent product ion spectral acquisition and data mining tools. J Am Soc Mass Spectrom. 2012;23:964–76.
41. Joly N, Vaillant C, Cohen AM, Martin P, El Essassi M, Massoui M, Banoub J. Structural determination of the novel fragmentation routes of zwitteronic morphine opiate antagonists naloxonazine and naloxone hydrochlorides using electrospray ionization tandem mass spectrometry. Rapid Commun Mass Spectrom. 2007;21:1062–74.
42. Bijlsma L, Sancho JV, Hernández F, Niessen WMA. Fragmentation pathways of drugs of abuse and their metabolites based on QTOF MS/MS and MSE accurate-mass spectra. J Mass Spectrom. 2011;46:865–75.
43. Mansoori BA, Dyer EW, Lock CM, Bateman K, Boyd RK, Thomson BA. Analytical performance of a high-pressure radiofrequency-only quadrupole collision cell with a axial field applied by using conical rods. J Am Soc Mass Spectrom. 1998;9:775–88.
44. Giles K, Pringle SD, Worthington KR, Little D, Wildgoose JL, Bateman RH. Applications of a travelling wave-based radio-frequency-only stacked ring ion guide. Rapid Commun Mass Spectrom. 2004;18:2401–14.
45. Chernushevich IV, Loboda AV, Thomson BA. An introduction to quadrupole–time-of-flight mass spectrometry. J Mass Spectrom. 2001;36:849–65.
46. Xia YQ, Miller JD, Bakhtiar R, Franklin RB, Liu DQ. Use of a quadrupole linear ion trap mass spectrometer in metabolite identification and bioanalysis. Rapid Commun Mass Spectrom. 2003;17:1137–45.

47. Hopfgartner G, Varesio E, Tschäppät V, Grivet C, Bourgogne E, Leuthold LA. Triple quadrupole linear ion trap mass spectrometer for the analysis of small molecules and macromolecules. J Mass Spectrom. 2004;39:845–55.
48. Tanner SD, Baranov VI, Bandua DR. Reaction cells and collision cells for ICP-MS: a tutorial review. Spectrochim Acta B. 2002;57:1361–452.
49. Yip Y-C, Sham W-C. Applications of collision/reaction cell technology in isotope dilution mass spectrometry. Trends Anal Chem. 2007;26:727–43.
50. Schmitt JP, Dawson PH, Beaulieu N. Chemical synthesis inside the collision cell of a MS/MS system: 1—Formation of adduct ions between protonated esters and ammonia. Org Mass Spectrom. 1985;20:269–75.
51. Pachuta RR, Kenttämaa HI, Cooks RG, Zennie TM, Ping C, Chang C-J, Cassady JM. Analysis of natural products by tandem mass spectrometry employing reactive collisions with ethyl vinyl ether. Org Mass Spectrom. 1988;23:10–15.
52. Cole MJ, Enke CG. Fast atom bombardment tandem mass spectrometry employing ion–molecule reactions for the differentiation of phospholipid classes. J Am Soc Mass Spectrom. 1991;2:470–5.
53. Moraes LAB, Mendes MA, Sparrapan R, Eberlin MN. Transacetalization with gaseous carboxonium and carbosulfonium ions. J Am Soc Mass Spectrom. 2001;12:14–22.
54. Chen H, Zheng X, Cooks RG. Ketalization of phosphonium ions by 1,4-dioxane: selective detection of the chemical warfare agent simulant DMMP in mixtures using ion/molecule reactions. J Am Soc Mass Spectrom. 2003;14:182–8.
55. Kostiainen R, Auriola S. Isomer-specific determination of tetrachlorodibenzo-*p*-dioxins by tandem mass spectrometry using low-energy reactive collisions between oxygen and negative molecular ions. Rapid Commun Mass Spectrom. 1988;2:135–7.
56. Jalonen J. Application of reactive collisions for differentiation of isomeric organic ions in the gas phase. J Chem Soc Chem Commun. 1985:872–4.
57. Meyerhoffer WJ, Bursey MM. Differentiation of the isomeric 1,2-cyclopentanediols by ion–molecule reactions in a triple quadrupole mass spectrometer. Org Mass Spectrom. 1989;24:169–75.
58. Schwartz JC, Schey KL, Cooks RG. A penta-quadrupole instrument for reaction intermediate scans and other MS–MS–MS experiments. Int J Mass Spectrom Ion Proc. 1990;101:1–20.
59. Eberlin MN. Triple-stage pentaquadrupole (QqQqQ) mass spectrometry and ion–molecule reactions. Mass Spectrom Rev. 1997;16:113–44.
60. Jonscher KR, Yates JR 3rd. The quadrupole ion trap mass spectrometer—a small solution to a big challenge. Anal Biochem. 1997;244:1–15.
61. March RE. An introduction to quadrupole ion trap mass spectrometry. J Mass Spectrom. 1997;32:351–69.
62. Schwartz JC, Senko MW, Syka JEP. A two-dimensional quadrupole ion trap mass spectrometer. J Am Soc Mass Spectrom. 2002;13:659–69.
63. Douglas DJ, Frank AJ, Mao D. Linear ion traps in mass spectrometry. Mass Spectrom Rev. 2005;24:1–29.
64. Wu SL, Jardine I, Hancock WS, Karger BL. A new and sensitive on-line liquid chromatography/mass spectrometric approach for top-down protein analysis: the comprehensive analysis of human growth hormone in an *E. coli* lysate using a hybrid linear ion trap/Fourier transform ion cyclotron resonance mass spectrometer. Rapid Commun Mass Spectrom. 2004;18:2201–7.
65. Olsen JV, Schwartz JC, Griep-Raming J, Nielsen ML, Damoc E, Denisov E, Lange O, Remes P, Taylor D, Splendore M, Wouters ER, Senko M, Makarov A, Mann M, Horning S. A dual pressure linear ion trap orbitrap instrument with very high sequencing speed. Mol Cell Proteomics. 2009;8:2759–69.
66. Kind T, Fiehn O. Advances in structure elucidation of small molecules using mass spectrometry. Bioanal Rev. 2010;2:23–60.

67. van der Hooft JJ, Vervoort J, Bino RJ, Beekwilder J, de Vos RC. Polyphenol identification based on systematic and robust high-resolution accurate mass spectrometry fragmentation. Anal Chem. 2011;83:409–16.
68. Gronert S. Quadrupole ion trap studies of fundamental organic reactions. Mass Spectrom Rev. 2005;24:100–120.
69. McLuckey SA, Glish GL, Asano KG, Van Berkel GJ. Self chemical ionization in an ion trap mass spectrometer. Anal Chem. 1988;60:2312–4.
70. Brodbelt JS, Louris JN, Cooks RG. Chemical ionization in an ion trap mass spectrometer. Anal Chem. 1987;59:1278.
71. Pons A, Lavigne V, Darriet P, Dubourdieu D. Determination of 3-methyl-2,4-nonanedione in red wines using methanol chemical ionization ion trap mass spectrometry. J Chromatogr A. 2011;1218:7023–30.
72. Libong D, Bouchonnet S, Ricordel I. A selective and sensitive method for quantitation of lysergic acid diethylamide (LSD) in whole blood by gas chromatography-ion trap tandem mass spectrometry. J Anal Toxicol. 2003;27:24–9.
73. Tzing SH, Ghule A, Chang JY, Ling YC. Selective adduct formation by furan chemical ionization reagent in gas chromatography ion trap mass spectrometry. J Mass Spectrom. 2003;38:401–8.
74. Bouchonnet S, Kinani S, Sablier M. Does the reagent gas influence collisional activation when performing in situ chemical ionization with an ion trap mass spectrometer? Eur J Mass Spectrom. 2007;13:223–6.
75. Ryzhov V, Sunderlin SS, Keller LMM, Gaillard ER. Measuring gas-phase basicities of amino acids using an ion trap mass spectrometer. A physical chemistry laboratory experiment. J Chem Educ. 2005;82:1071–3.
76. Nourse BD, Cooks RG. Proton affinity determination using the kinetic method in an ion trap mass spectrometer. Int J Mass Spectrom Ion Proc. 1991;106:249–72.
77. McCarley TD, Brodbelt J. Structurally diagnostic ion–molecule reactions and collisionally activated dissociation of 1,4-benzodiazepines in a quadrupole ion trap mass spectrometer. Anal Chem. 1993;65:2380–8.
78. Abirami S, Wong CC, Tsang CW, Ma NL. Dissociation of alkaliated alanine in the gas phase: the role of the metal cation. Chemistry. 2005;11:5289–301.
79. Fu M, Duan P, Li S, Habicht SC, Pinkston DS, Vinueza NR, Kenttämaa HI. Regioselective ion–molecule reactions for the mass spectrometric differentiation of protonated isomeric aromatic diamines. Analyst. 2008;133:452–4.
80. O'Hair RA. The 3D quadrupole ion trap mass spectrometer as a complete chemical laboratory for fundamental gas-phase studies of metal mediated chemistry. Chem Commun (Camb). 2006;14:1469–81.
81. Pitterl SJ, McLuckey SA. Recent developments in the ion-ion chemistry of high-mass multiply charged ions. Mass Spectrom Rev. 2005;24:931–58.
82. Xia Y, McLuckey SA. Evolution of instrumentation for the study of gas-phase ion-ion chemistry via mass spectrometry. J Am Soc Mass Spectrom. 2008;19:173–89.
83. Zaia J. Mass spectrometry of oligosaccharides. Mass Spectrom Rev. 2004;23:161–227.
84. Hsu FF, Turk J. Electrospray ionization with low-energy collisionally activated dissociation tandem mass spectrometry of glycerophospholipids: mechanisms of fragmentation and structural characterization. J Chromatogr B. 2009;877:2673–95.
85. Hsu FF, Turk J. Structural characterization of triacylglycerols as lithiated adducts by electrospray ionization mass spectrometry using low-energy collisionally activated dissociation on a triple stage quadrupole instrument. J Am Soc Mass Spectrom. 1999;10:587–99.
86. Cheng C, Gross ML. Applications and mechanisms of charge-remote fragmentation. Mass Spectrom Rev. 2000;19:398–420.
87. Michael SM, Chien BM, Lubman DM. An ion trap storage/time-of-flight mass spectrometer. Rev Sci Instrum. 1992;63:4277–84.

88. Michael SM, Chien BM, Lubman DM. Detection of electrospray ionization using a quadrupole ion trap storage/reflectron time-of-flight mass spectrometer. Anal Chem. 1993;65:2614–20.
89. Liu ZY. An introduction to hybrid ion trap/time-of-flight mass spectrometry coupled with liquid chromatography applied to drug metabolism studies. J Mass Spectrom. 2012;47:1627–42.
90. Römpp A, Taban IM, Mihalca R, Duursma MC, Mize TH, McDonnel LA, Heeren RM. Examples of Fourier transform ion cyclotron resonance mass spectrometry developments: from ion physics to remote access biochemical mass spectrometry. Eur J Mass Spectrom. 2005;11:443–56.
91. Kelleher NL, Lin HY, Valaskovicm GA, Aserud DJ, Fridriksson EK, McLafferty FW. Top-down versus bottom-up protein characterization by tandem high-resolution mass spectrometry, J Am Chem Soc. 1999;121:806–12.
92. Bogdanov B, Smith RD. Proteomics by FTICR mass spectrometry: top down and bottom up. Mass Spectrom Rev. 2005;24:168–200.
93. Patrie SM, Charlebois JP, Whipple D, Kelleher NL, Hendrickson CL, Quinn JP, Marshall AG, Mukhopadhyay B. Construction of a hybrid quadrupole–Fourier transform ion cyclotron resonance mass spectrometer for versatile MS–MS above 10 kDa. J Am Soc Mass Spectrom. 2004;15:1099–108.
94. Syka JE, Marto JA, Bai DL, Horning S, Senko MW, Schwartz JC, Ueberheide B, Garcia B, Busby S, Muratore T, Shabanowitz J, Hunt DF. Novel linear quadrupole ion trap/FT mass spectrometer: performance characterization and use in the comparative analysis of histone H3 post-translational modifications. J Proteome Res. 2004;3:621–6.
95. Wu R, McMahon TB. Structures, energetics, and dynamics of gas phase ions studied by FTICR and HPMS. Mass Spectrom Rev. 2009;28:546–85.
96. Nibbering NMM. Gas-phase reactions of organic anions. Adv Phys Org Chem. 1988;24:1–55.
97. Raczyńska ED, Gal JF, Maria PC, Zientara K, Szelag M. Application of FT-ICR-MS for the study of proton-transfer reactions involving biomolecules. Anal Bioanal Chem. 2007;389:1365–80.
98. Somogyi A. Probing peptide fragment ion structures by combining sustained off-resonance collision-induced dissociation and gas-phase H/D exchange (SORI-HDX) in Fourier transform ion-cyclotron resonance (FT-ICR) instruments. J Am Soc Mass Spectrom. 2008;19:1771–5.
99. Bou-Assaf GM, Chamoun JE, Emmett MR, Fajer PG, Marshall AG. Complexation and calcium-induced conformational changes in the cardiac troponin complex monitored by hydrogen/deuterium exchange and FT-ICR mass spectrometry. Int J Mass Spectrom. 2011;302:116–24.
100. Olsen JV, Macek B, Lange O, Makarov A, Horning S, Mann M. Higher-energy C-trap dissociation for peptide modification analysis. Nat Methods. 2007;4:709–12.
101. Michalski A, Damoc E, Hauschild JP, Lange O, Wieghaus A, Makarov A, Nagaraj N, Cox J, Mann, M, Horning S. Mass spectrometry-based proteomics using Q Exactive, a high-performance benchtop quadrupole orbitrap mass spectrometer. Mol Cell Proteomics. 2011;10:1–12 M111.011015.
102. Senko MW, Remes PM, Canterbury JD, Mathur R, Song Q, Eliuk SM, Mullen C, Earley L, Hardman M, Blethrow JD, Bui H, Specht A, Lange O, Denisov E, Makarov A, Horning S, Zabrouskov V. Novel parallelized quadrupole/linear ion trap/orbitrap tribrid mass spectrometer improving proteome coverage and peptide identification rates. Anal Chem. 2013;85:11710–4.
103. Fehsenfeld FC, Schmeltekopf AL, Goldan PD, Schiff HI, Ferguson EE. Thermal energy ion-neutral reaction rates. I. Some reactions of helium ions. J Chem Phys. 1966;44:4087–94.
104. Graul ST, Squires RR. Advances in flow reactor techniques for the study of gas-phase ion chemistry. Mass Spectrom Rev. 1988;7:263–358.

105. Smith D, Španěl P, Holland TA, al Singari W, Elder JB. Selected ion flow tube mass spectrometry of urine headspace. Rapid Commun Mass Spectrom. 1999;13:724–9.
106. Španěl P, Smith D. Progress in SIFT-MS: breath analysis and other applications. Mass Spectrom Rev. 2011;30:236–67.
107. Andrade FJ, Wetzel WC, Chan GCY, Webb MR, Gamez G, Ray SJ, Hieftje GM. A new, versatile, direct-current helium atmospheric-pressure glow discharge. J Anal At Spectrom. 2006;21:1175–84.
108. Rowe BR, Viggiano AA, Fehsenfeld FC, Fahey DW, Ferguson EE. Reactions between neutrals clustered on ions. J Chem Phys. 1982;76:742–3.
109. Castleman AW Jr, Weil KG, Sigsworth SW, Leuchtner RE, Keesee RG. Considerations of the rates and lifetimes of intermediate complexes for the association of various ligands to metal ions: Ag^+ and Cu^+. J Chem Phys. 1987;86:3829–35.
110. Poutsma JC, Seburg RA, Chyall LJ, Sunderlin LS, Hill BT, Hu J, Squires RR. Combining electrospray ionization and the flowing afterglow method. Rapid Commun Mass Spectrom. 1997;11:489–93.
111. Ferguson EE. A Personal history of the early development of the flowing afterglow technique for ion–molecule reaction studies. J Am Soc Mass Spectrom. 1992;3:479–86.
112. Ferguson EE. Mass spectrometry in ionospheric research. Mass Spectrom Rev. 2007;26:142–9.
113. Fournier JA, Shuman NS, Melko JJ, Ard SG, Viggiano AA. A novel technique for measurement of thermal rate constants and temperature dependences of dissociative recombination: CO_2^+, CF_3^+, N_2O^+, $C_7H_8^+$, $C_7H_7^+$, $C_6H_6^+$, $C_6H_5^+$, $C_5H_6^+$, $C_4H_4^+$, and $C_3H_3^+$. J Chem Phys. 2013;138:154201.
114. Ichino T, Andrews DH, Rathbone GJ, Misaizu F, Calvi RM, Wren SW, Kato S, Bierbaum VM, Lineberger WC. Ion chemistry of 1H-1,2,3-triazole. J Phys Chem B. 2008;112:545–57.
115. Shuman NS, Friedman JF, Miller TM, Viggiano AA. Electron attachment to 14 halogenated alkenes and alkanes, 300–600 K. J Chem Phys. 2012;137:164306.
116. Shuman NS, Miller TM, Viggiano AA, Troe J. Electron attachment to CF_3 and CF_3Br at temperatures up to 890 K: experimental test of the kinetic modeling approach. J Chem Phys. 2013;138:204316.
117. Shuman NS, Miller TM, Friedman JF, Viggiano AA. Electron attachment to $Fe(CO)_n$ (n=0–5). J Phys Chem A. 2013;117:1102–9.
118. Miller TM, Friedman JF, Caples CM, Shuman NS, Van Doren JM, Bardaro MF Jr, Nguyen P, Zweiben C, Campbell MJ, Viggiano AA. Electron attachment to sulfur oxyhalides: SOF_2, $SOCl_2$, SO_2F_2, SO_2Cl_2, and SO_2FCl attachment rate coefficients, 300–900 K. J Chem Phys. 2010;132:214302.
119. Garver JM, Yang Z, Kato S, Wren SW, Vogelhuber KM, Lineberger WC, Bierbaum VM. Gas phase reactions of 1,3,5-triazine: proton transfer, hydride transfer, and anionic σ-adduct formation. J Am Soc Mass Spectrom. 2011;22:1260–72.
120. Smith D, Španěl P. On-line determination of the deuterium abundance in breath water vapour by flowing afterglow mass spectrometry with applications to measurements of total body water. Rapid Commun Mass Spectrom. 2001;15:25–32.
121. Španěl P, Smith D. Accuracy and precision of flowing afterglow mass spectrometry for the determination of the deuterium abundance in the headspace of aqueous liquids and exhaled breath water. Rapid Commun Mass Spectrom. 2001;15:867–72.
122. Davies S, Španěl P, Smith D. Rapid measurement of deuterium content of breath following oral ingestion to determine body water. Physiol Meas. 2001;22:651–9.
123. Jecklin MC, Gamez G, Touboul D, Zenobi R. Atmospheric pressure glow discharge desorption mass spectrometry for rapid screening of pesticides in food. Rapid Commun Mass Spectrom. 2008;22:2791–8.
124. Tan BK, Smith D, Španěl P, Davies SJ. Dispersal kinetics of deuterated water in the lungs and airways following mouth inhalation: real-time breath analysis by flowing afterglow mass spectrometry (FA-MS). J Breath Res. 2010;4:017109.

125. Tan BK, Davies SJ, Španěl P, Smith D. Injection of deuterated water into the pulmonary/alveolar circulation; measurement of HDO in exhaled breath and implications to breath analysis. J Breath Res. 2012;6:036005.
126. Hansel A, Jordan A, Holzinger R, Prazeller P, Vogel W, Lindinger W. Proton transfer reaction mass spectrometry: on-line trace gas analysis at ppb level. Int J Mass Spectrom Ion Proc. 1995;149/150:609–19.
127. Blake RS, Monks PS, Ellis AM. Proton-transfer reaction mass spectrometry. Chem Rev. 2009;109:861–96.
128. Jordan A, Haidacher S, Hanel G, Hartungen E, Herbig J, Maerk L, Schottkowsky R, Seehauser H, Sulzer P, Maerk TD. An online ultra-high sensitivity proton-transfer-reaction mass-spectrometer combined with switchable reagent ion capability (PTR + SRI-MS). Int J Mass Spectrom. 2009;286:32–8.
129. Creaser CS, Griffiths JR, Bramwell CJ, Noreen S, Hill CA, Thomas CLP. Ion mobility spectrometry: a review. Part 1. Structural analysis by mobility measurement. Analyst. 2004;129:984–994.
130. Hill HH Jr, Siems WF, St Louis RH, McMinn DG. Ion mobility spectrometry. Anal Chem. 1990;62:1201A–9A.
131. Lindinger W, Albritton DL. Mobilities of various mass-identified positive ions in helium and argon. J Chem Phys. 1975;62:3517–22.
132. Shvartsburg AA, Jarrold MF. An exact hard-spheres scattering model for the mobilities of polyatomic ions. Chem Phys Lett. 1996;261:86–91.
133. Wyttenbach T, von Helden G, Batka JJ Jr, Carlat D, Bowers MT. Effect of the long-range potential on ion mobility measurements. J Am Soc Mass Spectrom. 1997;8:275–82.
134. Revercomb HE, Mason EA. Theory of plasma chromatography/gaseous electrophoresis—A review. Anal Chem. 1975;47:970–83.
135. Bowers MT, Kemper PR, von Helden G, van Koppen PA. Gas-phase ion chromatography: transition metal state selection and carbon cluster formation. Science. 1993;260(5113):1446–51.
136. Oxley JC, Smith JL, Kirschenbaum LJ, Marimganti S, Vadlamannati S. Detection of explosives in hair using ion mobility spectrometry. J Forensic Sci. 2008;53:690–3.
137. Zimmermann S, Barth S, Baether WK, Ringer J. Miniaturized low-cost ion mobility spectrometer for fast detection of chemical warfare agents. Anal Chem. 2008;80:6671–6.
138. Wyttenbach T, von Helden G, Bowers MT. Gas-phase conformation of biological molecules: Bradykinin. J Am Chem Soc. 1996;118:8355–64.
139. Clemmer DE, Jarrold MF. Ion mobility measurements and their applications to clusters and biomolecules. J Mass Spectrom. 1997;32:577–92.
140. Srebalus CA, Li J, Marshall WS, Clemmer DE. Gas-phase separations of electrosprayed peptide libraries. Anal Chem. 1999;71:3918–27.
141. Lapthorn C, Pullen F, Chowdhry BZ. Ion mobility spectrometry-mass spectrometry (IMS-MS) of small molecules: separating and assigning structures to ions. Mass Spectrom Rev. 2013;32:43–71.
142. Kanu AB, Dwivedi P, Tam M, Matz L, Hill HH Jr. Ion mobility-mass spectrometry. J Mass Spectrom. 2008;43:1–22.
143. Williams DM, Pukala TL. Novel insights into protein misfolding diseases revealed by ion mobility-mass spectrometry. Mass Spectrom Rev. 2013;32:169–87.
144. Wyttenbach T, Pierson NA, Clemmer DE, Bowers MT. Ion mobility analysis of molecular dynamics. Annu Rev Phys Chem. 2014;65:175–96.
145. Guevremont R. High-field asymmetric waveform ion mobility spectrometry: a new tool for mass spectrometry. J Chromatogr A. 2004;1058:3–19.
146. Kolakowski BM, Mester Z. Review of applications of high-field asymmetric waveform ion mobility spectrometry (FAIMS) and differential mobility spectrometry (DMS). Analyst. 2007;132:842–64.

147. Tsai CW, Yost RA, Garrett TJ. High-field asymmetric waveform ion mobility spectrometry with solvent vapor addition: a potential greener bioanalytical technique. Bioanalysis. 2012;4:1363–75.
148. Pringle SD, Giles K, Wildgoose JL, Williams JP, Slade SE, Thalassinos K, Bateman RH, Bowers MT, Scrivens JH. An investigation of the mobility separation of some peptide and protein ions using a new hybrid quadrupole/travelling wave IMS/oa-ToF instrument. Int J Mass Spectrom. 2007;261:1–12.
149. Koeniger SL, Merenbloom SI, Valentine SJ, Jarrold MF, Udseth HR, Smith RD, Clemmer DE. An IMS–IMS analogue of MS–MS. Anal Chem. 2006;78:4161–74.
150. Zhong Y, Hyung SJ, Ruotolo BT. Ion mobility-mass spectrometry for structural proteomics. Expert Rev Proteomics. 2012;9:47–58.
151. Jurneczko E, Barran PE. How useful is ion mobility mass spectrometry for structural biology? The relationship between protein crystal structures and their collision cross sections in the gas phase. Analyst. 2011;136:20–8.
152. Uetrecht C, Rose RJ, van Duijn E, Lorenzen K, Heck AJ. Ion mobility mass spectrometry of proteins and protein assemblies. Chem Soc Rev. 2010;39:1633–55.
153. Enders JR, McLean JA. Chiral and structural analysis of biomolecules using mass spectrometry and ion mobility-mass spectrometry. Chirality. 2009;21:E253–264.
154. Xia YQ, Wu ST, Jemal M. LC-FAIMS-MS/MS for quantification of a peptide in plasma and evaluation of FAIMS global selectivity from plasma components. Anal Chem. 2008;80:7137–43.
155. Olivova P, Chen W, Chakraborty AB, Gebler JC. Determination of N-glycosylation sites and site heterogeneity in a monoclonal antibody by electrospray quadrupole ion-mobility time-of-flight mass spectrometry. Rapid Commun Mass Spectrom. 2008;22:29–40.
156. Shimizu A, Ohe T, Chiba M. A novel method for the determination of the site of glucuronidation by ion mobility spectrometry-mass spectrometry. Drug Metab Dispos. 2012;40:1456–9.
157. Dear GJ, Munoz-Muriedas J, Beaumont C, Roberts A, Kirk J, Williams JP, Campuzano I. Sites of metabolic substitution: investigating metabolite structures utilising ion mobility and molecular modelling. Rapid Commun Mass Spectrom. 2010;24:3157–62.
158. Shimizu A, Chiba M. Ion mobility spectrometry-mass spectrometry analysis for the site of aromatic hydroxylation. Drug Metab Dispos. 2013;41:1295–9.
159. Badman ER, Cooks RG. Miniature mass analyzers. J Mass Spectrom. 2000;35:659–71.
160. Peng Y, Austin DE. New approaches to miniaturizing ion trap mass analyzers. Trends Anal Chem. 2011;30:1560–7.
161. Whitten WB, Reilly PT, Ramsey JM. High-pressure ion trap mass spectrometry. Rapid Commun Mass Spectrom. 2004;18:1749–52.
162. Misharin A, Novoselov K, Laiko V, Doroshenko VM. Development and characterization of a field-deployable ion-trap mass spectrometer with an atmospheric pressure interface. Anal Chem. 2012;84:10105–12.
163. Sparkman OD, Bierbaum VM. Focus on field-portable and miniature mass spectrometers. Presentations from the 12th Sanibel conference on mass spectrometry. J Am Soc Mass Spectrom. 2001;12:617–8.
164. Munson MSB, Franklin JL, Field FH. High pressure mass spectrometric study of alkanes. J Phys Chem. 1964;68:3098–107.
165. Munson MSB, Field FH. Chemical ionization mass spectrometry. I. General introduction. J Am Chem Soc. 1966;88:2621–30.
166. Kebarle P. Ion thermochemistry and solvation from gas-phase ion equilibria. Ann Rev Phys Chem. 1977;28:445–76.
167. Cunningham AJ, Payzant JD, Kebarle P. A kinetic study of the proton hydrate $H^+ (H_2O)_n$ equilibria in the gas phase. J Am Chem Soc. 1972;94:7627–32.
168. Searles SK, Kebarle P. Hydration of the potassium ion in the gas phase: enthalpies and entropies of hydration reactions: $K^+ (H_2O)_{n-1} + H_2O = K^+ (H_2O)_n$ for $n = 1$ to $n = 6$. Can J Chem. 1969;47:2619–27.

169. Kebarle P, Verkerk UH. Electrospray: from ions in solution to ions in the gas phase, what we know now. Mass Spectrom Rev. 2009;28:898–917.
170. Klassen JS, Blades AT, Kebarle P. Determinations of ion–molecule equilibria involving ions produced by electrospray. Hydration of protonated amines, diamines, and some small peptides. J Phys Chem. 1995;99:15509–17.
171. Tang IN, Castleman AW Jr. Mass spectrometric study of the gas-phase hydration of the monovalent lead ion. J Chem Phys. 1972;57:3638–44.
172. Tang IN, Castleman AW Jr. Mass spectrometric study of gas-phase clustering reactions: hydration of the monovalent bismuth ion. J Chem Phys. 1974;60:3981–6.
173. Raspopov SA, McMahon TB. A high-pressure mass spectrometric and density functional theory investigation of the thermochemical properties and structure of protonated dimers and trimers of glycine. J Mass Spectrom. 2005;40:1536–45.
174. Wu R, McMahon TB. Stabilization of zwitterionic structures of amino acids (Gly, Ala, Val, Leu, Ile, Ser and Pro) by ammonium ions in the gas phase. J Am Chem Soc. 2008;130:3065–78.
175. Vestal ML, Blakley CR, Ryan PW, Futrell JH. New crossed-beam apparatus for the study of ion–molecule collision processes. Rev Sci Instrum. 1976;47:15–26.
176. Wesdemiotis C, McLafferty FW. Neutralization–reionization mass spectrometry. Chem Rev. 1987;87:485–500.
177. Saykally RJ, Woods RC. High resolution spectroscopy of molecular ions. Ann Rev Phys Chem. 1981;32:403–31.
178. Polfer NC, Oomens J. Vibrational spectroscopy of bare and solvated ionic complexes of biological relevance. Mass Spectrom Rev. 2009;28:468–94.
179. Polfer NC, Oomens J. Reaction products in mass spectrometry elucidated with infrared spectroscopy. Phys Chem Chem Phys. 2007;9:3804–17.

Chapter 5
Applications of Association Reactions in the Gas Phase

Toshihiro Fujii

5.1 Alkali Metal Ions in Electron Ionization (EI) and Chemical Ionization (CI) Mass Spectrometry (MS)

Alkali metal ion/molecule association reactions have been exploited in various mass spectrometric methods since the early 1970s. Initial studies were to determine metal ion affinities of simple compounds, with use of Fourier transform ion cyclotron resonance (FTICR) mass spectrometers. Alkali ions offer unique and interesting potential in analytical chemistry and studies of chemical dynamics. Consequently, the use of metal ions as reagents for CI mass spectrometry has been developed. The major advances in the application of MS as a routine analytical instrument using the complex ion/molecule chemistry came around in the past decades. They include atmospheric pressure chemical ionization (APCI), proton transfer reaction mass spectrometry (PTR-MS), selected ion flow tube mass spectrometry (SIFT-MS), ion attachment mass spectrometry (IAMS) and ion molecule reaction mass spectrometry (IMM-MS).

APCI allows instantaneous air analysis, with use of ambient air as a carrier and reagent gas. The trace atmospheric gas analyzer (TAGA) [1] is well known as an early version of APCI. The potential uses for APCI are expanding continuously as the understanding of the complex ion-molecule chemistry grows due to the instrumental advance of TANDEM MS. The applications are not restricted by the use of ambient air as a reagent gas; addition to the air carrier of thermionic alkali metal ions allows using specific reagent ions such as Li^+ or Na^+ as specific reagent ions. PTR-MS makes use of proton-transfer reactions, while SIFT-MS and IMR-MS instruments can choose ion association reactions, in addition to proton-transfer, hydride transfer, hydroxide transfer reactions for the CI of analytes.

T. Fujii (✉)
C & V Technix Co., Ltd., 3-6-1 Higashi, Akishima, Tokyo 196-0033, Japan
e-mail: fujii.toshihiro@c-vtechnix.com

T. Fujii (ed.), *Ion/Molecule Attachment Reactions: Mass Spectrometry*,
DOI 10.1007/978-1-4899-7588-1_5

5.1.1 Alkali Metal Ions in CI MS

Hodges and Beauchamp [2] reported a technique for obtaining mass spectra consisting solely of quasi-molecular ions formed by addition of an alkali ion to the sample molecule. The use of alkali lithium ions as reagent species for CI MS alkali metal ion attachment chemical ionization mass spectrometry (ACIMS) has been explored for the first time. Alkali ions are generated by thermionic emission externally to the enclosed ion source and injected into a chamber containing a reagent gas with trace amount of sample. An ion source consists of a thermionic lithium ion emitter, a reaction chamber (RC) and appropriate electrostatic lenses. Alkali ions initially bind to the reagent molecules and then are transferred to the sample in bimolecular reactions. Classes of molecules that have intrinsically high attachment chemical ionization (ACI) sensitivities are molecules that are polar species or polarizable species. ACIMS is a useful addition to the field of analytical MS. For example, it can be used to detect a trace amount of alkene in the presence of the corresponding alkane. They anticipate that ACIMS will find numerous applications in the analysis of mixtures. In the area of chemical reactivity, ACI, along with high-pressure MS techniques, makes it possible to determine the alkali affinities of a large number of molecules and to study the reactivity of alkali ions in the gas phase.

Bombick et al. [3] presented a simple, low cost method for producing thermal potassium metal ions for use as CI reagents. All studies were performed on a commercial gas chromatography–mass spectrometry (GC-MS) system. Thermionic emitters of a mixture of silica gel and potassium salts were mounted on a fabricated probe assembly and inserted into the CI volume of the ion source through the direct insertion probe inlet. Since adduct ions (also referred to as "cationized molecular ions" or "pseudomolecular ion") of the type $(M + K)^+$ have been observed, molecular weight information is easily obtained. The method is adaptable to any mass spectrometer with a CI source and direct inlet probe (DIP). In addition, the technique is compatible with chromatographic inlet systems, i.e., GC-MS modes, which will provide additional dimensions of mass spectral information.

5.1.2 EI, CI and In-Beam, with Flash Rapid Heating

Another development of methodology leading to cationization of organic molecules as a method of ionization is due to the work of Roellgen and coworkers [4]. They developed a technique using a two-filament design. In a specially constructed ion source, two filaments are in close proximity. The first filament is loaded with a mixture of silica gel and alkali salt (LiCl, NaI, and/or KI), which, when heated, emits alkali ions. Thermally labile compounds are present on the second filament. These molecules thermally desorb and form adducts with the alkali ions from the first filament. This method produces simple mass spectra from which molecular weight information can be easily derived.

Rapid-heating (or flash evaporation) techniques have been explored with ionization methods such as EI and CI in the analysis of thermally labile compounds [5, 6]. Their principles are based upon that if the analyte is rapidly heated, intact molecules may evaporate before decomposition takes place. Davis et al. reports that, by this technique, alkali metal attachment (cationization) of sodium benzoate and sodium acetate occurs, giving rise to cationized molecular ions (e.g., $[M + Na]^+$) and other cluster ions similar to those produced by desorption ionization processes [7, 8]. In these experiments, the molecular ions were formed by electron impact of salt molecules or clusters of sample molecules and salts in the gas phase.

The Rollgen's group [9] reports the thermal desorption of $[M + Alkali]^+$ quasi-molecular ions from a electrically heated metal surface (wires or ribbons) for sodium alkali salts of carboxylic acids and mixtures of alkali halides with a crown ether, glucose and adenosine. With benzo-15-crown-5 the desorption of $[M + Na]^+$ ions takes place even below the threshold temperature for thermionic emission of alkali ions. Alkali ion attachment has also been performed by thermal desorption of thermally labile analytes such as saccharides, pharmaceuticals, peptides, steroids and their mixtures. Bombick and Allison [8] describe a desorption/ionization method where samples are deposited directly on thermionic emission materials ($1K_2O{:}1Al_2O_3{:}2SiO_2$) and heated within the mass spectrometer's source. Ions representative of the sample are formed. The proposed mechanism involves the gas-phase addition of emitted potassium ions to neutrals desorbed from the surface.

5.1.3 Ion Attachment Mass Spectrometry (IAMS)

In the late1980s, Fujii et al. developed a novel method [10, 11] for the detection of radical species in the gas phase by using Li^+ ion attachment. It is called IAMS, which is a technique where a sample is ionized by a primary ion in an ion–molecule association reaction. This approach is similar to ion association (cationization) for detection of stable molecules [2, 3]. They explore the mechanism of alkali-metal ion/molecule association reaction and develop both the instrumentation for the attachment to free radical intermediate and to other species (which are sometimes not found under ordinary conditions). The newly developed instrumentation exhibits several advantages over conventional mass spectrometers [12, 13]. Section 5.2 details IAMS, summarizing analytical applications.

5.2 Li^+ IAMS

5.2.1 Origin

As already described in the previous section, Hodges and Beauchamp [2] proposed to a technique for obtaining mass spectra consisting solely of quasi-molecular ions formed by the addition of Li^+ ions to the sample molecules in 1975. It is known as ACIMS, in which alkali metal ions such as Li^+, Na^+, and K^+ form adduct ions (also referred to as cationized molecules) with molecular species through association reactions. Several research groups have actively attempted to detect many organic complexes by attaching alkali metal ions to the sample molecules [3–9]. Rollgen and Schulten [14] explored this approach for achieving cationization using the field desorption technique. However, neither study made an intensive attempt to provide a complementary tool to identify the chemical species.

In 1989, the Fujii group developed the method for detection of chemical species in the gas phase with use of Li ion attachment to chemical species [11–13]. The principle is based upon a phenomenon that Li^+ ions get attached to chemical species (M) by means of intermolecular association reactions to produce $(M + Li)^+$ adduct ions. Then they are transferred to a mass spectrometer for mass analysis. This approach is exactly the same to cationization for detection of molecular species. Since the potential of Li ion attachment in MS has not yet been realized, they attempted to reveal and explore some of the unique properties of Li IAMS. Li ions have been chosen as reactant ions, because the affinity of the species is the highest among all the alkali metal ions. This technique provides mass spectra of quasi-molecular ions formed by lithium ion attachment to the chemical species (M) under high pressure. Results are obtained in the form of trace of Li^+ adduct ions $(M + Li)^+$ (also referred as cationized molecules). The newly developed IAMS [10], manufactured by instrumental maker (Canon Anelva Corp., Japan), exhibits several advantages over conventional mass spectrometers. Currently, ion association MS is available commercially in a various form. Recently, some reviews have been published on the principles, instrumental techniques, unique characteristics, and applications of IAMS [15–20].

In this section, the development of a Li^+ IAMS is described. Section 5.2.2 summarizes the MS instrumentation for the alkali-metal ion/molecule association reaction; a quadrupole mass spectrometer (QMS), combined with the Li^+ ion attachment ionization source (model L-241G-IA, Canon Anelva Corp, Tokyo). The performance and response of this equipment and unique features are also presented (Sect. 5.2.3). Finally, the interesting applications are reviewed. There are some interesting and unusual applications of IAMS that broaden MS: (i) intermediary free radical species detection, in the gas phase reactions, diagnosis of plasma and diamond film chemical vapor deposition (CVD), (ii) detection of atmospheric and interstellar species, (iii) detection of environmentally important species, and (iv) identification of unfamiliar or unstable species. Restriction of Hazardous Substances Detectives (RoHS) is also discussed.

5.2.2 Instrumentation

5.2.2.1 Basic of Instrumental Design

In Fig. 5.1, a schematic diagram of the IAMS apparatus is shown [12, 13, 15, 21]. The instrument is composed of four major blocks: the sample inlet system (possibly

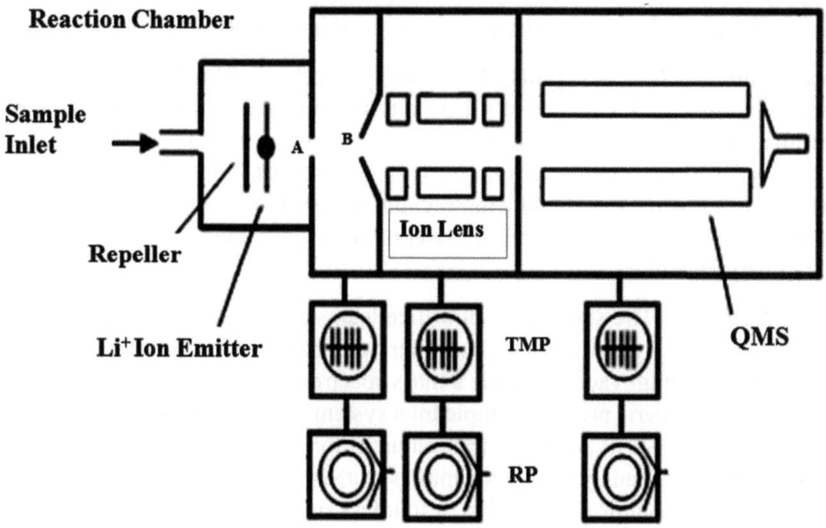

Fig. 5.1 Schematics of a quadrupole mass spectrometer with a Li^+ ion attachment reaction chamber. Any stream from sample inlet system whose components one want to analyze is directed to the reaction chamber. *RP* rotary pump, *TMP* turbomolecular pump, *QMS* quadrupole mass spectrometer

operated at atmospheric pressure), the Li^+ ion attachment reaction chamber (RC) with a Li^+ ion source and aperture (to provide for differential pumping), the ion focusing system (ion lenses), and the mass analyzer (e.g., QMS). Three vacuum pumps are generally employed to evacuate various sections of the system: the RC is maintained at l-760 Torr and a high-speed turbomolecular pump (TMP), producing for the ion focusing system and mass analyzer regions pressures less than 2×10^{-4} and 5×10^{-7} Torr, respectively.

Ionization takes place in the RC that contains a Li^+ ion source, which consists of a Li^+ ion emitter and repeller. The primary Li^+ ions ionize the target sample species by adduct formation to give $[M + Li]^+$ by termolecular association reactions. The adduct ions are focused by ion lens and transferred to the QMS chamber. A QMS is often employed, and detection is by a channeltron electron multiplier detector. Other mass spectrometers can be used.

5.2.2.2 Sample Inlet Systems

Samples can be introduced into the IAMS systems by several different methods, such as a leak for samples in a gas/liquid inlet, from a DIP for solids, and by on-line combinations with chromatography (see 6.1, application to GC-MS mode). A controlled leak is a simple method, but its use is restricted to samples with high vapor pressure. Solid and liquid samples can be introduced with a heated DIP (or an insertion probe). Either gas chromatography or liquid column chromatography can be easily applied as sample inlet systems. In principle, any gaseous sample whose components are to be identified can be introduced into the RC of IAMS.

Leaks, Capillary Tube A capillary leak inlet system is used for the introduction of gaseous samples [22]. One end of a 1-m-long fused-silica capillary tube (0.04 mm inner diameter, internally inert) is open to the atmosphere, and the other end is fixed on the flange of the vacuum envelope. This inlet makes it possible to effectively introduce gases while maintaining the vacuum required for IAMS. For instance, the sample gas from the diffusion cell [23, 24] or permeation tube [25] is allowed into the capillary leak inlet.

Orifice Leak Inlet The stainless steel orifice leak sample inlet is a simple, rugged and efficient inlet system to act as an interface between samples at atmospheric pressure and the high vacuum inside a mass spectrometer [26]. A schematic drawing of this atmospheric pressure sample inlet system and the part of IAMS is shown in Fig. 5.2. It consists of an orifice leak set up as follows to ensure a consistent and streamlined entry of analytes into the mass spectrometer. The sample inlet consists of two concentric stainless steel tubes; a stainless steel disk with a 30-μm orifice at its center is fastened to the end of this probe with a nut system. This arrangement ensures that the analyte leaks in a constant, unperturbed stream of excellent consistency from the atmosphere into the mass spectrometer. The inlet system is relatively maintenance free and is able to introduce high volumes of sample gas into the mass spectrometer. These characteristics result in near-real-time sample analysis. The sample inlet in combination with IAMS has a number of features that make it useful as an extremely sensitive instrument for continuous monitoring of trace organic species and small quantities of environmentally important analytes, such as radical

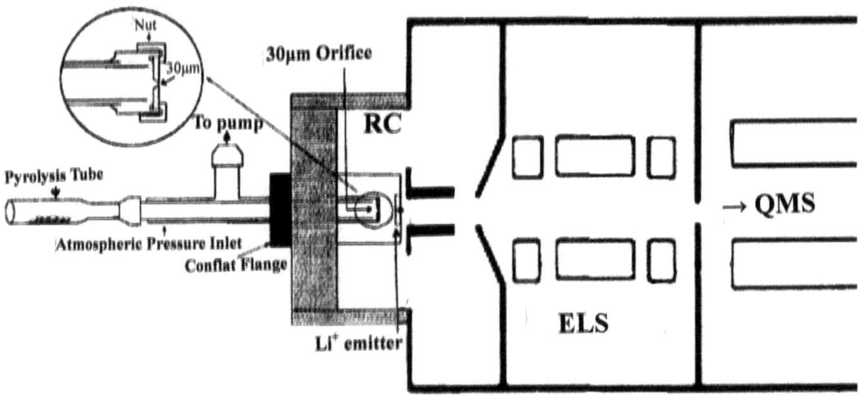

Fig. 5.2 Schematics of the atmospheric pressure orifice leak inlet system attached to the lithium ion attachment mass spectrometer. The sampling of the inlet probe is shown for polymer pyrolysis. *RC* reaction chamber, *ELS* electrostatic lens system, *QMS* quadrupole mass spectrometer. (Reprinted with permission from [26]. ©2001, Elsevier)

intermediates including toxic radicals present in ambient air. It is anticipated that a wide range of practical applications that will employ this inlet coupled with IAMS. For instance, the need for real-time trace analysis of air streams for environmental monitoring and process control has increased. Hence, there is a real need for instrumentation for the continuous in situ analysis of process gases for process control and for meeting stringent regulations. Further applications may include in breath analysis, emission control and monitoring of volatile organic compounds (VOCs) in urban and clean rural environments, monitoring catalytic processes, cigarette smoke, and material production in plasma reactors. For these volatile samples, the diffusion tube or permeation tube is used to prepare sample concentrations [23–25].

Direct Inlet Probe (DIP) The DIP is one of the earliest techniques used to introduce nonvolatile or highly insoluble samples into mass spectrometers. This technique is still popular today because it provides a means to perform rapid sample analysis with minimal or no sample preparation [27, 28]. Applications of the direct-probe technique include quality-control analyses and the screening of drug, polymer, and synthesis samples. It has also been used to study the temperature programmed decomposition of synthetic polymers and inorganic materials, to characterize the molecular properties of materials under thermal degradation conditions. The same is also true of the studies with IAMS (see Sect. 6.4).

Evolved gas analysis (EGA) Evolved gas analysis-mass spectrometry (EGA-MS), which can be considered a second generation of DIP-MS combination, has been developed and found to be useful in mass spectrometric analysis, particularly in the characterization of thermal decomposition processes [29, 30]. When this technique is used together with a temperature-programmed heating probe and with total ion monitoring (TIM) or selected ion monitoring (SIM), a program (thermogram) can also be obtained. The EGA probe can also serve as an isothermal or temperature-programmed flow reactor for homogeneous, heterogeneous, or thermal decomposition kinetic studies. The nonisothermal method has the advantage of using only one sample for the entire experiment.

Recently a new EGA system has been coupled with the IAMS technique (EGA-IAMS) to study involatile, untreated, and complex samples, such as polybrominated biphenyls (PBBs) and polybrominated diphenyl ethers (PBDEs) of the RoHS (see later, Sect. 5.8). It has been shown that by using the EGA-IAMS technique, temperature-programmed decomposition of a number of toxic solid materials can be studied in detail by directly detecting any chemical species in real-time and in situ. To vaporize samples, this EGA probe is provided with an infrared (IR) lamp for the DIP. The IR lamp heats the sample in the sample cup of the DIP to vaporize the components of the sample (see later, Sect. 5.2.8). The inset of Fig. 5.3 shows a schematic of the DIP assembly that allows the solid sample to volatile species.

Chromatographic Inlet Systems Also, IAMS is compatible with chromatographic inlet systems (i.e., gas chromatography, liquid chromatography, super fluid chromatography). Thermogravimetry IAMS or IR image furnace IAMS with a sampling

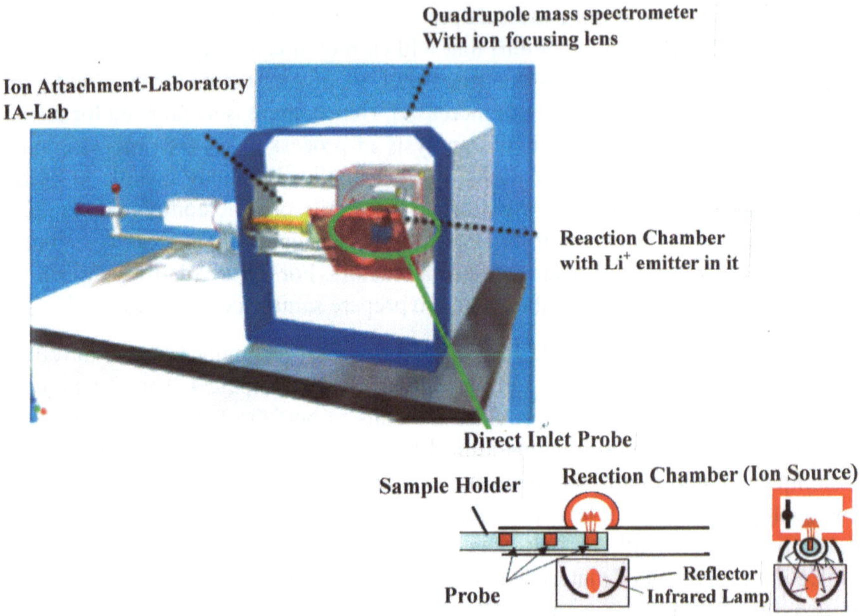

Fig. 5.3 Overview of commercial instrument, IA-Lab (model; L-250G-IA, manufactured by Canon Anelva, Kanagawa, JAPAN). This equipment consists of direct inlet probe (DIP), Li^+ attachment reaction chamber with Li^+ emitter in it, the ion focusing system, and the quadrupole mass spectrometer. Li^+ is supplied by a Li^+ emitter. There are two types of interface for sample introduction: an introduction tube for gases and a direct inlet probe for solids and liquids. (©2009, Canon Anelva, Data sheet)

inlet at atmospheric pressure seems promising for identifying and resolving complex coevolving thermal products. These techniques have been discussed in-considerable detail elsewhere (see Sect. 6.1 and 6.4).

5.2.2.3 Reaction Chamber (RC)

The RC is a cylindrical tube with a lithium ion emitter centered along one side (Fig. 5.1). The chamber can be evacuated through a cone aperture, A, with a vacuum pump and can be maintained at high pressure by altering the pumping speed or the carrier gas flow rate. The pressure at the center of the RC zone is measured with a Pirani gauge.

Li^+ Ion Emitter The lithium ion source consists of a lithium ion emitter and a repeller. The Li^+ emitter is a small mineral bead (about 0.2 cm in diameter) fused to a 0.25-mm-diam (IR) wire; the bead is prepared by thoroughly grinding a mixture of Li_2O:Al_2O_3:SiO_2 (1:1:1 mol ratio), and primary Li^+ ions are produced by heating the bead [31, 32]. The emission of contaminants along with the Li^+ ions is unavoidable as emission begins but in due course decreases to less than 0.1 % of the total

emission. The repeller electrode, consisting of a 1.5-cm-diam stainless steel disk, was placed 1 cm behind the emitter bead; its voltage was maintained at the same potential as the emitter. The performance of the Li^+ ion emissions was tested at operational temperatures. At a fixed pressure (0.2–1.2 Torr), the Li^+ ion current increased linearly, or nearly so, with emitter current up to 3.8 A and saturates after that. Maximum Li^+ ion emission current is around 6×10^{-7} A at a nitrogen carrier gas flow of 53 mL/min and pressure of 1.1 Torr; emitter current and gain of the secondary electron multiplier were 3.8 and 3.8×10^3A, respectively. Under these operating conditions, the lifetime of a filament is over 100 h. For the first few hours of operation, emission of the contaminants of other alkali-metal ions may be significant. These subsequently decrease to less than 0.1 % of the total emission.

Reactions in RC Ionization is initiated by formation of alkali-metal ions on the emitter. A successful method for obtaining lithium ion cationization mass spectra must provide a means for binding lithium ion to the sample molecules. Lithium ions can be injected into an inert gas, such as nitrogen, that contains a trace amount of sample, and $(M + Li)^+$ complexes are formed by termolecular association reactions. The ionic products observed are a reflection of association reaction rates. The formation of an adduct ion is commonly assumed to be a three-body process, in which a neutral molecule collides with an ion-molecule complex and removes an amount of energy, stabilizing the $(M + Li)^+$ complex. Based on this model, the cationization process would be suitable in a high-pressure environment, since the added gas molecules serve as the third body in the analyte addition reaction with alkali-metal ions.

The abundance of attachment products was greatly influenced by the voltage applied to the repeller, which provides the react ion, Li^+, with kinetic energy. The attachment probability decreases with increasing repeller voltage. At the higher ion source pressure, more adduct ions were produced at the higher repeller voltage. Low repeller voltage (E) and high ion source pressure (P) favor adduct formation; as E/P is raised, the probability of this process decreases. This ion energy dependence is consistent with the reported mechanism of the ion attachment reaction (see Sect. 2.1). The initial combination step of an ion–molecule association reaction is brought about by ion/dipole attraction between the reactants without any activation energy.

5.2.2.4 Interface, Mass Analyzer, and Detector

Adduct ions from the RC pass through an aperture, A, and are directed into the differentially pumped lens region through a 1-mm-diam orifice, B, drilled into the tip of a skimmer. No potential was applied to the skimmer. The ion focusing system is an electrostatic lens system consisting of an extractor, deflectors, and four sets of coaxial sequential cylinders, each biased at a particular direct current (DC) voltage. The mass-analyzing chamber is equipped with a QMS with rods biased below ground by connecting the DC rod driven circuit to separate DC supplies. Two possible modes of ion detection, analog and pulse counting, are usually employed using a Channeltron electron multiplier detector. In the pulsed mode, the channeled output from the multiplier was fed to the preamplifier-discriminator-counter system operated in the multichannel scaling mode. The system is evacuated differentially by three evacuation pumps. The RC into which the sample gas is introduced through the carrier gas flow line is maintained at 1–760 Torr by a pump. The lens region ahead of the QMS is held at less than 2×10^{-4} Torr. The QMS region is evacuated around 5×10^{-7} Torr.

5.2.2.5 Commercial Equipment, IA-Lab

This type of mass spectrometer has recently become available commercially in a complete form [33]. It allows the acquisition of the mass spectrum of pseudomolecular ions. Figure 5.3 shows schematics of this instrument, IA-Lab (ion attachment Lab). Li^+ is supplied by a Li^+ emitter. Li^+ attaches to the sample gas in the attachment area. The ion $[M + Li]^+$ is then introduced to the mass analyzer operating to 1000 m/z after passing through focusing devices (lenses) in the differential pumping chamber. In addition, use of the EI source can furnish complementary fragment information.

The IA-Lab can analyze not only liquid/solid samples but also adsorptive gases via a heating entry port (or introduction port). There are two types of interface for sample introduction: an introduction port for gases and a DIP for solids and liquids. The optimal utilization of "direct measurements" is facilitated by the constitution of this DIP. In DIP, the sample is heated, vaporized with the IR lamp, and ionized in the ion source, and its constituents are measured in the mass spectrometer. The heating temperature is controlled by the program. The range is room temperature to 500 °C. An easy linkup with GC connection via a heating entry port is possible, too. After sample introduction, all operations are performed on a PC. In a measurement screen of standard mode, the SIM and a mass spectrum are displayed at the same time. The mass spectrometer is fully equipped with the software necessary for quantitative analysis calculations.

5.2.3 Features and Performance

The developed (manufactured) device, a QMS (M-400-QA-M, Canon Anelva Corporation, Tokyo, Japan), with the Li^+ ion attachment ionization source (model L-241G-IA), exhibits several advantages over conventional mass spectrometers. The response characteristics of this equipment and unique features are presented.

5.2.3.1 Response Characteristics

Sensitivity Consideration Sensitivity is influenced by a large number of factors [13], a few of which on optimization adjustment can give a significant gain in sensitivity:

(i) kinetic energy of the Li^+ ion and (ii) Li^+ affinity of the sample molecule. For stabilization of the ionic adduct, the cationization process should be carried out in a high-pressure environment to provide necessary gas molecules that serve as third body in the addition reaction with the alkali-metal ions. The voltage applied to the repeller controls the kinetic energy of the reactant ion. The relationship between the Li^+ ion intensity and the repeller voltage should be adjusted for ion source pressures; at high ion source pressure, more adduct ions are produced at a higher repeller voltage. Low repeller voltage (E) and high ion source pressure (P) favor adduct formation; as E/P is raised, the probability for adduct formation decreases. This ion energy dependence is consistent with a mechanism of ion attachment reaction in which the initial step is brought about by ion-dipole attraction between the reactants without any activation energy.

Li^+ Ion Affinity The production rate of the Li^+ ion adduct in the $(M + Li)^+$ ion formation depends strongly on the Li^+ affinity of the molecule, which ranges up to 50 kcal/mol. The mass spectrometric observation of molecular ions formed by the attachment reaction of Li^+ ions with molecules may establish a relative order for the sensitivity and an empirical relationship between the Li^+ ion affinities and the intensity of the adduct ions. The relative sensitivity can then be estimated for the organic compound of interest when the Li^+ ion affinity is known.

To determine experimentally the relationship between the $(M + Li)^+$ intensity and the Li^+ ion affinities of chemical compounds and to investigate to which extent the empirical relationship can be used for the determination of the relative sensitivity for the chemical species of interest, the relative ion currents of the $(M + Li)^+$ ions formed by the attachment reaction were examined for 16 compounds with Li^+ affinities ranging from 0.3 to 2 eV (7 ~ 50 kcal/mol). Some species were chosen partly because of their heavy involvement in the global warming process.

A plot of $I(M + Li)^+$ ion intensities versus Li^+ ion affinities is shown in Fig. 5.4 for examined compounds. Intensity of $(M + Li)^+$ is the Li^+ adduct ion current obtained experimentally. Li^+ ion affinities (in eV) of several species were calculated at the B3LYP/6-311 + G(3df) level [34]. Disregarding from H_2O and HCHO, it is apparent that $I(M + Li)^+$ increases with increasing affinity values. $I(M + Li)^+$ generally correlates well with Li^+ ion affinities over the energy region. Li^+ ion affinities, in turn, are heavily dependent on the polarity or polarizability of the target species. In other words, the chemical species capture the Li^+ ions and yield the $(M + Li)^+$ adducts if the Li^+ affinities are sufficiently high. Under conventional Li^+ IAMS experimental conditions, compounds with Li^+ affinities bigger than 1 eV (23 kcal/mol) were highly detectable, even if they were present in small abundance. Their production rate reaches bimolecular Langevin collision rate. Li^+ affinities of molecular species may be helpful in predicting whether the species can be detected using a minimum sample concentration. Unfortunately, there is not yet a body of literature about many molecules. However, the wealth of information about the polarity and the polarizability is useful in predicting which molecule can be detected at a low-level concentration, because it is known that the affinity increases with the polarity and to a lesser degree with the polarizability of the molecules [35]. Finally, it should

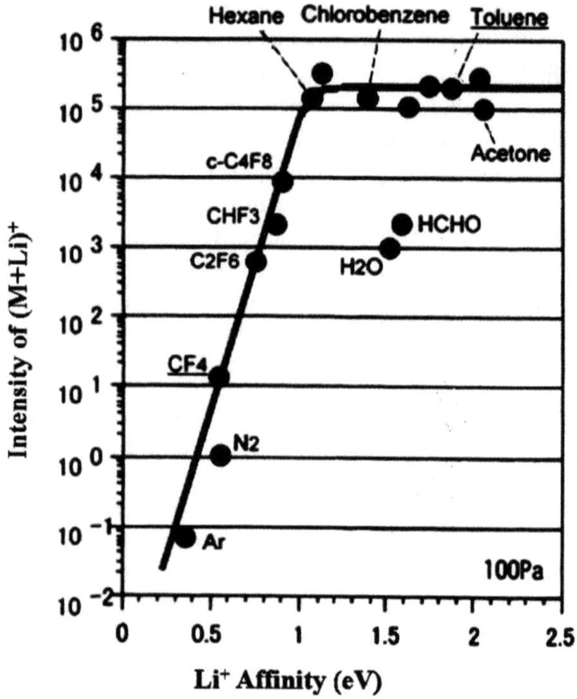

Fig. 5.4 Plot of $I(M+Li)^+$ vs Li^+ ion affinity for chemical compound (M) having different Li^+ ion affinities under the normalized condition that the pressure of the ion source is 100 Pa. The vertical axis is logarithm. Intensity of $(N_2+Li)^+$ is normalized to 1. (©2009, Canon Anelva, Data sheet)

be noted that many other alkali-metal species have been used as reactant ions. The affinity of the species, which decreases as the metal is varied from Li^+ to Cs^+, is the variable that can be adjusted to provide a selective ionization method.

Experimental Sensitivity The practical sensitivity [21] is determined as follows. When an acetone sample was introduced at a rate of 2.6×10^{-10} g/s from the permeation tube at 25 °C, measurement of the adduct ion current was 2.9×10^{-10} A (with an ion multiplier gain of 2×10^3); hence, the sensitivity was calculated as 1.12 A/(g/s). The minimum detectable amount (MDA) can be estimated using the actual noise level of the system as the ultimate limitation on the minimum detectable sample size. A signal-to-noise ratio of 10 is a practical and conservative lower limit for quantitative detection; in the analog detection mode, the MDA was calculated as 8.93×10^{-15} g/s (with a signal-to-noise ratio of 3), on the assumption that the detection limit of the ion detection system was around 1×10^{-14} A.

Linear Response Range (Dynamic Range) Linear response range is examined using ethyl acetate as a standard sample. Figure 5.5 shows log–log calibration, in which the logarithm of the response intensity (y-axis) is plotted against the logarithm of concentration (x-axis) and the calibration data are fit to a linear line. Signal response was linear over the chosen range of 10^{-4}–10^2 ppm (v/v) for the test sample. A plot of 10 introduction rates of ethyl acetate against their corresponding peak heights produced an essentially straight line up to 10^2 ppm, where more than

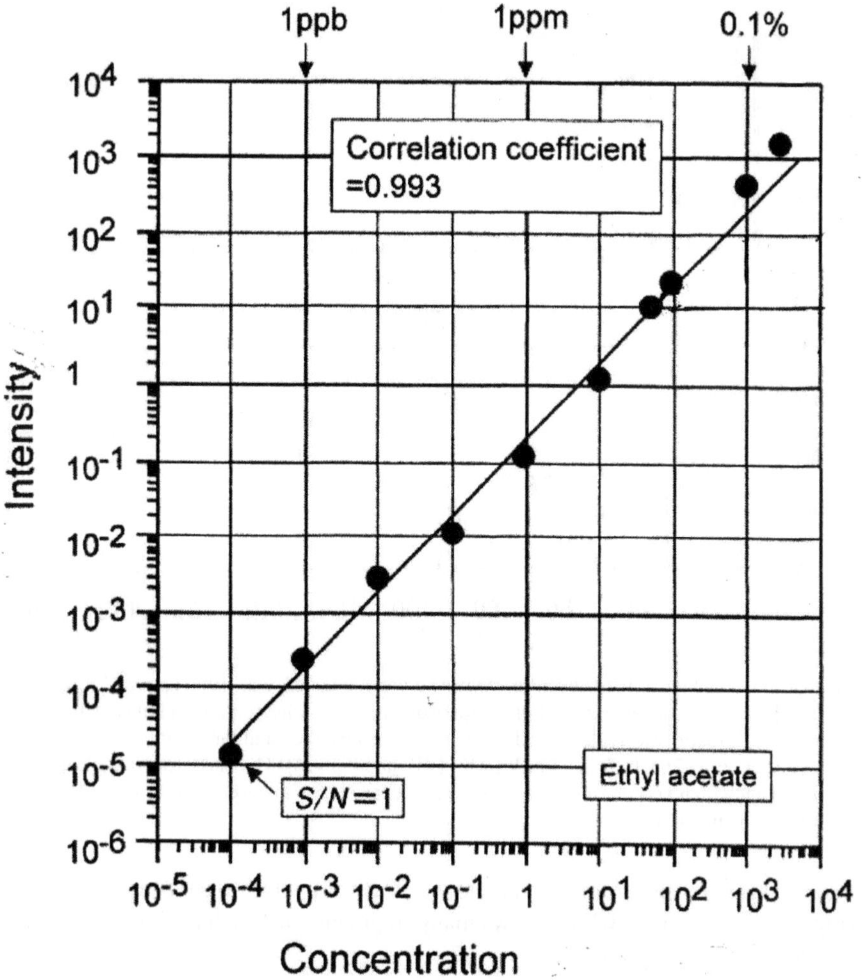

Fig. 5.5 Calibration plot of signal intensity vs concentration (v/v) for ethyl acetate as a testing sample. The signal intensity is normalized with N_2 signal at 100 Pa. Both axes are logarithm. (©2009, Canon Anelva, Data sheet)

3 % of the total Li^+ reactant ions is used for the attachment. In a manner identical with that of the electron capture detector the response follows Beer's law because the sample, as it is cationized, depletes the alkali-metal ion density. Therefore, for a linear response, the sample size must ensure a large excess of reactant ions in the cationization chamber (reaction chamber).

Typical Mass Spectrum A cyclic-C_4F_8 example shows the performance and response possible with this analytical instrument in an optimal case. Figure 5.6 shows a typical mass spectrum in the mass range up to m/z 210 [15]. Only the Li^+ adduct molecular ion (no fragmentation) is produced for this test compound (Fig. 5.6, upper

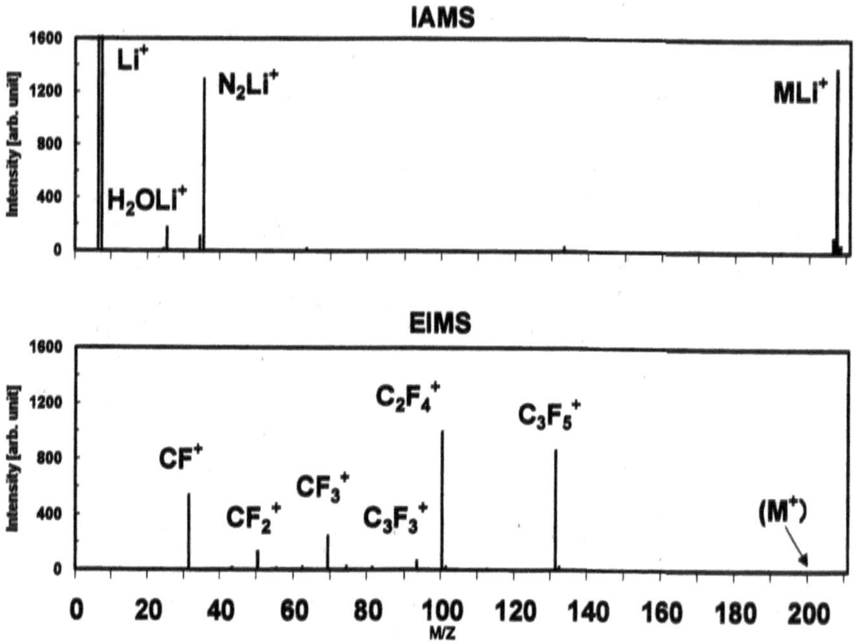

Fig. 5.6 Ion attachment mass spectrum *(upper column)* obtained from 1000 ppm c-C_4F_8 in the N_2 gas, showing c-$C_4F_8Li^+$ (m/z 207). For comparison, the 70 eV electron ionization (EI) mass spectrum *(lower column)* was measured; the abundant ions observed and their relative intensities are C_2F_4 (100), C_3F_5 (87), CF (54), CF_3 (25), CF_2 (13), and C_3F_3 (6). The background mass spectrum of EI is taken with no c-C_4F_8 in the N_2 carrier gas. (Reprinted with permission from [117]. ©2001, American Chemical Society)

trace). This mass spectrum is completely different from the 70 eV EI mass spectrum (Fig. 5.6, lower trace), which shows many fragment ions, such as $C_2F_4^+$, $C_3F_5^+$, and CF^+ ions, but no molecular ions.

5.2.3.2 Features

Some of the essential advantages of the IAMS system are the following: Li ion attachment MS is characterized by the following properties of importance in MS [21]:

1. No dissociative ionization takes place. There is little or no possibility of fragmentation of the adduct ions; IAMS provides only molecular ions and hence can be exploited for the determination of molecular weight and the analysis of mixtures when no fragmentation is desired.
2. Its sensitivity is high, especially for polar molecules, because ion-molecule reactions are fast.
3. Direct continuous measurements on a real-time basis of many radical species and stable molecules in a dynamic system are feasible. Mass spectra tend to be

simple and hence, it is easy to determine which form of which radical species is reacting. It also makes possible the analysis of a plasma via identification of both ionic and neutral radical species, and provides data so that correlations can be established between them. Such an instrument would extend significantly the use of IAMS in detecting any radical intermediates present in small quantities in the gas phase at atmospheric pressure and hence in directly identifying toxic radicals (e.g., from cigarette smoke, diesel engine exhaust, and smoke plume from combustion) present in the environment.

4. The detection of interstellar species and the identification of unfamiliar species, which would be immensely useful in atmospheric, astrophysical, and astrochemical studies, are possible with this IAMS method.

In addition to these appreciable advantages, the present technique also has certain limitations. Because identification of radical species or molecules is based only on the mass number, the peak assignment is not unambiguous (i.e., no direct structural information is afforded by the ionization). Unsatisfactory results have been obtained with nonpolar molecules or molecules with low polarizability, such as alkanes; these molecules have weak Li^+ affinities and are not easily attached. Also, the IAMS technique gives no direct structural information, and so the correct isomer of the detected molecule cannot be directly determined.

5.2.4 *Intermediary Free Radical Species*

The field of free radical chemistry using MS has been recently developed for its application in the region of heterogeneous and homogeneous thermal decomposition reactions, pyrolytic process, combustion and flame processes, electric discharges, photolysis reaction, CVD plasma processes, reaction with excited atoms, direct photolysis reactions, radicals produced by dissociation ionization and atmospheric reactions in which NO, NO, HNO_2, and N_2O figure prominently as intermediate products. Toxic radicals (e.g., from cigarette smoke, diesel engine exhaust, and smoke plumes from incomplete combustion) that are released into the environment are also interesting targets. The widespread occurrence of free radical intermediates in these chemical environments always provides an impetus to the study of kinetics [36, 37]. However, the experimental difficulties involved in detecting intermediate species (free radicals), identifying them and measuring their concentrations have proved to be great. The combination of low concentration and short lifetime of intermediate radicals has provided a real challenge to scientists.

Many methods, which include absorption spectroscopy [38], Fourier-transform spectroscopy [39], electron spin resonance [40], laser magnetic resonance [41], resonance fluorescence [37, 38], MS [42], and other chemical means, require special conditions of radical concentration or are applicable only to certain radicals with special properties [43]. In fact, many radicals have been detected to date, by MS with low-energy electrons [42], radical trapping in a CI source [44], or photons [45]. However, a number of interfering factors which are encountered in practice are unavoidable. So far, no completely satisfactory method has been developed for measurements on free radical intermediates and molecular and atomic species in ordinary gas-phase chemical reactions at the low, steady-state concentrations at which they usually occur.

This section describes a new method for detection and identification of intermediate radical species in the gas phase. This method is based upon the establishment of an alkali ion attachment to radicals through termolecular reaction. It provides mass spectra of quasi-molecular ions formed by lithium ion attachment to the radical species (R) under high pressure. Results are obtained in the form of a mass spectrometric trace of Li^+ adduct radicals. The advantages of the present method are: (i) a measure of mass is a guide to radical identity, (ii) adaptability to a condition of higher pressures, and (iii) direct continuous measurements of any species in dynamic systems can be made. The method is applied to the study of the microwave (MW) discharge in CH_4, C_2H_4, C_2H_2, CH_4/H_2, CH_4/O_2, or CH_4/N_2 and was successfully used for the qualitative analysis of the compositions of hydrocarbon (HC) radicals in the discharge. Various kinds of HC radicals are identified. In the following, the scheme, its operational procedure, and considerations of its sensitivity are detailed. Although the detection of radicals of greater instability awaits further studies in the technique, this method is of general applicability to problems involving free radicals and short-lived intermediates. Its contribution to radical chemistry is certain, especially in the dynamic studies of individual free radical species.

5.2.4.1 Diagnosis of Plasma

Various HC intermediate radicals can be directly identified in CH_4 [46, 47] and isobutane/He [11] MW plasmas [49–51]. The chemistry occurring in CH_4/O_2 [52, 53] and CH_4/N_2, [54] discharge plasmas shows that many o-containing and n-containing neutral chemical species are generated. The formation of many polycarbons, along with some aromatic radicals and interstellar species such as C_3H_2 and C_4H_2, was detected for the first time in the plasma of C_2H_2 [49], C_2H_4 [50], and CF_4 [51].

Although the use of plasma formed in the electric discharge has received considerable attention, the knowledge of radical species occurring in steady state plasmas has not advanced satisfactorily enough. Typically, in electric discharge plasmas, the radical concentration is about 50–100 times higher than the ion concentration [55] and, as a result, fast radical/molecule reactions may occur. Therefore, the role of radicals in reactive plasma must be great and hence, should be studied for the development of plasma processing [56]. In fact, the detection and evaluation of radicals in discharge plasma have been investigated by a variety of measurements. Mass spectrometric investigations of the CH_4 plasma were carried out by Rudolph and Moore. They measured mass spectra [57] of free radicals as well as stable molecules from a MW discharge in a fast flow of methane in argon and in hydrogen and reported many HC radicals. However, the main purpose of their work was not to determine the HC radicals. Electron impact energy of 15 eV was chosen for their work. Presumably, abundant formation of fragment ions from many stable molecules unavoidably hampered the determination of the radical species. IR tunable diode laser absorption spectroscopy is useful in the diagnostics of plasma. It is capable of detection of a wide variety of radicals and stable molecules found in the CH_4 system. Wormhoudt [58] reported the absolute concentration level, for CH_3 and C_2H_2 in CH_4 plasma.

Primarily because of the lack of spectral information for the majority of HC radicals formed in the reaction leading to the formation of higher HCs, no sufficient studies of the CH_4 plasma were performed with any of the methods; particularly in terms of HC polymer radical detection. Fujii and his group have shown [46–48] that the IAMS method was found to be especially suitable to detect free radicals in the plasma because the neutral radicals in the MW plasma are cationized by Li^+ in the RC. As a continuation of these and earlier studies on the detection of gaseous HC radicals, they report the identification of neutral radical species in a methane MW plasma discharge flow tube, indicating that the nature of the plasma can best be studied after free radicals have been fully identified. The setup used for the study of the CH_4 MW discharge plasma is reviewed briefly. The MW discharge tube forms part of a conventional flow system. The CH_4 gas is controlled so as to flow at a constant rate of 10 ml/min through a discharge tube. The discharge tube made of quartz glass tubing, with having an internal diameter of 3 mm, is mounted to the RC. The CH_4 gas enters the discharge tube where it is excited at 2450 MHz by a magnetron oscillator (model MR-301, Ewig Corp., Tokyo) having a nominal output power of 200 W. The formation and involvement of free radicals in the MW plasma was directly confirmed with Li^+ ion attachment to them. The experimental procedure consisted of admitting CH_4 gas with the gas flow rate at 10 ml/min. As a result, the gas pressure of the flow tube and the RC was ca. 1.7 and 0.12 Torr, respectively. The product analysis subtracts from each plasma spectrum its spectrum with no plasma; thereby, enabling the analysis of the influence on the whole spectrum which is caused by a change in any plasma parameter. Thus, the intensity of any mass peak in the condition of no plasma is subtracted from the corresponding mass peak under the plasma condition. The result shows the particular product peak due to plasma activation. The analysis of the mass spectra was made in terms of mass identification, peak height determination and the normalization of each peak considered. The relative abundance of the peaks represents the monoisotopic intensity (including contributions from all the isotopes for a given ions) with the relative intensities normalized to 100 units for the base peak of the C_2H_3 species.

The analysis of the individual spectral peaks was made in terms of mass identification and classification by their formulas. Table 5.1 shows the identified species which appeared. As far as the assignment of the mass spectral peaks goes, validity is based principally on the mass number. Therefore, some species show uncertainty with regard to their assignment. For example, the m/z 36 which was assigned to $C_2H_5Li^+$ may partially belong to $CHOLi^+$. With the MW discharge on, the following radicals and molecules were detected: C_nH2_{n+1}, C_nH_{2n-1}, C_4H_3, C_5H_5, C_6H_6 etc. A remarkable feature is the appearance of polymer HC radicals. No CH_4Li^+ was observed under the plasma activation conditions, indicating that almost all the CH_4 changed to other species in the MW discharge plasma. The occurrence of the

Table 5.1 Analysis of radical products formed from the MW discharge plasma of pure CH_4 in terms of mass identification, classification by their formula, and the relative intensities (percent values). (Adapted from [47] with permission)

Type	Chemical species
C_nH_{2n-1} [a]	C_2H_3(100), C_3H_5(68), C_4H_7(36), C_5H_9(28), C_6H_{11}(16), C_7H_{13}(13), C_8H_{15}(12), C_9H_{17}(6), $C_{10}H_{19}$(4), $C_{11}H_{21}$(3)
C_nH_{2n+1} [a]	C_2H_5(48), C_3H_7(42), C_4H_9(32), C_5H_{11}(29), C_6H_{13}(20), C_7H_{15}(12), C_8H_{17}(9), C_9H_{19}(6), $C_{10}H_{21}$(3), $C_{11}H_{23}$(2)
Others[a]	C_4H_3(13), C_4H_5(12), C_5H_5(10), C_5H_7(8), C_6H_5(8), C_6H_7(6), C_6H_9(6)
Molecules[b]	C_2H_2(10), C_4H_2(11), C_4H_4(7), C_4H_6(8), C_5H_6(11), C_6H_4(9), C_6H_6(12), C_6H_8(5), C_6H_{10}(3)

[a] Species possessing an unpaired electron; that is, open-shell species

[b] These closed-shell molecules may have high enough Li^+ affinities to be attached by Li^+ ions

polymerization processes taking place is also shown in Table 5.1 where the relative heights of the m/z peaks representing per cent values are normalized with the highest peak of $C_2H_3Li^+$. If one considers that the major radical constituents are those having relative heights higher than 5 %, it is interesting to note that 14 kinds of species with carbon numbers higher than 5 were obtained.

In the following section, the results are compared with the literature results from optical spectroscopy [58] and MS [43, 57, 59]. They are discussed separately for each classified species to evaluate the presence of the various chemical species. From the results given in Table 5.1, it can be concluded that (1) the C_nH_{2n+1} and C_nH_{2n-1} radicals, where $n=2$ to 11, have been detected. The intensity trend is clear; species with the higher n number decrease in peak intensities, (2) the plasma of CH_4 from the MW discharge leads to a polymerization process [60] of higher HCs, at least, partially through the radical/molecule reactions [61]. Rudolph and Moore [57] reported that the MW CH_4 discharge provided many HC radicals such as C_2H_5, C_3H_3, C_3H_7, C_4H_5, C_5H_7, C_5H_9, and C_5H_{11}. We have detected all these species except C_3H_3. However, C_2H_3 and C_3H_5, which exhibited very intense peaks in our spectrum, are absent in their spectrum. These disagreements may be partly because of the different experimental conditions.

The CH_3 radical, commonly observed in the CH_4 plasma [58, 59], was not detected in this system. The absence of this species is interpreted tentatively as follows: The present measurements have been made in a flow system in which the radical is continuously produced and detected through the attachment reaction in the afterglow region of the MW plasma. The distance between the MW cavity and the Li^+ ion emitter is about 3.2 cm. The chemical species formed in the cavity have to travel longer than this distance without conversion in order to be detected. A distance of 3.2 cm is too great to prevent conversion and therefore allow measurement of CH_3. The species, CH_2 and CH also did not appear, even though their production rates may be significant. These results may also be interpreted by the same token. This interpretation is consistent with Klein's finding [62] that CH_2 is quickly converted to CH in reactions with H atoms and the CH produces C_2H_5 by reacting more rapidly with CH_4.

Other radical species with carbon numbers *n* up to 6 may also be produced in the plasma process and detected by the present method. These are: C_4H_3, C_4H_5, C_5H_3, C_5H_5, C_5H_7, C_6H_5, C_6H_7, and C_6H_9. Also some molecules such as C_2H_2 and C_4H_4 appeared. Benzene and substituted benzene are also present. Unsatisfactory results are that nonpolar molecules or molecules with low polarizability, such as alkanes, have not been detected. They have weak Li^+ affinities and are not easily attached by the Li^+ ions. Therefore, even if abundant alkanes are produced in the plasma, they are not observed with the present method.

There appeared OH radicals in the spectrum when the discharge was on. The generation of H_2O concentration in the MW plasma, as well as the presence of H_2O in CH_4 as impurities, may be attributed partly to the formation of OH, because the presence of H_2O is associated with the production of OH radical [63] in the plasma.

5.2.4.2 Simultaneous Detection of Ionic and Neutral Species of Plasma

The understanding of the reactions in the plasma [64] is largely impeded due to the complex nature of plasmas. The most difficult task is to determine the relative importance of ions and neutral species in the reactions of plasma. To explain the process many different models have been postulated from the analysis of chemical products observed in the plasma. Models for plasma reactions have invoked ion–molecule reactions, radical–radical reactions, and radical–molecule reaction. However, the extent of their participation is still unclear. The mechanism of polymer production in gas discharge has also been the subject of some controversy [65]. Many reports of mass spectrometric sampling of electric discharges have appeared. Few attempts [57, 66] have been made to obtain correlations between the ions and neutral species.

C_2H_2 For many years there has been considerable interest in studying C_2H_2 chemistry in low-pressure electrical discharges at the molecular level because of its application in plasma polymerization for depositing thin films [65, 67, 68]. Fundamental investigations began in the mid 1960s and have been developed into a research field that includes a host of optical spectroscopic and mass spectrometric product analyses. The determination of the ionic and neutral chemical components of the flowing MW discharge plasma of acetylene was performed by Li ion attachment MS [49].

Plasma composition can be investigated as a function of controllable parameters such as C_2H_2 flow rate (pressure), MW-induced power input, and the MW cavity position. For these studies, the accelerating potential grid (APG) system [69] was mounted in the RC at the same axial position as the sampling orifice. This was done in order to study both the neutral species and the ionic species sampled from the MW plasma with any gaseous system, such as C_2H_2. All three grids are placed 3 mm apart. The stainless steel grid is disk-shaped with a 10-mm hole covered by platinum (Pt) wire mesh. To maximize the ionic signal, independently variable potentials are applied to the grids. With the APG at ground (in the neutral mode), negligibly small ions above the background signal could be detected, but with the

APG on (in the ionic mode) the characteristic ionic spectrum was obtained. In the neutral mode, the products were monitored by measuring Li^+ adduct ion mass spectrum. The neutral species flow over the point of Li^+ emission. After Li^+ ion attachment in the RC at approximately room temperature, the adduct ions are then analyzed mass spectrometrically. In the ion mode, the APG system is employed; approximately −390 V on the first grid, −80 V on the second, and −100 V on the third. Presumably, incoming positive ions were accelerated electrostatically. Therefore, this system serves to efficiently transfer the charged species present in the plasma to the mass spectrometer analysis. The MW source was used, which was constructed from a straight quartz tube (4-mm inner diameter, 6-mm outer diameter, and 30-cm long). The C_2H_2 gas flows down the tube. Connecting a cavity to a 2.465 MHz MW generator through a matching network created the MW plasma.

Mass spectra were obtained with plasma activation in two modes; (i) in the presence of Li^+, denoted as A, and (ii) in the absence of Li^+, as B. Thus, the intensity of any mass peak in A is subtracted from the corresponding mass peak in B and the results (A-B) show the Li^+ adducts of particular neutral products (Fig. 5.7). The mass spectrometric measurements have been performed at the down-stream position with respect to the MW cavity and the direction of the gas flow. The formation and involvement of free radicals in the MW plasma was also confirmed. The reaction schemes for some products have been considered, particularly with regard to

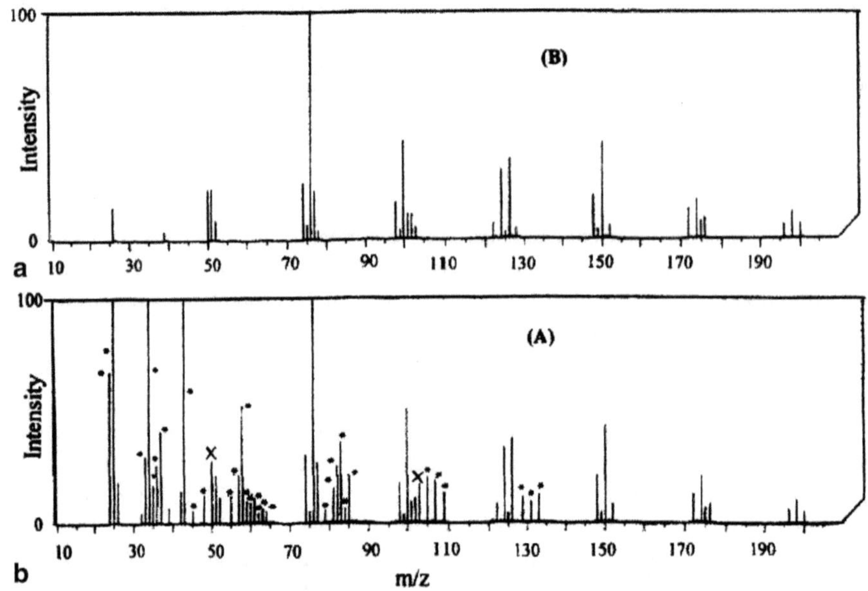

Fig. 5.7 Mass spectra of a 30-W, 840-Pa C2H2 MW discharge. **a** Sampled in the Li^+-on condition. **b** Sampled in the Li^+-off condition *(ionic species detection)*. Additional peaks in **a** due to Li^+ ion adducts are marked by *asterisk*. The significantly increased peaks at m/z 50 and 103 in **a** consist of ionic species and Li^+ ion adducts. These are marked by *x*. (Reprinted with permission from [49], ©1998, by The American Physical Society)

the role of radical and ionic condensation reaction processes. An attempt was made to link both the ionic and neutral species to a proposed mechanism. A model that explains the observed phenomena and is consistent with the dependence on plasma parameters is suggested. The model predictions indicate, for example, that the principal ionic condensation reaction can be summarized by the reaction sequence:

$$C_nH_2^+ + C_2H_2 \rightarrow C_{n+2}H_2^+ + H_2, \quad (n = 2, 4, 6, 8) \tag{5.1}$$

while the reactions involving a hydrogen atom may be those which lead to the formation of neutral C_mH_n $(n > m)$ species, including free radicals of C_2H_3, and C_4H_5. It can be concluded that a pure C_2H_2 MW discharge results in a high conversion of C_2H_2 to ionic and neutral polymer products. This process is explained by two types of general reactions: interaction between free radicals and HC molecules, and interaction between ion radicals and HC molecules. The observed ion chemistry is very much like that found by high-pressure MS: ionic species with even-numbered carbon atoms dominate. Almost all the significant neutral species coexist with the corresponding ionic species, except for C_4^+, C_6^+, C_8^+, and $C_mH_n^+$ $(n > m)$. An interaction between a hydrogen atom and HC molecules is the mechanism for the C_mH_n $(n > m)$ species.

CH_4/O_2 The ionic and neutral products of the CH_4/O_2 mixture MW plasmas in a discharge flow tube have been studied using MS in conjunction with the APG devices and the Li^+ ion attachment technique [52]. The APG devices provide a means of identifying the ionic products from the MW discharge.

The use of O_2 additive gases with the monomer CH_4 MW discharge leads to the production of various kinds of chemical species, demonstrating that the MW discharge is a method for generating an interesting oxygenated polymer species. Direct observation of these species is allowed by the measurement of the Li^+ adduct mass spectrum. Interestingly, the product species have the formulas $C_nH_{2n}O$ (presumably R-OH), $C_nH_{2n}O_2$, and $C_nH_{2n+2}O$(presumably R-CHO) with n continuing on to 8 and beyond. The most probable primary reaction is thought to be a reaction to yield a peroxy radical. It is well known that alkyl radicals R are rapidly converted into peroxy radicals in combination with molecular oxygen [70]:

$$R + O_2 \rightarrow R - OO \tag{5.2}$$

The peroxy radical is considered to be capable of a whole range of additional reactions, such as an abstraction reaction of oxygen [71]:

$$R - OO + R - OO \rightarrow R - CHO + R - OH + O_2 \tag{5.3}$$

The assorted products of these reactions would result in widespread peaks. This speculation is proved partially by the presence of the R-CHO and R-OH species.

The major features of the ion chemistry in CH_4/O_2 discharge can be summarized as follows: (i) the O_2 (or O atom) interacts with various HC species to produce oxygenated polymer HC, (ii) mass spectra show that these neutral species, thus produced, undergo the protonation ion-molecule reactions with proton transfer reagents, presumably CH_5^+, $C_2H_5^+$, and so on, including primary electron impact ionization of the corresponding neutral species. The distribution of ionic species from a discharge is, somehow, representative of the neutrals being protonated.

CH_4/N_2 MW discharge methods [72, 73] have been used to form various kinds of chemical substances. Among them, the use of CH_4/N_2 plasma formed in an electric discharge first received attention in the synthesis of hydrogen cyanide (HCN). Yields of HCN have been rather extensively studied in both DC and radio frequency (RF) thermal plasmas. A glow discharge of a mixture of CH_4 and N_2 has been used to generate the HNCN radical.

The CH_4/N_2 MW discharge led to the production of various kinds of chemical species [54], and effectively demonstrated that a MW discharge is a method for generating over 50 N-bearing products. They had the formulas C_mH_nCN, $NC(CH_2)_nCN$, $C_mH_nNH_2$, and $C_mH_nN_2H_3$, with m and n continuing on to 8 and 15, respectively. Some interesting products found in this study were $H_2C = NH$ and $NC(CH_2)_nCN$. These components are predicted to be possible uses as industrial products. Similar to the CH_4/O_2 study with IAMS, the features of the ion chemistry in CH_4/N_2 discharge can be summarized as follows: (a) the N_2 (or N atoms) interacts with various HC species to produce nitrogenated polymer hydrocarbons, and (b) mass spectra show that these neutral species undergo protonation ion-molecule reactions with proton transfer reagents, presumably CH_5^+, $C_2H_5^+$, and so on, in addition to primary electron impact ionization of the corresponding neutral species. The distribution of ionic species from a discharge is, somehow, representative of the neutral species being protonated.

CH_4/H_2 Despite the importance of understanding the properties of the CH_4/H_2 plasma and the high level of interest, the chemistry involved with this system is not yet well understood. The mass spectrometer-Li^+ reactor setup was successfully used also for qualitative and quantitative analysis of the neutral and ionic species in the CH_4/H_2 system [74]. Features of the chemistry of CH_4/H_2 discharge can be summarized as follows: (i) The reaction of CH_4/H_2 in MW plasma leads to a polymerization process [75]. The presence of neutral radicals suggests a radical-involved reaction that may be one of the mechanisms in the polymerization. (ii) An increase in the CH_4 component of the feed gas leads to greater formation of many neutral species, whose distribution hardly changes over the entire range of feed gas composition. (iii) Mass spectra of ionic products show that all these species, with even-numbered mass ions predominant, were species with corresponding neutrals observed in the plasma. However, no remarkable similarities in abundances exist between ionic and neutral product species. The distribution of ionic species from a discharge is not representative of the neutral being protonated.

N_2/O_2 An N_2/O_2 MW discharge leads to the conversion of N_2 and O_2 to ionic and neutral association products [76]. The presence of water as a contaminant resulted in the production of many cluster species, $NO^+(H_2O)_n$ (n=0–3), $H_2NO^+(H_2O)_n$ (n=0–2), $NO^+(NO)$, $NO^+(N_2O)$, $NO^+(NO_2)$, and $NO^+(HNO_2)$, in addition to the familiar nitrogen oxides. For the first time, the formation of gas-phase NO_2 and HNO_2 neutral

species in the plasma was confirmed, which was possible only through simultaneous assessment of the neutral and ionic species in the plasma.

H_2O A H_2O/He MW discharge results in a high conversion of H_2O to ionic and neutral association products [77–79]. The ion/molecule reaction chemistry plays an important role. Primary ion production in the H_2O MW plasma is observed under any plasma condition and is given by,

$$H_2O \rightarrow H_2O^+,\ H^+ + OH,\ H_2 + O^+,\ H + OH^+ \tag{5.4}$$

The study clearly showed the presence of various clustering compounds of water as well as their ionized species. Among compounds are the interesting ionic series $H_3O^+(H_2O)_nO$ (n=0–2) and $(H_2O)_n^+O$ (n=0–4), which have not been reported before. The reaction schemes for some products, particularly with regard to the role of radical and ionic condensation reaction processes, in order to link both the ionic and neutral species to a mechanism. One would expect OH radical chemistry in the H_2O discharge. It is well known that hydrogen atoms and hydroxy radicals are the primary products of the dissociation of water vapor in an electric discharge. A hydroxy radical, OH, was directly observed as a Li^+ ion adduct. A possible reaction for association may start with this species. The presence of (H_2O), Li^+OH adducts suggests that a radical-involved association reaction may be the formation mechanism.

$$(H_2O)_{n-1}\,Li^+OH + (H_2O) \rightarrow (H_2O)_n\,Li^+OH \tag{5.5}$$

It is summarized that the principal ionic condensation reaction can be occurred by the reaction sequence

$$H_3O^+(H_2O)_n\,X + H_2O \rightarrow H_3O^+(H_2O)_{n+1}\,X\ \ (X = O, OH) \tag{5.6}$$

while involving the OH radical may lead to the formation of $(H_2O)_nLi^+OH$ adduct species.

5.2.4.3 Diagnosis of Diamond Film Growth

In the past years there have been an increasing number of studies identifying the primary gas-phase diamond growth precursor [65, 67, 80–87] in CVD techniques, which have been the subject of controversy. The growth rate and film quality of materials synthesized by CVD processes depend critically on the composition of the species generated.

Many techniques exist for examining the composition of the species (including radical species) generated in CVD reactors: optical emission spectroscopy (OES) [80], FT-IR spectroscopy [81] laser induced fluorescence (LIF) spectroscopy [82], diode laser IR absorption spectroscopy [83], and MS [84–87]. Each has its own strengths and shortcomings. A major advantage of MS over other techniques is its ability to monitor a wide spectrum of species simultaneously. However, the identities of many reactive species are lost in the case conventional electron impact ionization techniques.

Several growth species have been proposed for the CVD environments, including C, CH*, CH_3*, C_2, C_2H*, C_2H_2, C_2H_3*, C_2H_4, and so on. The earliest studies on gas phase chemistry in MW plasma assisted diamond deposition systems were conducted with OES to detect electronically excited H, CH and C_2 [80]. Mitomo et al. [81] used FTIR spectroscopy to study the effect of using various carbon feed gases and confirmed that abundant quantities of C_2H_2 are formed in the plasma when methane is the HC feed gas. Celli et al. [83] detected gaseous compounds C_2H_2, C_2H_4, and CH_3* in CVD from a reactant mixture of 0.5 % CH_4 in H_2 by diode laser IR absorption spectroscopy. Harris et al. [86] measured the mole fractions of products in the hot filament CVD by MS. They indicated that detected compounds were CH_4, C_2H_4, and C_2H_2, while C_2H_2 was the major product.

A detailed chemical kinetic mechanism was composed to describe the evolution of reaction species [88–92] in the pyrolysis of a hydrogen-diluted CH_4 mixture, simulating the gas-phase conditions of diamond CVD in the MW reactor. Somewhat surprisingly atomic C is predicted [88] to have a large concentration at the surface. C atoms could conceivably play a role in diamond growth [90].

The IAMS technique is applied in a simulation study of a diamond CVD reactor [93]. It is most advantageous when the experimenter does not know a priori what species to expect. A second advantage is its adaptability to a wide variety of process environments. To simulate the reacting plasma environment, a sampling aperture, followed by the skimmer, is fixed at the center of the growth substrate (Fig. 5.8). Mass spectra were obtained (Fig. 5.9) with plasma activation in two modes: (i)

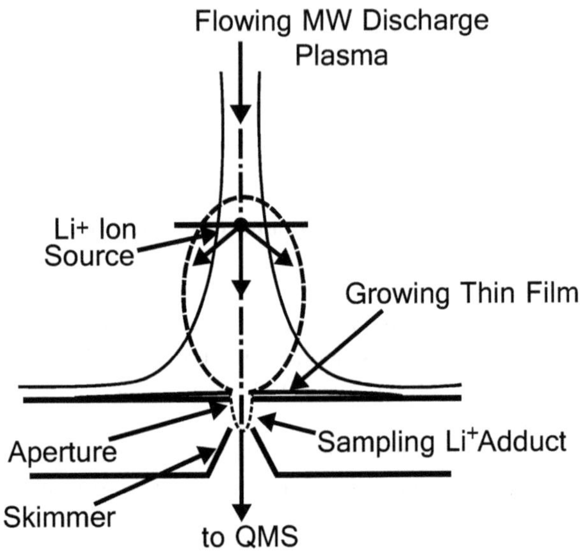

Fig. 5.8 Simulation system for the diagnosis of diamond deposition CVD with Li^+ ion attachment mass spectrometry. (Reprinted with permission from [93]. ©2001, American Institute of Physics)

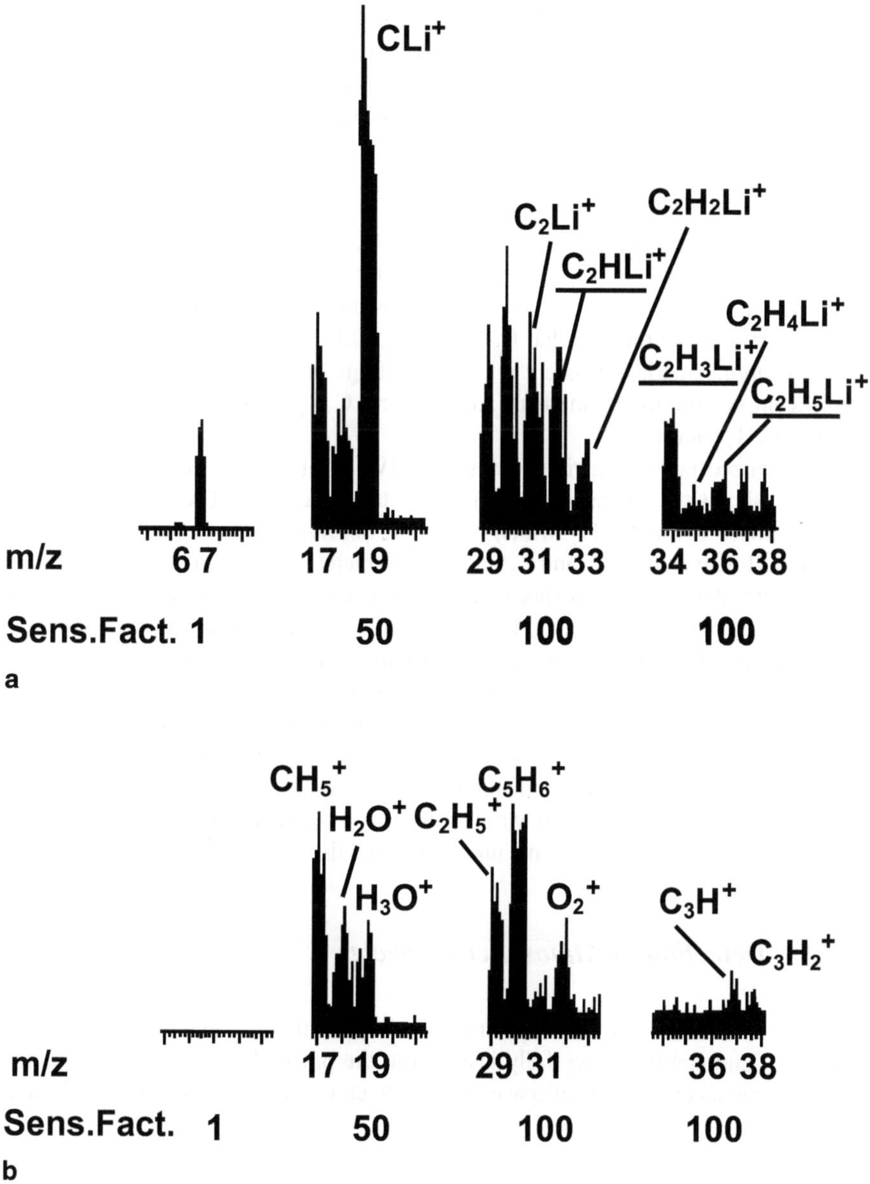

Fig. 5.9 Partial mass spectra of Li^+ ion adducts over the regions m/z 5–9, 17–21, 29–33 and 34–38. **a** The neutral mode (in the presence of Li^+) and **b** the ionic mode (in the absence of Li^+). The *underlined peaks* indicate the radicals observed. (Reprinted with permission [93]. ©2001, American Institute of Physics)

in the presence of Li^+, denoted as a, and (ii) in the absence of Li^+, as b. Thus, the intensity of any mass peak in a is subtracted from the corresponding mass peak in b and the results (a minus b) show the Li^+ adducts of particular neutral products. The mass spectrometric measurements have been performed for the 1: 100 CH_4: H_2 plasma at the downstream position with respect to the MW cavity and the direction of the gas flow. The formation and involvement of free radicals in the MW plasma was confirmed also.

The major species detected in (1: 100) CH_4/H_2 discharges are C, C_2, C_2H^*, C_2H_2, $C_2H_3^*$, C_2H_4, $C_2H_5^*$, and C_3. These are similar to those reported by others with some exceptions. C atoms were detected. However, we could not detect CH_3^* in the plasma, despite the fact that C_2H_2, $C_2H_3^*$, C_2H_4, and $C_2H_5^*$ were observed in the discharge. This is consistent with the findings of Stalder and Homsi [84] that significant conversion of methane to C_2H_2 and C_2H_4 is observed, as are radicals such as $C_2H_3^*$ and $C_2H_5^*$.

The ionized particles arriving from the MW plasma travel into the QMS. This spectrum shows two characteristic features in the ionic mode. The ions CH_4^+, CH_5^+, $C_2H_n^+$ with $n=1–7$ and $C_3H_n^+$ with $n=1–9$ were observed in the mass spectra in small amounts. However, both C^+ and C_2^+ ions are characteristic by their absence, while the neutrals are abundant. Mass spectrometric evidence is given that directly confirms the existence of C, C_2, C_2H^*, C_2H_2, $C_2H_3^*$, and C_2H_4 along with other polymer species. The chemistry of the diamond deposition plasma can best be studied after the component species have been fully identified. One of the most interesting results in this study is the observation of the abundant atomic carbon species, indicating that the incoming CH_4 decomposes to C through a loss of H in the gas phase and C species should be taken into account for diamond film growth. This finding is consistent with the Goodwin's calculations [75], which show that a significant amount of atomic carbon could be present in the plasma under diamond deposition conditions.

5.2.5 *Detection of Atmospheric Species*

The Li^+ ion attachment mass spectrometer is potentially relevant to atmospheric chemistry and can be successfully installed in the payload of a spacecraft to determine chemical composition of space, in order to enhance our understanding of space.

5.2.5.1 Titan Atmosphere

In the fields of chemical evolution, organic cosmochemistry, and exobiology, Saturn's satellite Titan has been considered as a very interesting object in the solar system [94, 95]. It should be noted that Titan's organic chemistry has received much attention due to its possible connection with the origin of life. Measurements performed by Voyager during a flyby of Titan [96] have revealed a dense atmosphere around Titan mainly composed of N_2 with some CH_4 and other minor components, including HCs and organic-nitrogen molecules. A great number of experiments have already been performed, simulating the evolution of gaseous mixtures containing N_2 and CH_4, when submitted to energy flux. Prior to the Voyager encounter, a few experiments had investigated the gas-phase products of radiation chemistry in N_2/CH_4 mixtures. Most of this work was done in the context of prebiotic chemistry on Earth.

Voyager with an IR interferometric spectrometer (IRIS) on it [96, 97] examined Titan's atmospheric chemistry. Six simple HCs (C_2H_2, C_2H_4, C_2H_6, C_3H_8, C_3H_4, C_4H_2) and three nitriles (HCN, NCCN, HC_3N) were found. The concentration of these compounds varied from the 10-ppm range for ethane down to the 1–10 ppb range for C_3H_4, C_4H_2, HC_3N, and C_2N_2. An interesting question comes up from the Voyager's IRIS whether these minor organic constituents are really produced from the major constituents (N_2 and CH_4) in the atmosphere of Titan. After Voyager, simulation experiments have been made with an acceptable set of conditions which might apply to the high altitude chemistry of Titan's atmosphere. Based on Voyager and some simulation data, it is desirable to examine that other complex organic-nitrogen molecules are formed in Titan's atmosphere.

The MW discharge of a 90% N_2–10% CH_4 gas mixture at 20 Torr was studied to simulate the Titan atmosphere with IAMS [98], because Thompson et al. [99] have suggested that such plasmas simulate the energetic process in the upper stratosphere of Titan. Many products are detected, which may be precursor molecules to the more complex organic compounds that are thought to exist in Titan's atmosphere, and which are of relevance to the study of the origins of life. The results are compared with the Voyager observations and other laboratory simulation systems producing organic compounds from Titan-like mixtures.

It was demonstrated that over 70 products could be detected, including HCs, nitriles, amines, possibly hydrazines, and two other N-bearing compounds of $H_2C=NH$ and H_2NCN. The hydrocarbons were notably saturated. Among the variety of products, molecules of highest mass include C_5H_9CN, $C_5H_{11}CN$ and so on up to (C+N) = 7. In comparing these results with the Voyager IRIS observations and with other laboratory simulations, it has been demonstrated that (i) all the gas-phase organics found by Voyager are detected in the simulation; the products provide an adequate explanation of the constituents detected by Voyager in the Titan atmosphere, (ii) additional products found from the simulation study are predicted as products for future, more sensitive, remote, and in situ observations. These results suggest that MW plasma may be the analogs to planetary lightning and plasmas associated with impacts of meteors; properly conducted laboratory experiments may be relevant to radiation-chemical (charged-particle–induced) processes in planetary atmospheres.

The results may also have implications for chemical evolution in the solar system. Our results suggest that nitriles, amines, hydrazines of higher molecular weight and their derivatives, such as cyano-acetylenes and dinitriles, might have been synthesized. The presence of these important compounds, as precursors of biologically important compounds on Titan, implies that the chemical reactions postulated as the formation processes of bases and amino acids on the primitive Earth may be commonplace in the planetary system.

5.2.5.2 Simulation of Earth's Atmosphere

Interest in NO*x* species is inspired by the desire to understand their complex spectroscopic characteristics on a fundamental level, as well as by their importance in the Earth's atmosphere, especially with regard to key processes relating to ozone (O_3), hydroxyl radicals, and acid formation [100]. Model calculations indicate that background concentrations of NO*x* in the troposphere may have risen by almost 100%, causing a 20% increase in O_3 levels during much of the year. Beck and Ehhalt [101, 102] reported significant evidence that 30–50% of the NO*x* in the upper troposphere is due to aircraft emissions.

Although nearly all (>90%) of the NO*x* in the atmosphere is emitted as NO, NO is quickly converted to a variety of other species such as NO_2, N_2O_5, HNO_2, and HNO_3 [103]. There are indications that all these species may be involved in a complex way in the behavior of NO*x* [104]. A full understanding of atmospheric NO*x* has not been achieved, and the existence of ionic NOx^+ has not yet been reported. Atmospheric NO*x* has been studied with the help of simulation experiments using plasma systems. The use of an electric discharge to partially decompose a chemical compound so that the various free radicals, reactive ions, excited molecules, and so on generated by the MW or RF discharge react almost immediately with the undecomposed parent compound to generate daughter compounds is a technique that should have general applicability.

Since MW discharges through N_2/O_2 in the presence of water are used for simulations [76] of atmospheric chemistry, it would be desirable to examine the effects of water in detail. Therefore, Li^+ IAMS with a flow system were employed to qualitatively detect the product species, paying special attention to the effects of water [105]. The products of an MW discharge in air-like gas mixtures (80% N_2, 20% O_2, and water) were measured. The products of the discharge are molecules that are thought to exist in the atmosphere and are relevant to the study of atmospheric chemistry.

The product species detected in this simulation study were compared (Table 5.2) with the species detected in field observations [106]. It is found that the neutral species observed in the laboratory resemble the constituents observed in the Earth's atmosphere. On the basis of this resemblance, it is concluded that the MW discharge-induced chemistry observed in the $N_2/O_2/H_2O$ plasma may be helpful for understanding atmospheric chemistry. Laboratory simulation experiments can yield results directly applicable to the Earth's atmosphere. It should be noted, however, that there are significant differences in the relative amounts of the products, especially for NO and HNO_2. The ionic products were compared with those detected by Eisele et al. [107] in a remote desert area of southeastern Arizona and on top of Mount Washington in New Hampshire. No agreement in terms of the products, except for one species at *m*/*z* 60. A number of ionic clusters with NO^+ and H_2O^+ produced by gas-phase ion–molecule reactions were detected, but the NH_4^+ products observed by Eisele were not observed. This difference remains unexplained at this time. The as-yet-unidentified ionic species at *m*/*z* 60 in both studies is probably $NO^+(NO)$, which is present with considerable intensity in the plasma at a H_2O partial pressure of 1 Pa. It is assumed that some of the ionic products found in the simulation study

Table 5.2 Analysis of neutral products formed the MW discharge of the mixture of N_2 (80 Pa), O_2 (20 Pa), H_2O (2 Pa) and comparison with literatures. Adapted from [105] with permission

m/z[a]	Simulation study (plasma) (relative intensity)[b]	Troposphere (field study)[c] (relative intensity)[d]
37	NO (trace)	NO (56)
51	N_2O (310)	N_2O (trace)
53	NO_2 (100)	NO_2 (320)
54	HNO_2 (550)	HNO_2 (trace)
69	NO_3 (27)	NO_3 (trace)
70	HNO_3 (18)	HNO_3 (750)
86	HO_2NO_2 (3.2)	
115	N_2O_5 (trace)	

[a] The observed m/z due to Li^+ -adduct

[b] The percent intensity relative to the peak m/z 53 from the $N_2/O_2/H_2O$ system is given for all the products of MW plasmas

[c] From [107]. Distribution of reactive nitrogen species over Colorado (continental) during August and September 1984

[d] Volume mixing ratio (ppt)

that have not been reported by field investigators may eventually be detected in more sensitive in situ observations, especially in polluted areas.

5.2.6 *Detection of Environmentally Important Species*

Many kinds of pollutants can coexist in the atmospheric environment. Real-time measurements of pollutants are essential for the conformation of hazards and for safeguarding against chemical disasters. In this section, IAMS has been used to measure formaldehydes, hydrogen sulfides and perfluorocarbons (PFCs) of volatile organic pollutants in the atmospheric environment.

5.2.6.1 Indoors Formaldehydes and Odorous Hydrogen sulfides

Formaldehyde (HCHO) Formaldehyde emitted from the furniture and walls in rooms injures the eyes, the nose, and the respiratory organs and causes allergies; this is the sick house syndrome. Formaldehyde was a suspected carcinogen. In 2004, the working group of the International Agency for Research on Cancer (IARC) classified it as a human carcinogen [108]. As formaldehyde is a very unstable compound, pretreatment for derivatization is essential for GC and liquid chromatography (LC), but not for IAMS. IAMS detects for formaldehyde itself in the mass spectrum and demonstrates changes in its concentration in real time in the selected ion monitor mode. The detection limit was 9.2 ppb and the correlation coefficient (R^2) was 0.999. The studies indicate that IAMS can simultaneously monitor other atmospheric odorous substances.

H_2S H_2S is responsible for most of the odor problems associated with brewing and food processing wastewater treatment. Heavier than air, colorless, corrosive, and extremely toxic, it raises serious workplace health and safety concerns [109]. H_2S was measured directly by IAMS, which yielded a calibration curve with 0.996 (R^2) and a detection limit of 0.05 ppm, while the limit of detection (LOD) by conventional MS with EI was 100 ppm, due to interference by the O_2 isotope. These results demonstrate the possibility not only of measuring odors but also of monitoring the quality control of indoors, foods, industrial facilities, and so on.

5.2.6.2 Perfluorocarbons (PFCs)

Global warming is one of the major threats humankind is presently facing. In addition to naturally available greenhouse gases, such as carbon dioxide, methane, and nitrous oxide, manmade greenhouse gases contribute significantly to the global warming process [110]. Hydrofluorocarbons (HFCs) and PFCs are potent greenhouse gases produced by industries. These gases, plus sulfur hexafluoride (the most potent greenhouse gas), play a significant role in the global-warming process. Furthermore, these compounds have extremely long atmospheric lifetimes, resulting in their essentially irreversible accumulation within the atmosphere. The commonly used PFCs in industries are CF_4, CHF_3, and C_2F_6. The global-warming potential (GWP) of these greenhouse gases is very high, and they have long lifetimes. The atmospheric half lives of CF_4, CHF_3, and C_2F_6 are 50,000, 250, and 10,000 years, respectively, and their GWPs over a 100-year period are 6300, 12,100, and 12,200, respectively, with reference to the absolute GWP for CO_2 [34].

PFCs, widely used as fluorine sources in semiconductor manufacturing processes such as CVD chamber cleaning and dry-wafer etching, can be emitted as gaseous byproducts, along with hazardous air pollutants (HAPs) and various other gases. Some of these by-product gases are corrosive, toxic, or long-lived greenhouse gases (CF_4, CHF_3, SiF_4, COF_2, HF). Reductions in the emissions of PFCs, greenhouse gases produced during integrated circuit (IC) manufacturing, are being actively pursued by semiconductor manufacturers. Hence, there is a need for reliable monitoring methods that are capable of measuring PFCs in real time, at or below the maximum permissible concentration, in the IC manufacturing facilities which must minimize or eliminate emissions of PFCs and HAPs from their vent streams.

PFC Monitoring Fourier transform infrared spectroscopy (FT-IR) and on-line QMS are two analytical techniques commonly used for developing PFC monitors [111]. Air and exhaust monitoring using FT-IR methods is a feasible, reliable, and cost effective way to support the environmental safety and health (ES&H) programs of a diversity of manufacturing companies. Zazzera [112] and Gubner [113] have shown that tool exhaust from semiconductor processes, in addition to indoor air and other emission sources, can be monitored using commercially available FT-IR equipment and published guidelines; however, FT-IR cannot measure concentrations of homonuclear diatomic species such as F_2. The ability to quantify F_2 emissions without on-site calibration would make FT-IR more versatile and valuable. A

QMS would be very helpful in determining the amount of specific PFCs present in air [114, 115]. In 1997, a QMS with an electron-impact ion source was used to measure PFC concentrations. The main disadvantage of this method may be the interference of ion signals due to the high probability of PFC fragmentation. This creates difficulty in quantitative analysis and the situation is further complicated as most of PFCs break into similar fragments. On-site calibration is essential for analysis with a QMS. Therefore Harnisch et al. [116] employed MS combined with a GC and/or a cryotrap technique for the measurements of CF_4, SF_6, and C_2F_6 in the atmosphere.

IAMS for a Direct, Real Time Monitoring In order to support the emission reduction objectives of the semiconductor industry, the development of standardized analytical methodologies is necessary. To meet this requirement, IAMS was applied to the development of a direct, real-time monitoring method for PFCs emitted from IC manufacturing facilities [117, 118]. This technology may significantly improve these areas of PFC measurement technology. The compounds studied were sulfur hexafluoride (SF_6), tetrafluoromethane (CF_4), trifluoromethane (CHF_3), perfluoroethane (C_2F_6), and perfluorocyclobutane (c-C_4F_8).

Effluents of Semiconductor Equipment (Dry Etching Tool) were characterized for their composition of unused process chemicals and the by-products to assess the environmental impact of its processes and operations. Li^+ ion attachment MS is the analytical methodology used for continuous measurement of PFCs found in the exhausted gases as a result of operating conditions with the intent of developing improved methods for PFCs analysis, particularly at the environmental level (ppb concentration range). A stream of gas from exhaust is directed into a Li^+ ion attachment RC. All sampling was performed downstream of the process tool vacuum system to characterize the by-products. The gas sampling system consisted of Teflon™ sample line, approximately 4 m long. The sample line was used to extract continuously process emissions from the exhaust of the dry vacuum pump, i.e., after the N_2 ballast dilution and near ambient pressure. The flow rate was approximately 12 sccm for all source sampling and standardization. Calibration gas cylinders were used for standardization.

The experimental results demonstrate the feasibility of measuring PFC concentrations in air by generating ions by Li^+ ion attachment and mass analysis. The analytical method exhibited the following features: (i) high sensitivity, with the ability to detect 7 ppb c-C_4F_8 molecules in air; (ii) the ability to accept high-capacity sample introduction from atmospheric pressure and to allow easy coupling of various sample introduction sources to the mass spectrometric analysis; (iii) the opportunity for real-time detection of any PFC species, including radical intermediates; (iv) the ability to identify compounds by the generation of ions that do not fragment, which is especially useful for the determination of molecular weight; and (v) analysis of mixtures where no fragmentation is desired.

Comparison with FT-IR Compared with FT-IR, which has quickly become established as a prime analytical method for tracking emission profiles from exhaust systems within the semiconductor industry, the present methodology has both merits and limitations. The MS method may be more sensitive in most cases; however, the most important consideration concerns the relative sensitivity for each species and byproduct gas, which is required to accurately characterize and quantify exhaust stream content. The relative sensitivity for some species, such as nonpolar molecules, is too low to be detected with this method. On the other hand, in FT-IR, interference from various gases could suppress the ability to analyze certain gases. FT-IR has a more limited dynamic range compared to the MS method; moreover, a few pollutants cannot be detected by FT-IR. Hence, it can be concluded from these considerations that MS and FT-IR are complementary analytical techniques for real-time PFC exhaust monitoring. These two methods are valuable to the semiconductor industry for a better understanding of the process and for supporting the PFC emission reduction programs; however, further investigation is needed to demonstrate the feasibility of lithium ion attachment to PFC concentrations in a "real" air sample, which may have a large amount of moisture and interfering molecules. The possible interference should be considered carefully.

5.2.7 Identification of Unfamiliar, Complex or Unstable Species

5.2.7.1 Cu Complex

As the electronics industry moves to higher-density IC devices, the circuits will have to be fabricated on an increasingly small scale. For this reason, copper will soon replace aluminum as the material for conductive interconnects because copper has higher conductivity and higher resistance to electron migration than aluminum [119]. CVD is a superior process for producing these microscopic metal features. CVD of metals such as Al and Cu, which are suitable for microelectronic applications, has been studied extensively [120, 121]. In the study of the CVD process, monitoring techniques can be useful for optimizing the process and associated tools.

Among the many Cu(I) precursors, Cu(hfac)(tmvs) [hexafluoroacetylacetonate (hfac) and trimethylvinylsilane (tmvs)], commercially known as CupraSelect, has been shown to be one of the best precursors for producing pure copper films with low resistivity at high growth rates by means of CVD. This Cu precursor is thermally labile, corrosive and condensable. The organometallic copper–olefin bond is relatively weak. Dubois and Zegarski [122] estimated qualitative trends in the bond strengths using temperature-programmed desorption to determine the bond strength's role in the deposition rate and selective deposition. According to their study, the Cu–(tmvs) bond strength is 12 kcal/mol. This figure is the activation energy (Ea) for the desorption of tmvs from the Cu(100) surface.

Accordingly, interest in accurate, rapid, and relatively low-cost monitoring methods for the determination of CVD precursors has greatly increased. The CVD processes used in the semiconductor industry usually involve precursors that dissociate and either deposit a thin film or etch away a thin film of metal [123]. During the deposition and etching processes, various byproduct gases are formed. For specifically determining these molecular species during CVD, the metal films may have to be formed at lower temperatures, at higher rates, and with higher qualities than is currently possible. Hence, there is a need for reliable monitoring methods that are capable of measuring qualities of precursors in real time, at or below the maximum permissible concentration, in IC manufacturing.

A number of methods have been developed to monitor CVD and etching processes, including QMS [124, 125] and FT-IR [126, 127]. QMS has high sensitivity, a large linear dynamic range, rapid response, and the ability to withstand the reactive, corrosive nature of the effluent streams being analyzed. For instance, Zheng et al. [124] studied the gas-phase evolution and decomposition pathways of $Cu(hfac)_2$ with an in situ EI MS and identified the most likely pathways for precursor decomposition. Substantial progress has been made in the interpretation of mass spectra. However, the severe fragmentation that occurs during EI MS hampers correct interpretation of the spectra. Naik emphasizes [128] that the absence of any copper-containing fragments is attributed to extensive ligand cleavage under the harsh EI conditions. These fragmentations are consistent with weak bond strength of the copper-olefin bond. FT-IR is a reliable and cost-effective way to monitor semiconductor processes, but it is not as sensitive as QMS. A new mass-spectrometric technique that provides high sensitivity and fast response time without extensive fragmentation is desirable.

Li^+ ion attachment MS was applied to detect [Cu(hfac)(tmvs)], a thermally labile precursor used in copper chemical vapor deposition (Cu-CVD), with the intent of developing improved methods for mass-spectrometric analysis of the CVD process [129–131]. A laboratory-built direct introduction system of the sample provided the molecular ion of Cu(hfac)(tmvs) as the Li^+ ion adduct for the first time.

Figure 5.10 shows a typical Li^+ adduct mass spectrum of the as-received, reagent Cu(hfac)(tmvs) in the mass range up to m/z 450. This spectrum was acquired by heating the precursor vaporizer and QMS inlet lines. This figure also includes the background peaks. One of the observed peaks is the Li^+ adduct molecular ion at m/z 387 for the test compound, Cu(hfac)(tmvs), without any fragmentation, which is completely different from that of EI mass spectrum (not shown). The 70 eV EI gives many fragment ions without molecular ions. In addition to the observation of the molecular ion, an important feature of IAMS spectrum is the presence of peaks at m/z = 107 (tmvs)Li^+, 125 (tmvs)$Li^+(H_2O)$, 207 $(tmvs)_2Li^+$, 215 H(hfac) Li^+, and 233 H(hfac)$Li^+(H_2O)$. Authors believe that these ions do not result from the fragmentation of Cu(tmvs)(hfac)Li^+, but are attributed instead to the presence of traces of the (tmvs) and H(hfac) ligands, the ingredients employed in the synthesis of Cu(tmvs)(hfac), in the as-received complex.

Since the feasibility of generating Li^+ ion adduct for real-time monitoring of products encountered in Cu-CVD process facilities is demonstrated by the use of IAMS system, the Fujii group [130] studied the reaction pathways for Cu-CVD by Cu(hfac)(tmvs) by analysis of the reaction products. H(hfac) and tmvs were identified as the main products when Cu(hfac)(tmvs) was heated at temperatures ranging from room temperature to 160 °C in the reactor. The rate constant for Cu deposition, k (1/s), was determined to be $1.6 \times 10^6 \exp(-10.2 \text{ kcal/mol } RT)$. This kinetic information will be useful for improving the growth rate uniformity of CVD processes. Volume balances for Cu-CVD reagent used during processing were also estimated.

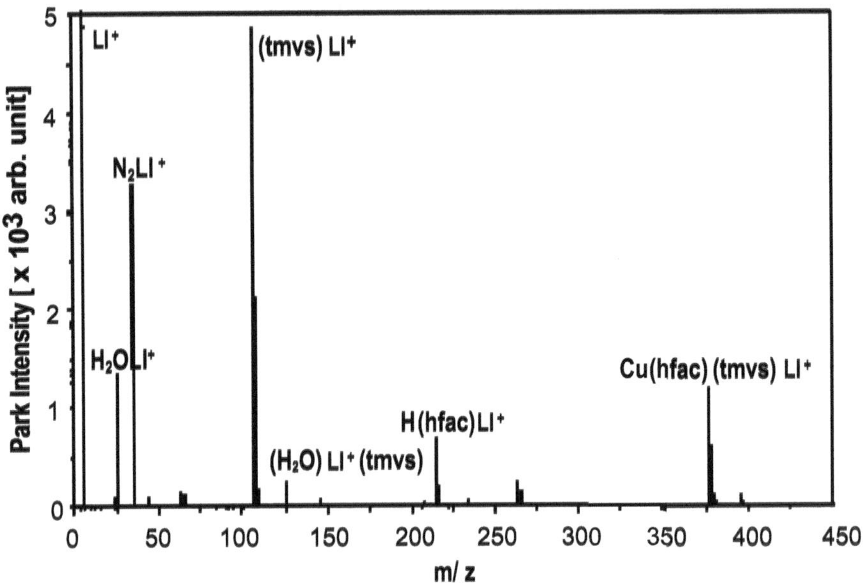

Fig. 5.10 Mass spectra of copper complex, Cu(hfac)(tmvs), obtained from the simulated Cu-CVD reactor at room temperature. (Reprinted with permission from [130]. ©2006, American Institute of Physics)

5.2.7.2 Polymerization in the C_2H_2 Plasma

The detection and evaluation of products in the C_2H_2 discharge plasma have been investigated from a phenomenological point of view by a number of investigators [66, 132–136]. The polymerization taking place in the plasma is one of the topics widely discussed. In most cases, the product has been measured using MS [66, 135, 136] and OES [132–134]. Kobayashi et al. [136], in a study of glow-discharge polymerization of C_2H_2, measured and identified that the polymer HC species are observed, which are most likely composed of highly branched oligomers. They suggested that the free radical mechanism is the key to plasma polymerization. Vasile and Smolinsky [66] in a series of mass spectrometric works succeeded in sampling positive ions and neutral species from RF plasma. An electron impact energy of 20 eV was chosen for their work of neutral species. They found hydrogen and diacetylene. But no free radical products presumably produced have been observed. Many astrophysicist have also measured the production in high-energy electrical discharges of organic molecules that are precursor molecules to the more complex organic compounds thought to exist in planetary atmospheres. For instance, Killian et al. [134] produced H_2CCCC and H_2CCC in the laboratory in a discharge through C_2H_2 and He in a molar ratio of 60:1. They determined them with a known MW spectrum.

The product analysis in MW C_2H_2 discharge by Li^+ ion attachment MS [48], has also brought forth a detection in the formation of many unfamiliar HC neutral products. The studies have been done in terms of MW discharge conditions that the ionic species leaving the MW C_2H_2 plasma were barely detectable intensities to confirm the possible presence of these neutral compounds in the gas phase.

Studies have clearly shown that the occurrence of the polymerization processes takes place. The identified polymer radicals and stable molecules include: C_nH_3 (n is up to 4), C_nH_5 (up to 6), C_nH_7 (up to 4), C_nH_9 (up to 4) and C_n (up to 8), C_nH_2 (up to 8), C_nH_4 (up to 8), C_nH_6 (up to 8), C_nH_8 (up to 7). This observation correlates qualitatively with a rapid deposition of products observed on the inside of the flow tube. It is interesting also to note that the most observed species contain even-numbered carbon atoms. The most conspicuous among the new features of the present spectra are the peaks at m/z 55, 79, 103, 33, 45, 57, 81, and 105, with many radical species peaks. These clear peaks observed incidentally must be assigned to C_n and C_nH_2, respectively. It can be concluded also that (1) the C_nH_{2n+1} ($n=2$–4) and C_nH_{2n-1} ($n=2$, 4) radicals have been detected. The intensity trend is clear; species with the higher n number decrease in peak intensities. (2) Other highly unsaturated radical species with carbon numbers n up to 6 may also be provided and detected by the MW C_2H_2 discharge. These are C_4H_3, C_4H_5, C_6H_5, which have been identified for the first time. (3) The plasma of C_2H_2 from the MW discharge leads to a polymerization process of higher HCs, at least, partially through the radical/molecule reactions [134]. There appeared OH radicals in the spectrum when the discharge was on. The presence of H_2O in C_2H_2 as impurities may be attributed partly to the formation of OH. The C_nH radical is observed in the C_2H_2 plasma [137]. Especially the ethynyl radical (C_2H) is of fundamental astrophysical interest and occurs in a wide variety of natural and artificial environment [138]. However, any the C_nH species in MW C_2H_2 discharge-IAMS system could not detected, even though C_n, C_nH_2, C_nH_3, C_nH_4 and so on were present.

C_2H_2/N_2 A MW discharge plasma is a prime method for the production of various kinds of novel compounds, since many products may be generated in complex ways [72]. In the search for materials with new physical properties, nitrogen compounds are a reasonable target since the nitrogen atom forms 3, 4, or 5 covalent bonds leading to unique structural characteristics. In studies related to C_2H_2 MW plasma, Bondybey et al. identified a variety of conjugated molecular products: cyanocarbon series C_nN_2 ($n=4$, 5) [139] and cyanopolyacetylene radical cations $NC(C{\equiv}C)_nCN^+$ [140]. These results suggested there would be considerable interest in seeking new products in a C_2H_2/N_2 MW discharge system. The production of gaseous HCs, nitriles, amines, and hydrazines in a continuous-flow MW plasma discharge excited in a 20 % of C_2H_2 + 80 % of N_2 mixture at a pressure of 20 Torr is also reported. The product analysis was made by Li^+ ion attachment MS [141]. A variety of N-containing organics (identified as HC_nN ($n=1$–7), $NC(C{\equiv}C)_nCN$ ($n=0$–2), $NC(CH_2)_nCN$ ($n=0$–6), $C_nH_{2n-1}NH_2$ ($n=0$–6), $C_nH_{2n+1}N(H)NH_2$ ($n=0$–5), and so on) were formed and these were tentatively assigned to nitriles, amines, and hydrazines. The mass-spectral analysis exhibited progressions differing by 14 mass units. Reaction schemes were proposed to explain the formation of some molecules. The observed peaks are classified to give three types of N-containing species: nitriles, amines, and hydrazines, and also possibly some HCs.

Nitriles Many products were formulated as $C_nH_{2n+1}CN$ (n=1–5), $C_nH_{2n-1}CN$ (n=5), $C_nH_{2n-3}CN$ (n=7), $C_nH_{2n-5}CN$ (n=3), $C_nH_{2n-7}CN$ (n=5), $C_nH_{2n-9}CN$ (n=3), and $C_nH_{2n-11}CN$ (n=3). As can be assumed from previous experimental results [98], which showed that that the dominant products from MW discharge in a CH_4/N_2 mixture were nitriles, these homologues (corresponding to additions of CH_2 units) may be saturated or unsaturated nitriles. Their abundance seems to vary smoothly with chain length, with CH_3CN being most abundant.

Among many the nitriles produced, the cyanopolyynes, HC_nN, are particularly interesting due to their linear conjugated structures, their biological importance, and their abundance in the interstellar medium. The well-known interstellar cyanopolyynes HC_nN (n=1, 3, 5, 7) [142] can be formed abundantly in the present system: HCN (relative intensity, 43), HCCCN (51), HCCCCCN (46), and HCCCCCCCN (19). In this study, the even numbers of the HC_nN series (with n at 2, 4, and 6) were also detected; however, their relative intensities were much less than those of the cyanopolyynes with odd carbon numbers.

Other significant peaks are observed at m/z 59, 83, and 107, which can be assigned to dicyanopolyacetylenes $NC(C{\equiv}C)_nCN$ with n at 0, 1, and 2. Cyanogen, NCCN, is well known for its production in high temperature environments [143] and its postulated presence in the interstellar medium. Since dicyanoacetylene, NCCCCN, was first reported in 1909, there have been surprisingly few studies of this species [144], despite being of interest as an unusual linear molecule which results in an extended π orbital (delocalization of the electrons along the whole molecule). The interest in these species can be expected to increase in view of the recent observations of Grosser and Hirsch [72]: They were able to produce large amounts of $NC(C{\equiv}C)_nCN$ molecules by vaporizing graphite under helium in a reactor designed for fullerene production in the presence of cyanogen, C_2N_2. Dicyanoalkanes, $NC(CH_2)_nCN$ (n=1–4), are interesting products found in the study that have not been reported before. The fraction was fairly constant, the observed intensity ratio of the mass lines corresponding to the compounds $NC(CH_2)_nCN$ (n=1–4) produced was 100:100:85:77, and the intensity was relatively constant over the n values. It is speculated that these compounds will be synthesized in the future.

Amines The products included alkyl amines with single, double, and triple bonds, with many homologous members often present in roughly equal yields (within a factor of 4). The higher homologues of $C_nH_{2n-1}NH_2$ (n=6–8) were found in higher fractions, mainly because the corresponding mass lines were possibly enhanced with the same mass signal as HCCCCCCC, $CH_3CCCCCCN$, and $C_2H_5CCCCCCCN$ species.

Hydrazines Some products were formulated as $C_nH_{2n+1}N(H)NH_2$, $C_nH_{2n+1}N{=}NH$, $C_nH_{2n-1}N{=}NH$, and $C_nH_{2n-3}N{=}NH$. These homologues (corresponding to additions of CH_2 units) may be hydrazines. Other significant peaks at m/z 36 are also observed, which can be assigned to $H_2C{=}NHLi^+$. It is suggested that hot NH_2 or CN radicals generated by dissociation of initial amine or nitrile products can react with each other and with HC species to produce these unfamiliar organic compounds by way of their recombination. In conclusion, the C_2H_2/N_2 MW discharge led to the production of various kinds of chemical species, and effectively demonstrated that the MW discharge is a method for generating over 70 N-bearing products. Some interesting products found in this study were $H_2C = NH$ and $NC(CH_2)_nCN$. These components are predicted to be possible future products. The results may also have implications for understanding of chemical evolution in the solar system. These results suggest that nitriles, amines, and hydrazines of higher molecular weight, and their derivatives, such as cyano-acetylenes and dinitriles, may have been synthesized as precursors of biologically important compounds in planetary lightning and plasmas associated with meteor impacts which may be acceptable analogs to MW plasmas.

C_3N_4 Crystalline β-C_3N_4 is a novel hypothetical compound predicted by Liu and Cohen [145]. Great interest has been motivated by its super hardness (maybe harder than that of diamond) and possibility of synthesis [146]. Many investigators have put great effort [147] into synthesizing the polycrystalline C_3N_4 thin films by using various technologies, such as RF sputtering, shock compression, pyrolysis, DC sputtering, implantation, electron beam evaporation, laser beam ablation, plasma arc, etc. However, no experimental results present evidence for the existence of C_3N_4 in the gas phase. The presence of a gas-phase C_3N_4 nitride as a marker could be invaluable in adjusting experimental conditions to optimize the crystalline production. Unexpectedly, the product analysis in MW C_2H_2–N_2 discharge by Li^+ ion attachment MS [141], has indicated the formation of the C_3N_4 species. The mass spectra exhibit the Li^+ adduct peak at m/z 99. This peak is assigned tentatively [148] to C_3N_4. The possibility of production of C_3N_4 was investigated. The variation of m/z 99 intensities with gas composition confirms the involvement of nitrogen. Furthermore, variation of nitrile species with gas composition and MW discharge power shows simple, clear correlation with that of C_3N_4. These findings, which are at least consistent with the C_3N_4 assignment, are additional, indirect evidence that the m/z 99 peak is $C_3N_4Li^+$.

5.2.7.3 N_3H_3 and N_4H_4

A study has been made of the production of gaseous hydronitrogen chemical species in a continuous flow MW plasma discharge excited in hydrazene (N_2H_4) gas and gas mixtures (N_2 and H_2) at a pressure of 10 Torr. The product analysis was made by Li ion attachment MS [149]. Hydrazine has been a subject of photochemical investigation since the 1960s. Hydrazine vapor is photolyzed at a variety of wavelengths [150]. Several researchers have reported the formation of unexpected species when N_2H_4 is subjected to a photolysis. For instance, Schurath et al. reported [151] the formation of many hydronitrogen species, including the free radicals NH_2, N_2H_3, and so on. Dissociation into two amino radicals was postulated [152] by researchers who used the flash photolysis technique (the amino radical is readily detected by absorption spectroscopy). In a study related to the N_2H_4 MW plasma, Willis and Back [153] identified diimide (N_2H_2).

Hydrazine in a fast-flow system was subjected to a MW discharge near the sampling orifice of an ion attachment mass spectrometer. A variety of products are formed and identified mass spectrometrically, assigned tentatively to N_2, H_2, NH_3, N_2H_2, N_3H_3*, N_4H_4, N_2H_3*, and so on. The mass spectral analysis exhibits progressions differing by 14 mass units. The interesting product found in the study was N_4H_2 with which chemical formula no species were listed in the National Institute of Standards and Technology (NIST) database. Mass spectrometric evidence is given for the elusive chemical compounds triazene (N_3H_3) and tetrazene (N_4H_4), and data on the free radicals NH_2* and N_2H_3*, effectively demonstrating that the MW discharge is a method for generating possible new products. Triazene and tetrazene have been postulated as intermediates in the oxidation of hydrazine in solution. In addition, the interesting neutral hydronitrogen species N_3H_5 and N_4H_6 were tentatively assigned. No species with these chemical formulas are listed in the NIST database. Formation of N_3H_5 and N_4H_6 may involve rearrangement reactions.

MW discharges through molecular N_2/H_2 gas mixtures are ubiquitously used for the synthesis of ammonia and other reagents as well as for plasma processing [154]. Most of the processing mixtures have aroused considerable attention as cleaning systems, while among the $HxNy$ products, the well characterized compounds diazene (HNNH) and isodiazene (H_2NN) [155] are widely used as synthetic reagents [156].

MW discharge through a N_2/H_2 mixture produced various kinds of chemical species. Ion attachment mass spectrometric evidence for the elusive chemical compound N_4H_2 is presented [157]. No species with chemical formula N_4H_2 is listed in the NIST database. Recent ab initio studies of the structure, vibrational frequencies, and intensities of the open-chain species HN = N–N = NH [158] indicate that there are a considerable number of possible conformations for HN_4H. This species can be assumed as a possible molecule. Production of triazene and tetrazene as intermediates in the N_2/H_2 MW plasma was expected, but neither species was directly observed in the Li^+ adduct mass spectrum. The ionic components observed were N^+, NH_3^+, NH_3H^+, $(NH_3)_nH^+$ (n = 1–3), N_2^+, N_2H^+, $N_2H_2^+$, and some clustered ions. N_2H and N_2H_2 were not observed in the neutral form. The yields of these species readily varied when the gas composition was changed, just as the yields of the neutral species did.

5.2.7.4 H_2O_2, $H_2O_2H^+$, and H_2O_3

The following topic is concerned with the presence production of H_2O_2, $H_2O_2H^+$, and H_2O_3 in the gas phase [78, 79].

$H_2O_2H^+$ The protonated hydrogen peroxide ($H_2O_2H^+$) has been long postulated as a possible intermediate. According to the NIST-ONLINE search, there are only five ionic chemical species which have 35 amu as the molecular weight: Cl^+, H_3S^+, H_4P^+, H_2FN^+, CH_4F^+. Surprisingly, $H_2O_2H^+$ is not found in the search results, suggesting that the validity of the $H_2O_2H^+$ presence has been questioned. During the studies on the identification of the principal species produced by the CH_4/O_2 MW discharge, Fujii et al. [78] have succeeded in determining the presence of the protonated hydrogen peroxides $H_2O_2H^+$ in the gas phase. The mass spectra of the discharge plasma, obtained by a direct sampling system, clearly show the peak m/z 35

($H_2O_2H^+$), giving the direct evidence that $H_2O_2H^+$ is formed in the plasma environment and detected mass spectrometrically. The influence of certain discharge variables on yield and concentration of $H_2O_2H^+$ was investigated. $H_2O_2H^+$ is only minor products, compared with other ionic products formed from the CH_4/O_2 discharge. They explain the small yield as follows. Probably the major part of the $H_2O_2H^+$ formed in the gas phase results from the protonation reaction

$$H_2O_2 + RH^+ \rightarrow H_2O_2H^+ + R, \tag{5.7}$$

where RH^+ is proton donor. In the present CH_4/O_2 system, RH^+ must be CH_5^+. The difference in proton affinity between H_2O_2 and CH_4 (corresponding base) [159] are ca. 40 kcal/mol. This value may not be small enough to avoid direct decomposition of $H_2O_2H^+$ into O and H_2OH^+. In conclusion, the direct proof of the presence of $H_2O_2H^+$ in the gas phase was provided by the direct sampling technique incorporated with QMS, which is obviously applicable to any reactive species and can be used to obtain the mass spectrum.

H_2O_2 The electric discharge of water vapor has been studies extensively in connection with the synthesis of H_2O_2 [160]. When water vapor, dissociated by an electric discharge, passed into a cold trap, it yielded products that at room temperature consisted of H_2O_2 and H_2O after the O_2 evolution. H_2O_2 is formed only on a sufficiently cold surface (below −150 °C), suggesting the involvement of surface properties. The reaction mechanism in electrically discharged water vapor is not very well understood: the necessary fundamental information is mostly lacking. For instance, the possible presence of H_2O_2 in the H_2O MW discharge plasma has been a matter of question from the viewpoint of the gas phase. Some attempts have been made to determine whether H_2O_2 is really present in the gas phase of H_2O discharge plasma, using MS with Li^+ ion attachment technique. The mass spectra show that no appreciable number of H_2O_2, $H_2O_2^+$, or $H_2O_2H^+$ has been observed, suggesting these species could neither form nor survive outside the plasma discharge. In the course of investigations on the product analysis of the CH_4/O_2 MW discharge plasma [52, 53], it was found that H_2O_2 may be produced. The Li^+ adduct mass spectra of the complex product mixture, obtained by MS with Li^+ ion attachment technique, clearly show the peak *m/z* 41 ($H_2O_2Li^+$) among many peaks, suggesting that H_2O_2 is formed in the gas phase of the CH_4/O_2 plasma. The unique interest of the H_2O_2 compound comes from the important role as intermediates of the chemical reactions [161, 162]. English chemists Fabian and Bryce suggested the formation of H_2O_2 during the discharge of CH_4/O_2 in the 7th symposium on combustion (1956). They used electron impact MS. Since this species is very labile and require special conditions for its formation, its detection was always difficult for a long time. The evidence was either indirect or doubtful. Therefore, it was desirable to secure further direct evidence from another method. The Li^+ mass spectra of the product species provide the first positive identification of stabilized H_2O_2 in the MW discharge CH_4/O_2 plasma. These are significant mainly from two viewpoints; mass spectrometric assignment of the peak at *m/z* 41 and isotopic consideration of the concerned peaks at *m/z* 40–41. It can be luckily supposed that, under the CH_4/O_2 plasma conditions, the peak at *m/z* 41 is $H_2O_2Li^+$ as the source of the possible compounds whose

molecular weights are 34 amu. Furthermore, the observation of the minor $H_2O_2H^+$ peak gives indirect evidence of H_2O_2 formation (see above paragraph).

H_2O_3 During the studies on the identification of the principal species produced by the CH_4–O_2 MW discharge [52, 53, 78, 79], the presence of hydrogen trioxide H_2O_3 was postulated in the gas phase. The Li^+ adduct mass spectra of the discharge plasma, obtained by the Li^+ ion attachment technique, clearly show the peak m/z 57 ($H_2O_3Li^+$), giving direct evidence that H_2O_3 is formed in the plasma environment.

The technique used is IAMS and thereby arrives at concrete evidence of the presence of the hydrogen trioxide. A most important advantage of the present study comes again from the fact that m/z 57, where the mass number of the Li^+ adduct complex happens, is virtually free from interference, in contrast with most m/z. The presence of a chemical species, whose molecular weight is 50 amu is denoted by an increase in the current of the Li^+ adduct ions (m/z 57). According to a CASONLINE search, there are, surprisingly, only 43 chemical species which have 50 amu as their molecular weight. Under the CH_4/O_2 plasma conditions, the possible compounds are either H_2O_3 or C_2H_3LiO. It is concluded that from the two candidates the peak at m/z 57, due to the Li^+ adduct ion, is assigned to $H_2O_3Li^+$, since the isotopic abundance consideration of the m/z 56 peak relative to the m/z 57 peak indicates containing one Li atom and hence, rules out the possibility of $C_2H_3LiOLi^+$. An ab initio molecular orbital calculation [163] predicts the collisional stabilization of H_2O_3 by the exothermic reaction

$$OH^* + HO_2^* \rightarrow H_2O_3 \tag{5.8}$$

In conclusion, direct proof of the presence of H_2O_3 in the gas phase was provided. Further investigation is desirable to determine the mechanism of production of H_2O_3 under the present conditions. It also appeared desirable to apply the present approach to the H_2O or H_2O_2 discharge system, which has been investigated for the production of solid H_2O_3 and related compounds over the last 60 years [164–166].

5.2.8 Restriction of Hazardous Substances, RoHS

RoHS Since the "Directive on the restriction of the use of certain hazardous substances in electrical and electronic equipment" (commonly referred to as the Restriction of Hazardous Substances directive of the RoHS directive) came into force in the European Union (EU) in July 2006 [167]. Some of the same regulations on chemical management have been adopted in other parts of the world. The RoHS directive restricts the amount of PBBs and PBDEs [168] contained in electrical and electronic equipment to less than 1000 ppm. No electronic electric apparatus can be marketed in an EU member state if it contains those materials in excess of a designated value of RoHS. PBBs and PBDEs are among the large variety of brominated flame retardants (BFRs) used in plastics and textiles. PBBs are no longer produced, but PBDEs were in widespread use before the RoHS directive came into effect, and various recycled polymers may contain PBDEs. Therefore, efforts are being made to realize efficient and reliable methods for the management of hazardous RoHS substances, including instrumental technologies.

The International Electrotechnical Commission (IEC) [169] is a worldwide organization for standardization. In 2008, IEC publishes the documents of IEC 62321 as an international standard for determination of levels of six regulated substances in electrotechnical products, such as, PBBs and PBDEs. The IEC 62321 method consists of two steps, screening and high-precision chemical analysis. Regarding PBDEs, the suggested method is screening by fluorescent X-ray analysis (XRF) [170], followed by precision analysis by solvent extraction GC-MS. However, this method is only included in the informative annex because its reliability has not been confirmed. Other precision analysis methods, such as HPLC-UV/MS [171], GC-ECD [172], and GC-MS [173, 174], have been previously reported for PBDEs. However, these methods have various problems. Some are time-consuming, while others require the use of toxic solvents for sample preparation for quality control. Therefore, a determination analysis method that does not need the use of a toxic solvent in the confirmatory analysis is required. Several spectroscopic methods [175, 176], such as atomic absorption spectroscopy, FT-IR spectroscopy and Raman spectroscopy are used for the rapid analysis of PBDEs in polymers. Although these methods require no solvent and are relatively quick, they cannot identify all PBDE congeners. Additionally, in the analysis of actual samples, the polymer matrix may interfere in the detection of PBDEs. When mixtures are analyzed by MS methods, it is usually necessary for each component of the mixture to be separated by chromatography (such as GC or HPLC), because peaks of fragment ions may be detected from one of the components. As a solution to these problems, Toshiba has developed a new quantitative analysis method [177] for BFRs in electrical and electronic products that utilizes IAMS, which is a very soft ionization method that requires no separation technique before analysis. This method makes it possible to measure the RoHS elements in the plastic materials efficiently and reliably.

DIP-IAMS The Toshiba group performed IAMS analysis with an L-250G-IA (IA-Lab, refer to Sect. 5.2.2.5) manufactured by Canon Anelva Inc. (Kanagawa, Japan). The IAMS coupled with DIP allows the direct mass spectrometric analysis of PBB and PBDE in plastic materials without prior pretreatment process [178]. Identification is based on their different mass number and isotope distribution pattern. The samples were put into aluminum pans of DIP. The plastic samples were cut into small pieces of around 1 mg using nippers or a box cutter before they were put into the pans (up to 5 pans can be placed on the DIP at one time). In the case of standard solution sample, around 2 ml using micro syringes were also put into the pans. The samples were sent into the RC one by one and were thermally desorbed by programming DIP. The total analysis time required with this method is generally 40 min, that is, very short. The IAMS method is faster than the GC-MS method and results in exact differentiation of Br compounds because there is no fragmentation.

Mass Spectrum The ability of the present methods to identify PBDEs and decabromodiphenyl ether (decaBDE) in the plastic samples that contain several BFRs.

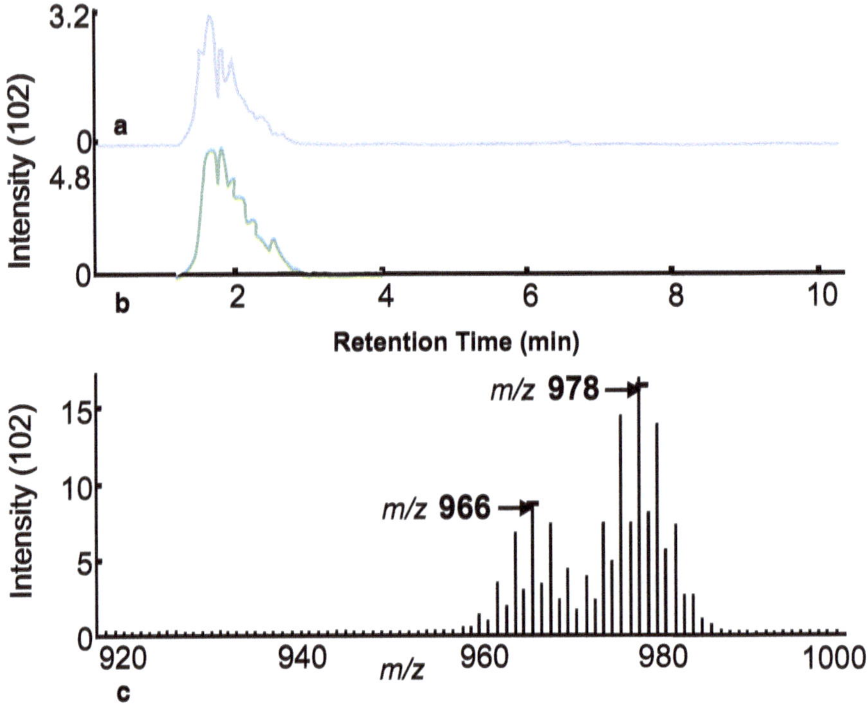

Fig. 5.11 An ion attachment mass spectrum and selected ion monitoring (SIM) of the CRM sample (No. 8, decaBDE of 3900 ppm in polystyrene resin). **a** SIM for m/z 966, **b** SIM for m/z 978, and **c** reconstructed mass spectrum for the retention time period of 1–3 min. (Reprinted with permission from [177]. ©2010, Royal Society of Chemistry)

Decabromodiphenyl ether, which was recently included in the list of compounds restricted by the RoHS directive, was used widely till a short time ago. The probability that decaBDE is present in recycled polymers is higher than other PBDEs. In addition, decaBDE has the largest molecular weight and the highest boiling point among the PBDEs. Therefore, decaBDE is the most difficult PBDE to measure. For this reason, the polystyrene resin sample was tested and the presence of decaBDE (m/z 966, (M + Li^+) and bis(pentabromophenyl)ethane (BPBPE, m/z 978, (M + Li)$^+$) was confirmed as can be seen in Fig. 5.11. The molecular weights of decaBDE and BPBPE are similar and their chromatograms overlap, but each compound could be identified based on their different mass spectrum and isotope distribution pattern.

Validation Validation study of the IAMS method was made. A calibration curve was generated by the least squares method in the range of 0.04–2.0 μg (0.02–1.0 μg ml^{-1}, injection volume 2 μl) using decaBDE solution within the range permitted by the RoHS directive in terms of concentration in polymers. The coefficient of determination for the linear regression was 0.99, indicating that the calibration curve for the IAMS method has linear dynamic range. The recovery of decaBDE from the polymer samples was evaluated using certified reference materials CRM 8108-a (317 ± 14 ppm) and CRM 8110-a (886 ± 28 ppm). The recovery and repeatability values for CRM 8108-a were 81.4 and 6.19 %, respectively, while those for CRM 8110-a were 85.4 and 5.86 %, respectively. The LOD and the limit of quantification (LOQ) were evaluated based on the standard deviation of the peak areas obtained from the analysis of a sample containing 0.04 μg of decaBDE, which was the lowest level of calibration curve. LOD was assumed to be three times the standard deviation, while LOQ was assumed to be ten times the standard deviation. The obtained LOD and LOQ values were 13.5 and 45.0 ppm, respectively (in the case of using 1-mg solid sample for analysis). These results indicate that the IAMS method is applicable to the qualitative analysis of samples to check whether their decaBDE levels satisfy the restrictions prescribed by regulations such as the RoHS directive.

Prospective IAMS is well suited for rapid qualitative analysis of PBDEs, and can be used for the quality control of materials and parts. Comparison study showed that decaBDE concentrations of around 1000 ppm, which is the minimum level required by the RoHS directive, could not be detected by FT-IR spectroscopy because of its low sensitivity. In fact, decaBDE concentrations of even around 100,000 ppm in industrial parts could not be identified by FT-IR spectroscopy, mostly due to interference by the matrix. In contrast, IAMS was able to detect a decaBDE concentration of approximately 300 ppm in polymer (analysis conditions: sample weight of 1 mg, extraction temperature of 100–300 °C with a temperature increase rate of 256 °C min^{-1}). IAMS could also identify other BFRs, such as ethylene (bis-tetrabromophthal) imide (EBTBPI) and bis(pentabromophenyl)ethane (BPBPE), and decaBDE could identify in mixtures of BFRs. These results indicate that IAMS can detect decaBDE in polymer with less interference from the matrix and with higher sensitivity than FT-IR spectroscopy. The validation results confirmed that IAMS could be used to detect lower concentrations of PBDEs than the maximum concentrations permitted by the RoHS directive, and has a possibility of application to quantitative analysis in the future in accordance with its improvement. The present DIP-IAMS combination system appears to hold great promise for the application to analysis of many nonvolatile and thermally labile compounds. It includes, for instance, quantification of phthalates (possible endocrine) in polyvinylchloride resin, organic tin compounds (environmental pollutant) in antifouling coatings on underwater structures, ships and other craft and polychlorinated-n-alkanes (PCAs) or chlorinated paraffins in adhesives, paints, rubber, and sealants. It should be noted, finally, that IEC technical committee111 has just evaluated IAMS [179] as international standard method for determination of certain substances in electrotechnical products: phthalates, PBBs, and PBDEs in polymer.

References

1. Thomson BA, Davidson WR, Lovett AM. Applications of a versatile technique for trace analysis: atmospheric pressure negative chemical ionization. Environ Health Persp. 1980;36:77–84.
2. Hodges RV, Beauchamp JL. Application of alkali ions in chemical ionization mass spectrometry. Anal Chem. 1976;48:825–8.

3. Bombick D, Pinkston JD, Allison J. Potassium ion chemical ionization and other uses of an alkali thermionic emitter in mass spectrometry. Anal Chem. 1984;56:396–402.
4. Schmelzeisen-Redeker G, Giessmam U, Rollgen FW. In-beam ionization by alkali ion attachment applying a two-filament ion source. Org Mass Spectrom. 1985;20:305–309.
5. Anderson WR Jr, Frick W, Daves CD Jr A direct technique for obtaining electron-impact mass spectra of polar, involatile compounds. Application to underivatized disaccharides. J Am Chem Soc. 1978;100:1974–75.
6. Daves GD Jr, Anderson WR Jr. Cationization in electron ionization mass spectrometry of polar organic molecules. Int J Mass Spectrom Ion Phys. 1979;30:385–388.
7. Cotter RJ, Yergey AL. Thermally produced ions in desorption mass spectrometry. Anal Chem. 1981;53:1306–07.
8. Bombick DD, Allison A. Desorption/Ionization mass spectrometric technique for the analysis of thermally labile compounds based on thermionic emission materials. Anal Chem. 1987;59:458–66.
9. Stoll R, Rollgen FW. Thermal desorption of quasimolecular ions org. Mass Spectrom. 1981;16:72–75.
10. Shiokawa Y, Nakamura M, Hirano Y, Fujii T. Ionization apparatus and ionization method for mass spectrometry. From U.S. Pat Appl Publ. 2002;US 20020053636 A1 20020509, Language: English.
11. Fujii T. A novel method for detection of radical species in the gas phase: usage of Lithium ion attachment to chemical species. Chem Phys Lett. 1992;191:162–8.
12. Fujii T, Ogura M, Jimba J. Chemical ionization mass spectrometry with use of alkali ion attachment to molecule. Anal Chem. 1989;61:1026–29.
13. Fujii T. Quadrupole mass spectrometry in combination with lithium ion attachment for sampling at atmospheric pressure: possible coupling to a superfluid critical chromatography. Anal Chem. 1992;64:775–8.
14. Rollgen FW, Schulten HR. Molecular weight determination by cationization. Org Mass Spectrom. 1975;10:660–8.
15. Fujii T. Ion attachment mass spectrometry, Vol 6: ionization methods. In: Gross M, editors. Encyclopedia of mass spectrometry. America society for mass spectrometry. New York: Elsevier; 2007. pp 327–34.
16. Sablier M, Fujii T. Mass spectrometry of free radicals. Chem Rev. 2002;102:2855–924.
17. Fujii T. Alkali-metal ion/molecule association reactions and their applications to mass spectrometry. Mass Spectrom Rev. 2000;19:111–38.
18. Sablier M, Fujii T. Mass spectrometry of free radicals: a methodological overview. In: Webb G, editor. Progress in chemistry, Sect. C (Phys Chem). Cambridge: Royal Society of Chemistry; 2005. pp 53–99.
19. Fujii T. A new method for thermal analysis: ion-attachment mass spectrometry (IAMS). J Therm Anal Calorim. 2012;110:17–25.
20. Kitahara Y, Takahashi S, Tsukagosi M, Juhász M, Fujii T. Ion attachment mass spectrometry for environmental analysis. In: Gauglitz G, Moore DS, editors. Handbook of spectroscopy, 2nd edn vol 3. London: Wiley-VCH;2014. pp. 1287–312.
21. Selvin PC, Fujii T. Lithium ion attachment mass spectrometry: Instrumentation and Features. Rev Sci Instrum. 2001;72:2248–52.
22. Bach HT, Meyer BA, Tuggle DG. Role of molecular diffusion in the theory of gas flow through crimped-capillary leaks. J Vac Sci Tech. 2003;A 21:806–13.
23. Miguel AH, Natusch DFS. Diffusion cell for the preparation of dilute vapor concentrations. Anal Chem. 1975;47:1705–7.
24. Altshuller AP, Cohen IR. Application of diffusion cells to the production of known concentrations of gaseous hydrocarbons. Anal Chem. 1960;32:802–10.
25. Scaringelli FP, O'Keeffe AE, Rosenberg E, Bell JP. Preparation of known concentrations of gases and vapors with permeation devices calibrated gravimetrically. Anal Chem. 1970;42:871–6.
26. Fujii T, Selvin PC, Sablier M, Iwase K. Lithium ion attachment mass spectrometry for on-line analysis of trace components in air: direct introduction. Int J Mass Spectrom. 2001;209:39–45.

27. Chan KC, Tse RS, Wong SC. A temperature programmed fractionation inlet system for mass spectrometers. Anal Chem. 1982;54:1238–40.
28. Yuen HK, Mappes GW, Grote WA. An automated system for simultaneous thermal analysis and mass spectrometry. Thermochim Acta. 1982;52:143–53.
29. Materazzi S, Gentili A, Curini R. Applications of evolved gas analysis part 2: EGA by mass spectrometry. Talanta. 2006;69:781–94.
30. Brown ME. Introduction to thermal analysis:techniques and applications. New York: Kluwer Academic Publishers; 2004. p. 264.
31. Blewett JP, Jones E. Filament sources of positive ions. Phys Rev. 1936;50:464–8.
32. Fujii T, Ohta M. Filament thermionic sources of Li^+ ions in low heating power. J Phys D: Appl Phys. 1995;28:1268–72.
33. C & V Technix Co., Ltd. http://c-vtechnix.com/. Accessed 14 April 2015.
34. Arulmozhiraja S, Fujii T. Li^+ ion affinities of global-warming perfluorocarbons. J Phys Chem. 2000;104:9613–8.
35. Aue DH, Bowers MT. Stabilities of positive ions from equilibrium gas-phase basicity measurements. In: Bowers MT, editor. Gas phase ion chemistry. Vol. 2. New York: Academic Press; 1979. pp 1–51.
36. Parsons AW. An introduction to free radical chemistry. New Jersey: Wiley; 2000. p. 252.
37. Alfassi ZB. General aspects of the chemistry of radicals (the chemistry of free radicals). New Jersey: Wiley; 1999. p. 578.
38. Sugden TM, Ashmore PG, Dainton FS. Photochemistry and reaction kinetics. Cambridge: Cambridge University Press; 2010. p. 404.
39. Smith BC. Fundamentals of fourier transform infrared spectroscopy. 2nd ed. Boca Raton: CRC Press; 2011. p. 207.
40. Rieger PH. Electron spin resonance: analysis and interpretation. Cambridge: Royal Society of Chemistry; 2007. p. 173.
41. Slichter CP. Principles of magnetic resonance (Springer series in solid-state sciences 1). Berlin: Springer; 1996. p. 658
42. Kojima H, Toyoda H, Sugai H. Observation of CH_2 radical and comparison with CH_3 radical in a RF methane discharge. Appl Phys Lett. 1989;55:1292–94.
43. Herzberg G. The spectra and structures of simple free radicals: an introduction to molecular spectroscopy. New York: Dover Publications; 2012. p. 240.
44. McEwen CN. Radicals in analytical mass spectrometry. Mass Spectrom Rev. 1986;5:521–47.
45. Jones ITN, Bayes KD. Detection of steady-state free-radical concentrations by photoionization. J Am Chem Soc. 1972;94:6869–71.
46. Fujii T, Syouji K. Identifications of intermediate radicals in the CH_4 microwave plasma by Li^+ ion attachment method. Phys Rev A. 1992;46:3555–7.
47. Fujii T, Syouji K. Mass spectrometric detections of neutral radicals in CH_4 microwave discharge by usage of Li^+ ion attachment techniques. J Appl Phys. 1993;74:3009–12.
48. Fujii T. Neutral product analysis of the microwave C_2H_2 plasma: C_n, C_nH_2, C_nH_3, C_nH_4, C_nH_5 and larger species. J Appl Phys. 1997;82:2056–9.
49. Fujii T. Diagnostics of microwave plasmas of C_2H_2: mass spectrometric investigations of ionic and neutral species. Phys Rev E. 1998;58:6495–02.
50. Fujii T, Kim HS. The mass spectrometric analysis of the neutral products of MW C_2H_4 plasma: carbon carbenes and aromatic compounds. Chem Phys Lett. 1997;268:229–34.
51. Sablier M, Iwase K, Sato G, Fujii T. Generation and observation of CHF_2, CF_2, and CF_3 in a CF_4/He microwave discharge system: a mass spectrometric method. Chem Phys Lett. 2005;409:342–8.
52. Fujii T, Syouji K. Mass spectrometric studies of the neutral and ionic products in a CH_4/O_2 microwave discharge plasma. J Phys Chem. 1993;97:11380–4.
53. Fujii T, Syouji K. Production of large O containing neutral hydrocarbon species by a CH_4–o_2 microwave discharge. Phys Rev E. 1994;49:657–62.
54. Kareev M, Sablier M, Fujii T. Diagnosis of a CH_4/N_2 microwave discharge: ionic and neutral species. J Phys Chem. 2000;104:7218–23.

55. Kushner MJ. On the balance between silylene and silyl radicals in rf glow discharges in silane: the effect on deposition of a-Si:H. J Appl Phys. 1987;62:2803–10.
56. Davies PB, Isaacs NA, Johnson SA, Resell DK. Detection of SiH(X 2) fundamental and hot band transitions by diode laser absorption spectroscopy. J Chem Phys. 1985;83:2060–3.
57. Rudolph RN, Moore JH. Plasma polymerization and a-C:H film ablation in microwave discharges in methane diluted with argon and hydrogen. Plasma Chem Plasma Process. 1990;10:451–71.
58. Wormhoudt J. Radical and molecular product concentration measurements in CF_4 and CH_4 radio frequency plasmas by infrared tunable diode laser absorption. J Vac Sci & Technol A. 1990;8:1722–5
59. Toyoda H, Kojima H, Sugai H. Mass spectroscopic investigation of the CH_3 radicals in a methane rf discharge. Appl Phys Lett. 1989;54:1507–9.
60. Havens MR, Biolsi ME, Mayhan KG. Survey of low temperature r.f. plasma polymerization and processing. J Vac Sci Technol. 1976;13:575–84.
61. Kobayashi H, Shen M, Bell AT. The role of halogens in the plasma polymerization of hydrocarbons. J Macromol Sci Chem A. 1974;8:1345–60.
62. Kline LE, Partlow WD, Bies WE. Electron and chemical kinetics in methane rf glow-discharge deposition plasmas. J Appl Phys. 1989;65:70–8.
63. Radford HE, Evenson KM, Howard CJ. HO_2 detected by laser magnetic resonance, J. Chem. Phys. 1974;60:3178–83.
64. Fridman A. Plasma Chemistry. London: Cambridge University Press; 2012. p. 1017.
65. Angus JL, Hayman CC. Low pressure metastable growth of diamond and diamondlike phase. Science. 1988;241:913–21.
66. Vasile MJ, Smolinsky G. The chemistry of radiofrequency discharges: acetylene and mixtures of acetylene with helium, argon and xenon. Int J Mass Spectrom Ion Phys. 1977;24:11–23.
67. Slovetsky DI. Mechanisms of decomposition of hydrocarbons in electrical discharges. Pure Appl Chem. 1988;60:753–68.
68. Smith AM, Agreiter J, Hartle M, Engel C, Bondybey VE. Rare gas matrix studies of absorption and fluorescence of reactive intermediates formed in discharges through acetylene. Chem Phys. 1994;189:315–34.
69. Franklin JL, Studniarz SA, Ghosh PK. Translational energy distribution of electrons and positive ions in the plasma of microwave and high frequency discharges of He, Ne, and Ar. J Appl Phys. 1968;39:2052–61.
70. Atkinson R, Baulch DL, Cox RA, Hampson RF Jr., Kerr JJ, Troe J. Evaluated kinetic and photochemical data for atmospheric chemistry. Supplement III. J Phys Chem Ref Data. 1988;18:881–1097.
71. Lightfoot PD, Veyret B, Lesclaux R. Flash photolysis study of the methylperoxy + hydroperoxy reaction between 248 and 573 K. J Phys Chem. 1990, 94,708–14.
72. Grosser T, Hirsch A. Dicyanopolyynes: formation of new rod-shaped molecules in a carbon plasma. Angew Chem Int Ed Engl. 1993;32:1340–42.
73. Niu C, Lu YZ, Lieber CM. Experimental realization of the covalent solid carbon nitride. Science. 1993;261:334–7.
74. Fujii T, Kareev M. Diagnostic studies of a CH_4/H_2 microwave plasma by mass spectrometry: ionic and neutral species, J Phys Chem A. 2001;105:4923–27.
75. Harris SJ, Goodwin DG. Growth on the reconstructed diamond (100) surface, J Phys Chem. 1993;97:23–28.
76. Selvin PC, Iwase K, Fujii T. Mass spectrometric analysis of ionic and neutral species produced in an N_2/O_2 microwave discharge plasma. J Phys D: Appl Phys. 2002;35:675–9.
77. Selvin PC, Iwase K, Fujii T. Determination of the ionic and neutral chemical components of an H_2O microwave discharge plasma. Chem Phys Lett. 2002;360:367–73.
78. Fujii T, Iijima S, Iwase I. Mass spectrometric detection of $H_2O_2H^+$: a CH_4/O_2 microwave discharge plasma. Chem Phys Lett. 2001;341:513–7.
79. Fujii T, Iijima S, Iwase K, Arulmozhiraja S. The production of H_2O_2 in the microwave discharge plasma of CH_4/O_2. J Phys Chem A. 2001;105:10089–2.

80. Wagner J, Wild Ch, Pohl F, Koidl P. Optical studies of hydrogenated amorphous carbon plasma deposition. Appl Phys Lett. 1986;48:106–8.
81. Mitomo T, Ohta T, Kondoh E, Ohtsuka K. An investigation of product distributions in microwave plasma for diamond growth. J Appl Phys. 1991;70:4532–39.
82. Raiche A, Jeffries JB. Laser-induced fluorescence temperature measurements in a dc-arcjet used for diamond deposition. Appl Opt. 1993;32:4629–35.
83. Celii FG, Pehrsson PE, Wang H. –t, Butler JE. Infrared detection of gaseous species during the filament-assisted growth of diamond. Appl Phys Lett. 1988;52:2043–5.
84. Stadler KR, Homsi W. Mass spectroscopic investigations of a hydrocarbon arcjet plasma operating under diamond deposition conditions. Appl Phys Lett. 1996;68:3710–2.
85. Harris SJ, Weiner AM, Perry TA. Measurement of stable species present during filament-assisted diamond growth. Appl Phys Lett. 1988;53:1605–7.
86. Celli PG, Butler JE. Hydrogen atom detection in the filament-assisted diamond deposition environment. Appl Phys Lett. 1989;54:1031–3.
87. Mitsuda Y, Tanaka Y, Yoshida Y. In situ emission and mass spectroscopic measurement of chemical species responsible for diamond growth in a microwave plasma jet. J Appl Phys. 1990;67:3604–08.
88. Goodwin DG. Simulations of high-rate diamond synthesis: methyl as growth species. Appl Phys Lett. 1991;59:277–9.
89. Brown RC, Roberts JT. Microstructure evolution in diamond CVD: computer simulations of 111 surface site formation on a growing diamond-100 surface. J Phys Chem B 2000;104:8420–9.
90. Cappelli MA, Paul PH. An investigation of diamond film deposition in a premixed oxy-acetylene flame. J Appl Phys. 1990;67:2596–02.
91. Netto A, Frenklach M. Kinetic Monte Carlo simulations of CVD diamond growth-Interlay among growth, etching, and migration. Diam Relat Mater. 2005;14:1630–46.
92. Richley JC, Harvey JN, Ashfold MNR. CH_2 group migration between H-Terminated 2 × 1 reconstructed {100} and {111} surfaces of diamond. J Phys Chem C. 2012;116:7810–16.
93. Fujii T, Kareev M. Mass spectrometric studies of a CH_4/H_2 microwave plasma under diamond deposition conditions. J Appl Phys. 2001;89:2543–6.
94. Chang S, Scattergood T, Aronowitz S, Flores J. Organic chemistry on titan. Rev Geophys Space Phys. 1979;17:1923–33.
95. Sagan C, Thompson WR, Khare BN. Titan: a laboratory for prebiological organic chemistry. Acc Chem Res. 1992;25:286–92.
96. Kunde VG, Aiken AC, Hanel RA, Jennings DE, Maguire WC, Samuelson RE. C_4H_2, HC_3N and C_2N_2 in Titan's atmosphere. Nature. 1981;292:686–8.
97. Hanel R. et al. Infrared observations of the saturnian system from voyager 1. Science. 1981;212:192–200.
98. Fujii T, Arai N. Analysis of N-containing hydrocarbon species produced by CH_4/N_2 microwave discarge. Astrophys J 1999;519:858–63.
99. Thompson WR, Henry TJ, Schwartz JMS, Khare BN, Sagan C. Plasma discharge in N_2+ $CH_{4\,at}$ low pressures: Experimental results and applications to Titan. Icarus. 1991;90:57–73.
100. Crutzen PJ. The role of NO and NO_2 in the chemistry of the troposphere and stratosphere. Ann Rev Earth Planet Sci. 1979;7:443–72.
101. Beck JC, de Leeuw RF, Penkett S. The effect of aircraft emissions on tropospheric ozone in the northern hemisphere. Atmos Environ A. 1992;26:17–29.
102. Ehhalt DH, Rohrer F, Wahner A. Sources and distribution of NO in the upper troposphere at northern mid-latitudes. J Geophys Res. 1992;97:3725–38.
103. Johnson C, Henshaw J, McInnes G. Impact of aircraft and surface emissions of nitrogen oxides on tropospheric ozone and global warming. Nature. 1992;355:69–71.
104. Singh HB. Reactive nitrogen in the troposphere. Environ Sci Technol. 1987;21:320–7.
105. Fujii T, Iwase K, Selvin PC. Observation of NO_x Species in a $N_2/O_2/H_2O$ microwave discharge plasma: a laboratory simulation of Earth's atmosphere. J Geophys Res Atmos. 2003;108:4148–51.

106. Folkins IA, Weinheimer AJ, Ridley BA, Walega JG, Anderson B, Collins JE, Sachse G, Pueschel RF, Blake DR. O_3, NO_y, and NO_x/NO_y in the upper troposphere of the equatorial pacific. J Geophys Res 1995;100:20913–26.
107. Eisele FL, McDaniel EW. Mass spectrometric study of tropospheric ions in the northeastern and southwestern United States. J Geophys Res. 1986;91:5183–8.
108. Suzuki Y, Nakano N, Suzuki K. Portable sick house syndrome gas monitoring system based on novel colorimetric reagents for the highly selective detection of formaldehyde. Environ Sci Technol. 2003, 37, 5695–700.
109. Endecott BR, Sanders DC, Arvind K, Chaturvedi AK. Simultaneous gas chromatographic determination of four toxic gases generally present in combustion atmospheres. J Anal Toxicol. 1996, 20, 189–94.
110. Ravishankara AR, Solomon S, Turnipseed AA, Warren RF. Atmospheric lifetimes of long-lived halogenated species. Science. 1993;259:194–9.
111. Langan J, Maroulis J, Ridgeway R. Strategies for greenhouse gas reduction. Solid State Technol. 1996;39:115–22.
112. Zazzera L, Reagen W, Cheng A. Infrared study of process emissions during C_3F_8/O_2 plasma cleaning of plasma enhanced chemical vapor deposition chambers. J Electrochem Soc. 1997;144:3597–601.
113. Gubner AE, Kohler U. FTIR spectroscopy for the analysis of selected exhaust gas flows in silicon technology. J Mol Struct. 1995;348:209–12.
114. Stoffels E, Stoffels WW, Tachibana K. Electron attachment mass spectrometry as a diagnostics for electronegative gases and plasmas. Rev Sci Instrum. 1998;69:116–22.
115. Stoffels W, Stoffels E, Tachibana K. Polymerization of fluorocarbons in reactive ion etching plasmas. J Vac Sci Technol A 1998;16:87–95.
116. Harnisch J, Borchers R, Fabian P, Maiss M. Tropospheric trends for CF_4 and C_2F_6 since 1982 derived from SF_6 dated stratospheric air, Geophys Res Lett. 1996;23:1099–102.
117. Fujii T, Arulmozhiraja S, Nakamura M, Shiokawa Y. Mass spectrometry for on-line monitoring of perfluoro compounds using Li^+ ion attachment techniques. Anal Chem. 2001;73:2937–40.
118. Fujii T, Nakamura M. On-line monitoring of perfluoro compounds in exhaust gases during semiconductor manufacture: use of Li^+ ion attachment mass spectrometry. J Appl Phys. 2001;90:2180–4.
119. Helneder H, Korner H, Mitchell A, Schwerd M, Seidel U. Comparison of copper damascene and aluminum RIE metallization in BICMOS technology. Microelectron. Eng. 2001;55:257–68.
120. Lakshmanan SK, Gill WN. A novel model of hydrogen plasma assisted chemical vapor deposition of copper. Thin Solid Films. 1999;338:24–39.
121. Kim D, Wentorf RH, Gill WN. Low pressure chemically vapor deposited copper films for advanced device metallization. J Electrochem Soc. 1993;140:3273–9.
122. Dubois LH, Zegarski BR. Selectivity and copper chemical vapor deposition. J Electrochem Soc. 1992;139:3295–9.
123. Girolami GS, Jeffries PM, Dubois LH. Mechanistic studies of copper thin film growth from cui and cuii face β- Diketonates. J Am Chem Soc. 1993;115:1015–24.
124. Zheng B, Goldberg C, Eisenbraun ET, Liu J, Kaloyeros AE, Toscano PJ, Murarka SP, Loan JF, Sullivan J. In situ quadrupole mass spectroscopy studies of water and solvent coordination to Copper(II) β- Diketonate precursors: implications for the chemical vapor deposition of Copper. Mater Chem Phys. 1995;41:173–81.
125. Lakshmanan SK, Gills WN. Experiments on the plasma assisted chemical vapor deposition of copper. J Vac Sci Technol A 1998;16:2187–97.
126. Farkas J, Hampden-Smith MJ, Kodas TT. FTIR studies of the adsorption/desorption behavior of copper chemical vapor deposition precursors on Silica. 2. (1,1,1,5,5,5-Hexafluoroacetylacetonato)(2-butyne)copper(I). J Phys Chem. 1994;98:6763–70.
127. Pinkas J, Huffman JC, Baxter DV, Chisholm MH, Caulton KG. Mechanistic role of H_2O and the ligand in the chemical vapor deposition of Cu, Cu_2O, CuO, and Cu_3N from bis(1,1,1,5,5,5-hexafluoropentane-2,4-dionato)copper(II). Chem Mater. 1995;7:1589–96.

128. Naik MB, Gill WN, Wentorf RH, Reeves RR. CVD of Copper by using Copper (I) and Copper (II) β-Diketonates. Thin Solid Films. 1995;262:60–6.
129. Fujii T, Arulmozhiraja S, Nakamura N, Shiokawa Y. Detection of Cu(hfac)(tmvs) by ion attachment mass spectrometry. Chem Phys Lett. 2006;425:134–7.
130. Fujii T, Arulmozhiraja S, Nakamura M, Shiokawa Y. Chemistry of Cu deposition by Cu(hfac)(tmvs) monitored by Li^+ ion attachment mass Spectrometry. J Appl Phys. 2006;100:084912.
131. Arulmozhiraja S, Fujii T. Is the (hfac)Cu(I)-(tmvs) bond intrinsically weak? Molecular Phys. 2005;103:3293–8.
132. Durrant SF, Mota RP, Bica de Moraes MA. Relationships between the plasma environment and the composition and optical properties of plasma-polymerized thin films produced in rf discharges of C_2H_2-SF_6 mixtures. J Appl Phys. 1992;71:448–55.
133. Pedersen JOP, Opansky BJ, Leone SR. Laboratory studies of low temperature reactions of C_2H with C_2H_2 and implications for atmospheric models of Titan. J Phys Chem. 1993;97:6822–9.
134. Killian TC, Vrtilek JM, Goottlieb CA, Gottlieb EW, Thaddeus P. Laboratory detection of a second carbon chain carbene—Butatrienylidene H_2CCCC. Astrophys J. 1990;365:L89–L92.
135. Benedikt J, Hecimovic A, Ellerweg D, von Keudell A. Quadrupole mass spectrometry of reactive plasmas. J Phys D: Appl Phys. 2012;45:403001 (23 pp).
136. Kobayashi H, Bell AT, Shen M. Plasma polymerization of saturated and unsaturated hydrocarbons. Macromolecules. 1974;7:277–83.
137. Gottlieb CA, Vrtilek JM, Gottlieb EW, Thaddeus P, Hjalmarson A. Laboratory detection of the C_3H radical. Astrophys J. 1985;294:L55–8.
138. Kaiser RI, Lee YT, Suits AG. Crossed beam reaction of $C(^3P_j)$ with $C_2H_2(^1S^+g)$: observation of Tricarbon-Hydride C_3H. J Chem Phys. 1995;103:10395–8.
139. Smith AM, Engel C, Thoma A, Schallmoser G, Wurfel BE, Bondybey VE. Tentative identification of C_5N_2 in rare gas matrices. Chem Phys. 1994;184:233–45.
140. Agreiter J, Smith AM, Bondybey, VE. Laser-induced fluorescence of matrix-isolated $C_6N_2^+$ and of $C_8N_2^+$. Chem Phys Lett. 1995;241:317–27.
141. Fujii T. Analysis of products from a C_2H_2/N_2 microwave discharge: new nitrile species. Chem Phys Lett. 1999;313:733–40.
142. Smith D. The ion chemistry of interstellar clouds. Chem Rev. 1992;92:1473–85.
143. Goede J., de Kanter FJJ, Bickelhaupt F. Investigations on doubly nitrogen- 1 5 labeled isocyanogen. (CNCN). J Am Chem Soc. 1991;113:6104–7.
144. Fomey D, Freivogel P, Fulara J, Maier JP. Electronic absorption spectra of cyano-substituted polyacetylene cations in neon matrices. J Chem Phys. 1995;102:1510–4.
145. Liu AY, Cohen ML. Prediction of new low compressibility solids. Science. 1989;245:841–2.
146. Zhang Y, Zhou Z, Li H. Crystalline carbon nitride films formation by chemical vapour deposition. Appl Phys Lett. 1996;68:634–6.
147. Dawei W, Dejun F, Huaixi G, Zhihong Z, Xianquan M, Xiangjun, F. Structure and characteristics of C_3N_4 thin films prepared by rf plasma-enhanced chemical vapor deposition. Phys Rev B. 1997;56:4949–54
148. Fujii T, Muraki J, Arulmozhiraja A, Kareev M. Possible production of C_3N_4 in the microwave discharge of C_2H_2/N_2. J Appl Phys. 2000;88:5592–6.
149. Fujii T, Selvin CP, Sablier M, Iwase K. Analysis of hydronitrogen species generated by a microwave discharge in (N_2H_4)/He. J Phys Chem A 2002;106:3102–5
150. Vaghjiani GL. Laser photolysis studies of hydrazine vapor: 193 and 222-nm H-atom primary quantum yields at 296 K, and the kinetics of $H + N_2H_4$ reaction over the temperature range 222–657 K. Int J Chem Kinet. 1995;27:777–90.
151. Schurath U, Schindler RN. The photolysis of hydrazine at 2062 A in the presence of ethylene. J Phys Chem. 1970;74:3188–94.
152. Arvis M, Devillers C, Gillois M, Curtat M. Isothermal flash photolysis of hydrazine. J Phys Chem. 1974;78:1356–60.
153. Willis C, Back RA. Di-imide: some physical and chemical properties, and the kinetics and stoichiometry of the gas-phase decomposition. Can J Chem. 1973;51:3605–19.

154. Nagapal R, Garscadden A. Dissociation of hydrogen in glow discharges in hydrogen-nitrogen mixtures. Chem Phys Lett. 1994;231:211–5.
155. Goldberg N, Holthausen MC, Hrusak J, Koch W, Schwarz H. Mass-Spectrometric and GAUSSIAN2 studies of the Diazene (HNNH) and Isodiazene (H_2NN) molecules and their radical cations. Chem Ber. 1993;126:2753–8.
156. Sylwester AP, Dervan PB. Low-temperature matrix isolation of the 1,1-diazene H_2NN. Electronic and infrared characterization. J Am Chem Soc. 1984;106:4648–50.
157. Fujii T, Iwase K, Selvin PC. Mass spectrometric analysis of a N_2/H_2 microwave discharge plasma. Int J Mass Spectrom. 2002;216:169–75.
158. Fu Y, Tyrrell J. Ab initio investigation of the structure, vibrational frequencies, and intensities of HN_nH, HN_nF, and FN_nF ($n=3, 4$). J Phys Chem. 1995;99:1909–12.
159. Walder R, Franklin JL. Proton affinities of neutral molecules. Int J Mass Spectrom Ion Phys. 1980;36:85–112.
160. Roychowdhury S, Roychowdhury UK, Venugopalan M. Effect of heating on dissociation of water vapor in high-frequency plasmas and formation of hydrogen peroxide in a cold trap downstream of the plasma plasma chem. Plasma Process. 1982,2:157–66.
161. Bufalini JJ, Gay BW Jr, Brubaker KL. Hydrogen peroxide formation from formaldehyde photooxidation and its presence in urban atmospheres. Environ Sci Technol. 1972;6:816–21.
162. Thiel WR. New routes to hydrogen peroxide: alternatives for established processes? Angew Chem Int Ed Engl. 1999;38:3157–8.
163. Fujii T, Yashiro M, Tokiwa H. Proton and Li^+ cation Interactions with H2O3 and H2O/O2: *ab Initio* molecular orbital study. J Am Chem Soc. 1997;119:12280–4.
164. Giguere PA, Herman K. Studies on hydrogen-oxygen systems in the electrical discharge. IV. Canad J Chem. 1970, 48, 3473–82.
165. Gonzalez C, Theisen, J., Schlegel, H. B, Hase WL, Kaiser EW. Kinetics of the reaction between hydroxyl and hydroperoxo on the triplet potential energy surface. J Phys Chem. 1992;96:1767–74.
166. Koller J, Plesnicar B. Mechanism of the participation of water in the decomposition of hydrogen trioxide (HOOOH). A theoretical study. J Am Chem Soc. 1996;118:2470–72.
167. Barnes JR. Directive 2002/95/EC of the European parliament and of the council of 27 january 2003 on the restriction of the use of certain hazardous substances in electrical and electronic equipment. Off J Eur Union. 2003;46:L37/19–23.
168. Environmental risk evaluation report: Decabromodiphenyl ether, Environment Agency, UK, 2009; p. 282 https://www.gov.uk/government/organisations/environment-agency. Accessed 15 April 2015.
169. The International Electrotechnical Commission (IEC); http://www.iec.ch/. Accessed 15 April 2015.
170. Samsonek J, Puype J, Uype F, YPE F, Vit DD. Rapid determination of certain BFRs in plastics by X-ray fluorescence spectrometry (XRF) and thermal desorption GC-MS (TD-GC-MS) for the RoHS directive. Organohalogen Compd. 2007;69:2789–92.
171. Guerra P, De La Torre A, Martinez MA, Eljarrat E, Barcelo D. Identification and trace level determination of brominated flame retardants by liquid chromatography/quadrupole linear ion trap mass spectrometry. Rapid Commun Mass Spectrom. 2008;27:916–24.
172. Wang D, Atkinson S, Hoover-Miller A, Shelver WL, Li QX. Simultaneous use of gas chromatography/ion trap mass spectrometry—electron capture detection to improve the analysis of bromodiphenyl ethers in biological and environmental samples. Rapid Commun Mass Spectrom. 2008;22:647–56.
173. Dirtu AC, Ravindra K, Roosens L, Van Grieken R, Neels H, Blust R, Covaci A. Fast analysis of decabrominated diphenyl ether using low pressure gas chromatography-electron-capture negative ionization mass spectrometry. J Chromatog A. 2008;1186:295–301.
174. Li Y, Wang T, Hashi Y, Li H, Lin JM. Determination of brominated flame retardants in electrical and electronic equipments with microwave-assisted extraction and gas chromatography-mass spectrometry. Talanta. 2009;78:1429–35.

175. Oppermann U, Schram J, Felkel D. Improved background compensation in atomic absorption spectrometry using the high speed self reversal method. Spectrochim Acta B. 2003;58:1567–72.
176. Kikuchi S, Kawauchi K, Ooki S, Kurosawa M, Honjho H, Yagishita T. Non-destructive rapid analysis of brominated flame retardants in electrical and electronic equipment using Raman spectroscopy. Anal Sci. 2004;20:1111–2.
177. Sato Y, Oki M, Kondo A, Takenaka M, Satake H. Rapid analysis of polybrominated diphenyl ethers by ion attachment mass spectrometry. Anal Methods. 2010, 2:701–6.
178. Maruyama H, Homma K, Wada A, Shiokawa Y. The rapid analysis of PBDEs (poly brominated diphenyl ethers) utilizing the ion attachment mass spectrometry. Organohalogen Compd. 2007; 69:2767–68.
179. IEC technical committee 111 http://www.iec.ch/tc111. Accessed 15 April 2015.

Chapter 6
Hybrid System with Ion Attachment Techniques

Toshihiro Fujii

6.1 Application to Gas Chromatography-Mass Spectrometry (GC-MS) Mode

The advantage of ion attachment (IA) techniques is due in part to its compatibility with most MS techniques [1–3]. Also, it is compatible with chromatographic inlet systems (i.e., gas chromatography (GC)/mass spectrometry (MS), liquid chromatography (LC)/MS) [4]. Thus, applications to GC-MS mode are promising analytically in view of the simple mass spectra obtained where pseudo-molecular (cationized) ions dominate. Application to GC-MS mode is evaluated as a method for qualitative and quantitative determination of several aroma compounds, which have very similar retention times. IA technique has the possibility of verifying the compound's identity through its mass spectrum. The co-eluting compounds can be separated by their spectra when overlap takes place. The work presented here represents the example that solves the interference problem.

A synthetic aroma compound of 2,6-Octadien-1-ol, 3,7-dimethyl ($MW = 154$) is studied. This compound is analyzed by glass capillary GC-MS by using ion attachment ionization in the mass spectrometer. Figure 6.1 shows ion attachment mass spectrum over the retention time, 39.5 min. An IA mass spectrum does show some abundance (~2 %) at m/z 205, indicating the co-eluting interfering compound. This is detected as possible interferences. Sablier et al. carry out a survey of detection limits by studying a mixture of phthalates (dimethyl, diethyl, di-n-octyl phthalate) and deriving the relative IA mass spectra for phthalates by ion trap quadrupole mass spectrometer with a gas chromatograph (see Sect. 6.5). They showed that the expected ability of this IA technique to characterize elusive complex samples is very promising. It is also suggested that the sensitivity for detection could be extended considerably under the ion trap conditions of operation via a gas chromatography inlet.

T. Fujii (✉)
C & V Technix Co., Ltd., 3-6-1 Higashi, Akishima, Tokyo 196-0033, Japan
e-mail: fujii.toshihiro@c-vtechnix.com

T. Fujii (ed.), *Ion/Molecule Attachment Reactions: Mass Spectrometry*,
DOI 10.1007/978-1-4899-7588-1_6

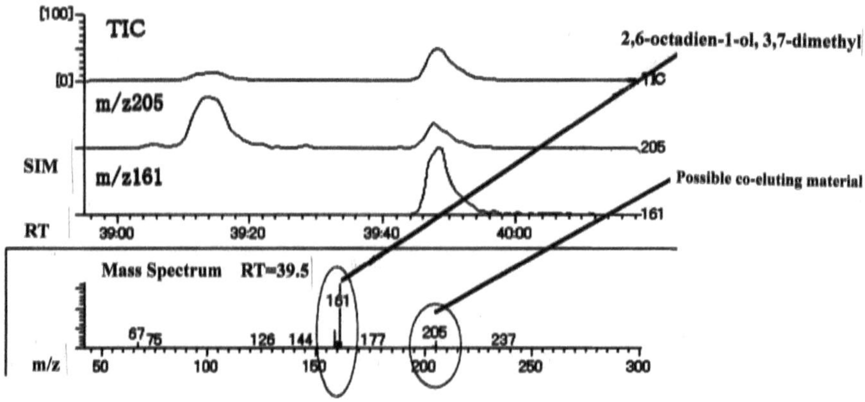

Fig. 6.1 An ion attachment mass spectrum of 2,6-octadien-1-ol, 3,7-dimethyl together with total ion chromatogram (TIC) and selected ion monitoring (SIM) for m/z 161 and 205. The mass peak at m/z 205 indicates the co-eluting interferences. GC separation column; Inert Cap CHIRAMAX 30 m x 0.25 mm ID, *df*=0.25 μm, Carrier gas; He 1.2 ml/min. (©2009, Canon Anelva, Data sheet)

6.2 Aerosol Mass Spectrometry

The application of mass spectrometric techniques to the real-time measurement and characterization of aerosols represent a significant advance in the field of atmospheric science. In 1990s, several groups had developed instrumentation to perform online analysis of individual aerosol particles by laser desorption (LD)/ionization. These methods are based upon an approach originally described by Sinha [5] and McKeown et al. [6] in which aerosol particles are drawn into a mass spectrometer and then ablated with a high-energy laser pulse as they pass through the source region. Online LD has been combined with a variety of optical particle sizing techniques to characterize ambient particles in the atmosphere over the approximate size range of 0.3–8 μm diameter. Online LD has also been applied to liquid chromatography/mass spectrometry (LC/MS) using an aerosol inlet [7].

This section focuses on the aerosol mass spectrometer (AMS), an instrument designed and developed at Aerodyne Research, Inc. that is the most widely used thermal vaporization AMS [8]. The AMS uses aerodynamic lens inlet technology together with thermal vaporization and mass spectrometry to measure the real-time volatile chemical speciation and mass loading as a function of particle size of fine aerosol particles with aerodynamic diameters between 50 and 1000 nm. The original AMS utilizes a quadrupole mass spectrometer with electron impact ionization and produces ensemble average data of particle properties. Later versions employ time-of-flight (TOF) mass spectrometers and can produce full mass spectral data for single particles. Instrumental developments, such as the incorporation of softer ionization techniques (vacuum ultraviolet (VUV) photo-ionization, Lithium ion attachment, and dissociative electron attachment) and high-resolution TOF mass

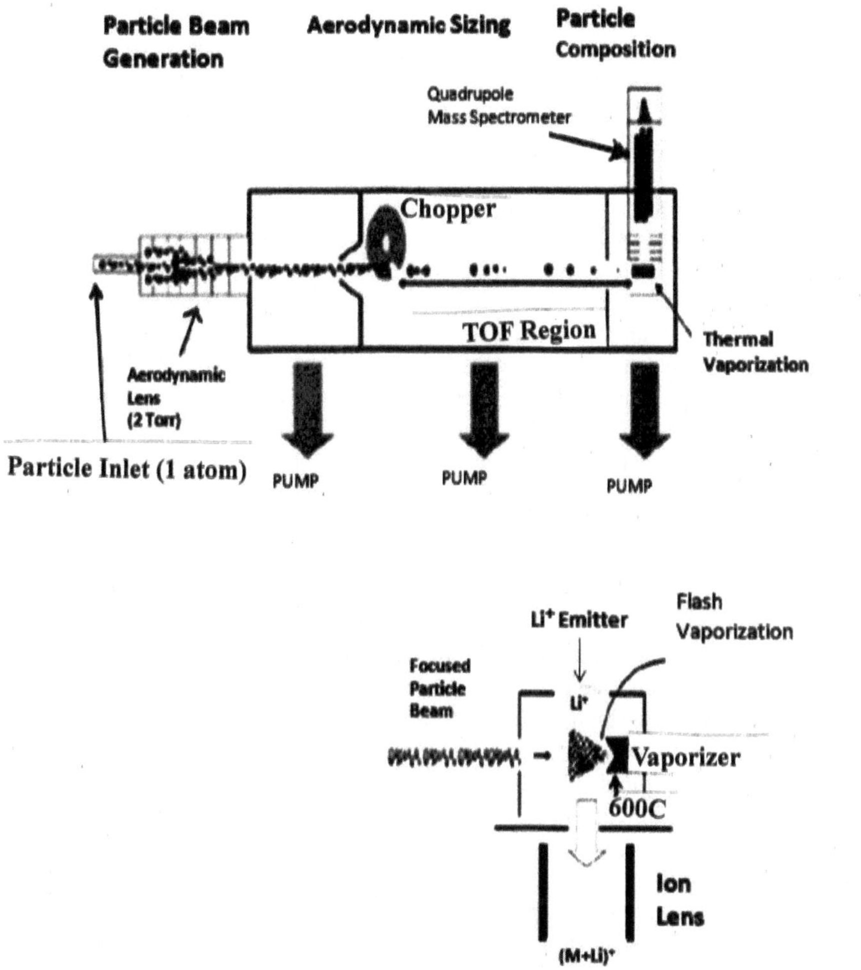

Fig. 6.2 Schematic of an Aerodyne aerosol mass spectrometer (AMS). Vaporized aerosol species are ionized and analyzed via mass spectrometry. This figure shows the ion attachment version of ionization methods. Other existing versions of the AMS that utilize several unique ionization methods and developments that are not shown in this schematic are discussed in the text. Extended drawing of flash vaporizer system shows that the particle beam first impacts on a vaporizer, and volatile aerosol components that vaporize are subsequently subjected to cationization. The unique feature of this detection scheme is the fact that a vaporizer is directly coupled into an ion attachment technique to enable a two-step particle vaporization and ionization process. The separation of the vaporization and ionization processes allows for quantitative detection of aerosol mass with the AMS. (Reprinted with permission from Ref. [8]. ©2007, John Wiley and Sons)

spectrometers, that yield more detailed information about the organic aerosol component are reported.

A schematic representation of the Aerodyne AMS is shown in Fig. 6.2. The instrument has three main sections: the aerosol inlet (particle inlet), the particle sizing chamber (aerodynamic sizing), and the particle composition (detection section). The aerosol inlet samples submicron aerosol particles into the AMS through an aerodynamic lens, forming a narrow particle beam that is transmitted into the detection chamber where volatile components are flash vaporized upon impact on a hot surface (~600 °C) under high vacuum (10^{-5} Pa) and chemically analyzed via MS (an Li^+ ion source integrated into the ionizer/vaporizer of the AMS is drawn below in this figure). Three versions of the AMS are currently in use. These versions vary in the type of mass spectrometer, using a quadrupole mass spectrometer (QMS), a time-of-flight mass spectrometer (TOF), or a high-resolution TOF mass spectrometer (HR-TOF). The three AMS versions are described in detail by Jayne et al. [9], Drewnick et al. [10] and DeCarlo et al. [11], respectively.

Detailed organic analysis is difficult to achieve with the electron impact ionization data because extensive fragmentation reduces the differences in mass spectra among species. One of the recent developments in AMS technology is the use of ionization methods that are softer than 70 eV EI ionization. Soft ionization methods would also enhance the specificity of the AMS technique to target specific groups of species that are preferentially ionized by a given technique. The use of single-photon ionization using VUV light from a krypton discharge lamp that emits radiation at 10 and10.6 eV has been demonstrated at Aerodyne Research, Inc. and deployed in field measurements at University of California, Riverside and in chamber studies at the Paul Scherrer Institute, Switzerland [12]. In this technique, a VUV resonance lamp is integrated into the AMS vaporization/ionization region and used in alternation with the standard EI ion source to optimize quantitative information. VUV ionization reduces fragmentation enough that the parent molecular ion is observed. A second soft ionization technique under development for AMS deployment is Lithium ion attachment MS [1, 4, 13, 14]. IAMS has the advantage that it generally occurs without fragmentation or the occurrence of side reactions and therefore, almost solely results in the formation of quasi-molecular adduct ions. The resulting mass spectra are relatively simple and easy to interpret. A Li^+ ion source has been integrated into the AMS instrument (see Fig. 6.2). A note; the high-resolution TOF-AMS will produce more detailed spectra, from which additional information about the organic aerosol components may be extracted.

6.3 Ion Mobility Spectroscopy

Ion mobility spectrometry (IMS) [15, 16] is basically an ion-separation technique at atmospheric pressure. IMS is a relatively inexpensive and useful method for the detection of many compounds such as explosives, chemical warfare agents, and drugs of abuse. Ions in IMS are separated according to their individual velocities as they drift through an inert gas driven by an electric field. For over 20 years, radioactive sources, especially ^{63}Ni, have been the ionization source of choice for IMS. Due to oxidation, a ^{63}Ni radioactive source must be periodically tested following nuclear regulatory commission (NRC) procedures. Furthermore, manufacturing, transport-

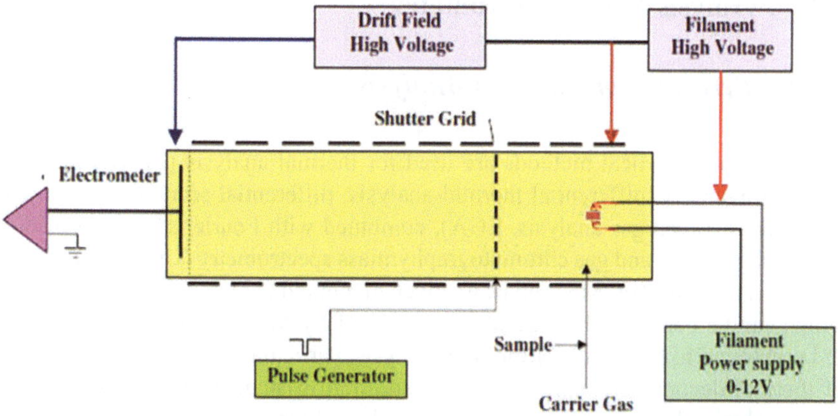

Fig. 6.3 Schematic of the ion mobility spectrometer with a thermionic source of graphite intercalation salt compounds. (Reprinted with permission from Ref. [20]. ©2008, Institute of Physics. IOP Science)

ing, storing, operating, and disposing of hardware containing a ^{63}Ni radioactive source must follow the NRC regulations. These problems and limitations encourage the use of a substitute nonradioactive ionization source for IMS. One of the alternate sources proposed for IMS is the alkali cation emissive source. Alkali ions of K^+, Na^+, Li^+, Rb^+,and Cs^+ are easily produced by heating chemical salts of different alkali metals on filaments [17, 18].

Alkali cation emitters were used in IMS by Tabrizchi et al. [19, 20]. The experimental setup is shown in Fig. 6.3. The IMS cell is made of a 4 cm OD glass tube on which 15 stainless steel guard rings are mounted. The ionization and drift regions are 5 and 15 cm wide, respectively. The shutter grid is mounted inside the glass tube. A graphite intercalation salts compound coated on the filament is used in this instrument as the ionization source. The filament is powered by a variable power supply isolated from the ground by a transformer. Another isolated high-voltage power supply is employed to adjust the potential of the filament (50–200 V) with respect to the first guard ring. Pure nitrogen is used as the drift and carrier gases at flow rates of 500 and 100 mL min^{-1}, respectively. In this work, graphite intercalation compounds are proved to have the capability of being used as an alkali ion source. A number of compounds including methyl isobutyl ketone, dimethyl sulfoxide, acetone, and tetrahydrofuran are successfully ionized by these alkali ion sources. Each substance produced a distinctive ion mobility pattern. The source also ionizes some alkali salts via cation attachment reaction [19]. Although the source is demonstrated to work for ion mobility spectrometry that operates under atmospheric pressure, it can also be used in vacuum for applications such as mass spectrometry or ion beams.

6.4 Pyrolysis Mass Spectrometry

6.4.1 Pyrolysis for Thermal Analysis

A number of analytical methods are used for thermal analysis [21, 22] including thermogravimetry, differential thermal analysis, differential scanning calorimetry, pyrolysis (evolved gas analysis, EGA), combined with Fourier transform infrared spectroscopy, MS, and gas chromatography/mass spectrometry (GC/MS). Pyrolysis mass spectrometry (Py-MS) can be a universal method [23]. In addition, EGA–MS, which can be considered a second generation of Py-MS, has been developed [24, 25] and found to be useful for thermal analysis, particularly in the characterization of thermal decomposition processes. In addition to standard mass spectra, mass spectra obtained in total ion monitoring or selected ion monitoring mode as a function of temperature (spectra that are equivalent to a pyrogram or thermogram) can provide information for kinetic studies of thermal decomposition processes.

In almost all MS studies, electron-impact ionization is used to analyze the degradation products. An electron impact spectrum consists of the molecular and fragment ions produced by sample pyrolysis. Some difficulties occur such as thermal decomposition products with energetic electrons cause further fragmentation, and fragmentation of the degradation products may depend strongly on the instrument conditions, especially in the ion source of the mass spectrometer. Therefore, the development of an efficient mass spectrometer for rapidly identifying species in thermal (chemical) processes is needed.

For this purpose, a system that combines an ion attachment quadrupole mass spectrometry (IAMS) [1–4] with a various kind of direct inlet probe [26] or an infrared image furnace (IIF) [27] unit for EGA has been developed [28]. EGA is most effectively performed by means of MS because of its specificity and sensitivity. The system enables us to introduce thermally decomposed analytes at atmospheric pressure directly into the IAMS system. One of the biggest advantages of IAMS is that it can be used to directly analyze gaseous compounds because it provides mass spectra only of the molecular ions formed by ion attachment to any chemical species introduced into the spectrometer, including free radicals. Direct continuous measurements are feasible on a real-time basis for many radical species and stable molecules in a dynamic system. These features can help us to expand the applications of IAMS into areas such as product analysis in plasma reactors, monitoring of catalytic processes, detection of cigarette smoke, detection of interstellar species, etc. The compatibility of thermogravimetry with IAMS also provides [29, 30] excellent temporal correspondence between thermogravimetry and mass spectral data, as well as the ability to identify and resolve complex coevolving products. This method appears to hold great promise for the analysis of nonvolatile d-metal complexes, and thermally labile medicinal compounds. Applications involving pyrolysis GC/IAMS are also promising analytically, in view of the simple mass spectra obtained where pseudo-molecular ions are predominant (see Sect. 6.1).

In this section, the technical details and operation of the EGA–IAMS system are described, together with several advantages that this system has over conventional mass spectrometers. The EGA–IAMS system work well for nonvolatile, untreated, and complex samples such as lacquers because the simplicity of the ion attachment spectrum permits the analysis of mixtures electron-impact spectra of which are difficult to interpret. Interesting applications for thermal analysis and detection of free radicals in thermal processes are reviewed.

6.4.2 *Temperature-Programmed Heating Probe for Evolved Gas Analysis (EGA)*

The group of Fujii designed a simple EGA system to act as a sampler between solid samples at atmospheric pressure and the high vacuum inside a mass spectrometer [26, 28]. The newly designed stainless steel EGA system is simple, small, and rugged, and it fulfills all the basic requirements for EGA. Temperature parameters are programmable with a maximum heating rate of 60 °C min^{-1} and a maximum temperature of 600 °C. With this system coupled with IAMS, it is possible to study the temperature-programmed decomposition of a number of solid materials by detecting any chemical species on a real-time basis. Another advantage is the ability to directly analyze gaseous constituents; because the ion-attachment process is non-dissociative, it generates $(M + Li)^+$ that do not fragment. The fragment-free measurement of chemical species permits the analysis of mixtures with electron-impact spectra that are difficult to interpret.

EGA Probe Figure 6.4 is a schematic drawing of the cutaway side view of the EGA-probe inlet system fabricated and part of the lithium ion attachment mass spectrometer [26]. The newly designed inlet system for the mass spectrometer consists of a gas inlet in the inner tube through which a buffer gas is introduced to carry the evolved gas. This arrangement ensures that the analyte flows in a constant stream from the sampler into the mass spectrometer. The EGA probe consists of two concentric stainless steel tubes (200 mm long with a 15.8 mm o. d., and 280 mm long with a 6.4 mm o. d.). A holder with the sampler at its center is fastened to the end of this probe. The EGA probe is placed in front of the reaction (ionization) chamber of the mass spectrometer with a conflat flange in such a way that the sampler is located 10 mm from the Li^+ emitter. This distance is adjustable, but 10 mm was found to be optimal in terms of sensitivity. The sample is introduced through a vacuum isolation valve into the MS. The purpose of the isolation valve is to permit the sampler, which is at atmospheric pressure, to be injected into the vacuum system while still keeping a good vacuum for the MS operation. The sampler (Fig. 6.4) is made of heat-resistant tantalum alloy. The heating element (a micro ceramic heater, 15 V, 15 W) is contacted on a holder body that holds the sampler on the upper end and a thermocouple at the bottom. An EGA probe controller controls the programmable heating, in which the goal temperature, the rate of heating, and the length of constant-temperature periods in each step can be set up. The maximum operating temperature is 600 °C. While the analytes are introduced directly into the mass spectrometer, the pressure of the RC is maintained between 60 and 133 Pa, which is the optimum pressure for the formation of Li^+ adducts through a termolecular asso-

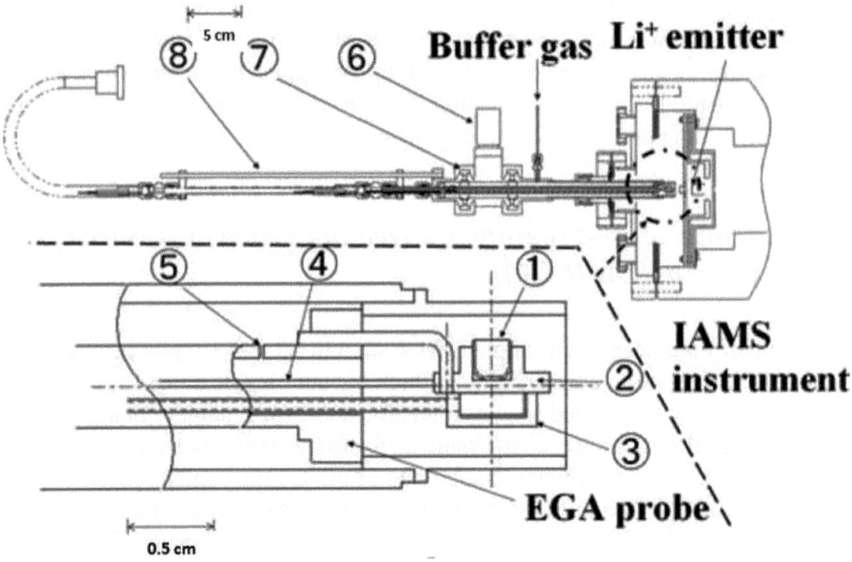

Fig. 6.4 General overview of the EGA probe attached to the lithium ion attachment mass spectrometer used; extended view of EGA probe showing (1) sampler, (2) sampler holder, (3) heater, (4) thermocouple, (5) gas inlet, and the whole system of EGA probe and IAMS, showing (6) sampler, (7) isolation valve, (8) detachable flange, (9) slide guide, and (10) Li^+ emitter. (Reprinted with permission from Ref. [26]. ©2010, John Wiley and Sons)

ciation reaction. A thermogram can be obtained with a temperature-programmed heating probe with total ion monitoring or selected ion monitoring. The probe can also serve as an isothermal or temperature-programmed flow reactor for homogeneous, heterogeneous, or thermal decomposition kinetic studies. The non-isothermal method has the advantage of using only one sample for the entire experiment.

Acrylamides With the EGA-IAMS system, the thermal decomposition of polyacrylamide (PAA) is investigated. The thermal decomposition of polyacrylamide (PAA) has received continued attention in the literature [31, 32] for the following reasons: (i) PAA is used as an important thermoplastic material in many industries; (ii) the acrylamide monomer, which is a neurotoxin and carcinogen, can be present in small amounts in polymerized acrylamide; (iii) PAA used in the food industry may contaminate food with acrylamide. Thermal decomposition of PAA under certain conditions has been reported to cause the release of acrylamide, and at cooking temperatures, degradation reactions are likely to occur. Therefore, a thorough understanding of its thermal behavior is of crucial importance for end-use applications. However, only a few reports about acrylamide production [33] or the thermo-oxidative behavior [34] and kinetics of acrylamide decomposition [35] have appeared in the literature. This information is essential to understand the thermal decomposition properties of PAA materials during their thermal destruction.

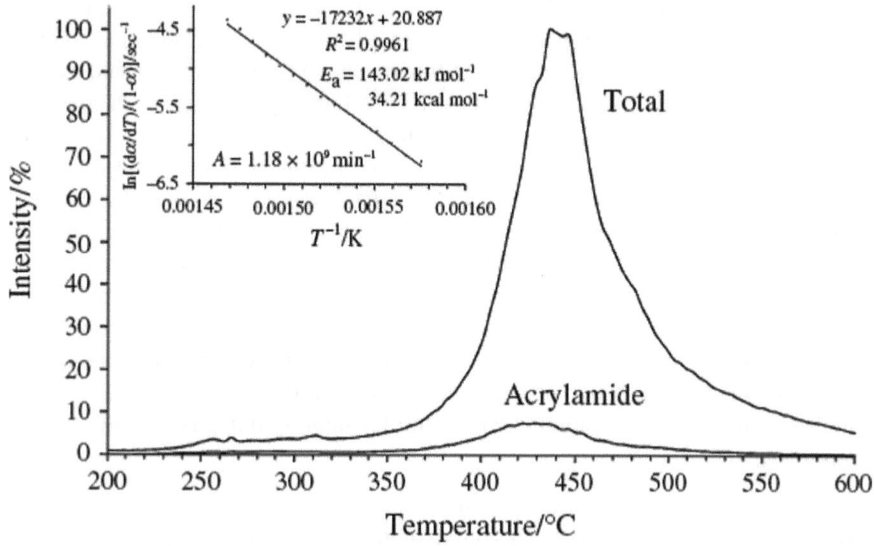

Fig. 6.5 Evolution profiles of thermal products during pyrolysis of polyacrylamide (PAA) in nitrogen, shown by total ion and selected ion chromatograms.*1*, total ion, *2*, acrylamide. Inset Arrhenius plot for acrylamide signals over the temperature range 350–400 °C. (Reprinted with permission from Ref. [36]. ©2012, Springer)

The ion attachment mass spectrum clearly indicates [36] that pyrolysis of PAA at around 450 °C produces many products (e.g., amides, imides, nitriles, ketones, aldehydes, and acrylamide oligomers). Monomer acrylamide, which is a possible carcinogen, is produced abundantly in the thermal decomposition process. The mass chromatogram (pyrogram) of the many products associated with reaction temperature can be informative for kinetic studies to show how acrylamide, a neurotoxin and animal carcinogen, is released. By selecting the specific ion chromatogram of m/z 78 and following it over the entire reaction temperature range, the evolution profile of the corresponding acrylamide monomer during decomposition can be seen in Fig. 6.5. The peak at m/z 78 appeared at 350 °C, increased to a maximum at 430 °C, and then almost disappeared at around 530 °C. This figure also illustrates the total ion monitoring (TIM) profiles of all the decomposed species, indicating that thermal decomposition started at above 350 °C.

The apparent activation energy (Ea) of a decomposed reaction is considered to be an energetic threshold for that reaction. Thus, Ea is commonly determined to evaluate the thermal stability of polymers. The integral method of Knumann et al. [37] is commonly used to evaluate Ea for the pyrolysis of solid materials. A pyrogram (thermogram, multi-ion detection) obtained from an EGA system gives the relative number of decomposition product molecules and indicates their production rates. The degree of conversion at any temperature T can be obtained from the area (determined by integration) under the program curve between the temperature at the start of the signal, T_0, and T. The ionic signal (i) acquired from real-time multiple ion detection of chemicals released from thermally decomposed specimens, such as acrylamide, has been used to obtain the functional (α) form of kinetic rate expressions. From plots of $\ln[(d\alpha/dT)/(1-\alpha)]$ vs. 1/T, Arrhenius parameters such as Ea and the pre-exponential factor (A) can be determined. The intensities of the Li^+ adduct signals at m/z 78 are measured over the temperature range 350–400 °C in a nitrogen atmosphere to obtain the rate expressions for PAA degradation, with the assumptions that (i) a simple correlation exists between the adduct ion signal of the acrylamide product and acrylamide concentrations, and (ii) the decomposition reaction is first order (unimolecular). The slopes of the plots of temperature versus signal intensity in the region are constant (inset of Fig. 6.5). An Ea of 143.02 kJ mol^{-1} with an A of 1.18×10^9 from the plots for acrylamide is calculated. The activation energy of the thermal decomposition of acrylamide exhibits different behaviors in different environments. This information will be helpful in understanding acrylamide release from PAA.

Vitamin C and B_6 Ion attachment mass spectrometry with a temperature-programmed direct probe allows the detection of intact pyrolysis products. It, therefore, offers the opportunity to monitor directly thermal byproducts on a real-time basis and potentially to detect thermally unstable products. EGA-IAMS is used to study the real-time, non-isothermal decomposition of vitamin C [30]. The results were compared with those obtained in a similar study on thermal decomposition of vitamin C using pyrolysis GC/MS. Significant differences were found between the two techniques, in terms of the nature and relative amounts of products formed. A major difference between the two techniques was in the transportation time of the pyrolysis products out of the pyrolysis chamber (or hot zone). The time was significantly shorter in EGA–IAMS than in pyrolysis GC/MS, which reduces the occurrence of secondary reactions of the primary pyrolysis products. Some decomposition products formed in the EGA–IAMS system were not detected in the previous pyrolysis GC/MS study [38] and thus were detected for the first time. For instance, dehydro-L-ascorbic acid was observed as a decomposition product. This compound was the main degradation product detected by means of EGA–IAMS. While it is an important compound because it possesses some biological activity, dehydro-L-ascorbic acid is difficult to measure due to its chemical instability.

Thermal decomposition of pyridoxine (Vitamin B_6 group) is also investigated [39]. Pyridoxine is an important vitamer in food and pharmaceutical products. Heat treatments applied during preparation or storage of the products cause the decomposition of pyridoxine [40, 41]. Identification and understanding of the degradation products of pyridoxine and studying its decomposition kinetics are essential in the preparation and preservation of pyridoxine-containing foods and pharmaceutical. Real-time, non-isothermal decomposition of pyridoxine was studied. Arrhenius parameters for the thermal decomposition of pyridoxine were obtained via the TIM curve. The results demonstrate that most of the pyridoxine evaporated in molecular form, but the formation of pyridoxal and o-quinone methide, both biologically important species, was also observed from the solid-phase degradation of pyridoxine. The observation of o-quinone methide, a species possessing anticancer activity, was particularly noteworthy due to its chemical instability. The activation energy (Ea) for pyridoxine decomposition determined by EGA-IAMS was found to be 20.0 kcal mol^{-1}, and the pre-exponential factor (A) was 5.7×10^9 min^{-1}. The calculated kinetic parameters are important for predicting the thermal stability of pyridoxine vitamer. The estimated lifetime (t_{90} %,25 °C) of 1.7×10^{-2} years in nitrogen was also obtained from the EGA-IAMS experiment.

$Ti(C_5H_5)_2Cl_2$ Characterization of the compound $Ti(C_5H_5)_2Cl_2$ was studied using Li^+ ion attachment mass spectrometry (IAMS) as an analytical methodology [42]. Since this target compound is used as an anticancer drug in the treatment of leukemia [43], accurate and rapid monitoring methods for the determination of titanium drugs in a hospital environment are desirable. An EGA probe was used to study the temperature-resolved behavior of this compound. The slope of the plot of signal intensity of $Ti(C_5H_5)_2Cl_2Li^+$ versus temperature for $Ti(C_5H_5)_2Cl_2$ sublimation from 60 to 100 °C was used to determine an apparent activation energy (Ea) of 124.43 kJ/mol for the sublimation of $Ti(C_5H_5)_2Cl_2$. However, Δ(PV) corrections should be considered for the calculation of enthalpy. The enthalpy of sublimation is calculated using the following relationship

$$\Delta H = \Delta E_{\text{internal energy}} + \Delta(PV) \tag{6.1}$$

where Δ(PV) = RT = −2.5 kJ/mol at 298.15 K. Since the sublimation process is very endothermic, Ea for sublimation can be assumed (approximated) as the internal energy difference, $\Delta E_{\text{internal energy}}$. When the above-calculated Ea values were used for $\Delta E_{\text{internal energy}}$ and Δ(PV) correction was made, the enthalpy of sublimation was calculated as 121.93 kJ/mol. This value should be closely associated with the molar enthalpy of sublimation of $Ti(C_5H_5)_2Cl_2$. In fact, it agrees well with the reported enthalpy value of 118.8 kJ/mol obtained by combustion calorimetry of $Ti(C_5H_5)_2Cl_2$ [44]. These results demonstrate that the IAMS methodology can be used to study the enthalpy of sublimation for d-metal complex materials.

6.4.3 *Infrared Image Furnace (IIF)*

To permit thermal stability studies of a wide range of nonvolatile materials under atmospheric conditions or in a flowing stream, an IAMS system coupled with an infrared image furnace (IIF) was developed (Fig. 6.6) [27]. An orifice interface system to be placed between the sample IIF at atmospheric pressure and the high vacuum inside a mass spectrometer was designed. The IIF used as the heat source consists of two tungsten lamps. The lamps are placed within gold-plated, parabolic reflecting surfaces, which can heat a sample up to 1100 °C at a heating rate of 10 °C sec^{-1}. The buffer gas, which also acts as the cooling gas, is drawn at a rate of 30 mL min^{-1} from the gas cylinder into the IIF. To ensure entry of the evolved gas

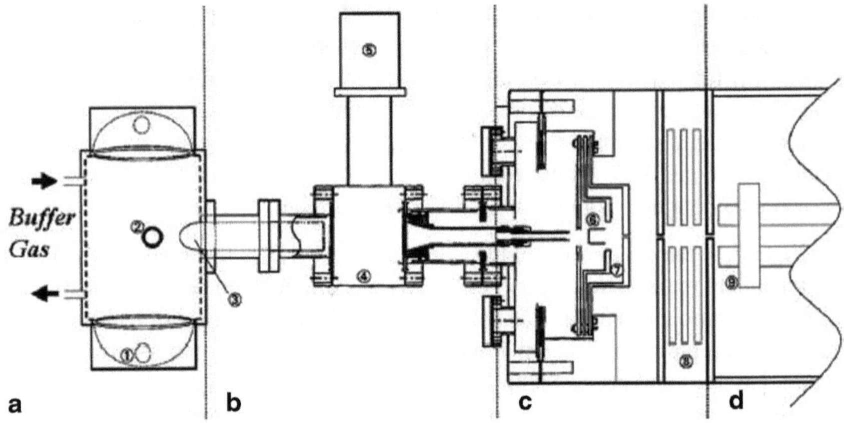

Fig. 6.6 General scheme of the IIF attached to the Li$^+$IAMS: **a** the infrared image furnace (IIF) at atmospheric pressure, showing the IR lamp unit (1) and position of the sample holder (2), **b** the interface, which has a concentric quartz tube with a 70 μm orifice at its center (3) and exit stainless steel tube (4), both of which are fixed to the flange of the gate valve (5), **c** the reaction (ionization) chamber at the pressure of 100 Pa, with the Li ion emitter (30 V) bead fused onto the Ir wire (6), electrode (40 V) (7), and electrostatic lens (0,-150, 0 V, respectively) (8), **d** the envelope for quadrupole MS (9). (Reprinted with permission from Ref. [27]. ©2009, American Chemical Society)

into the IAMS, an orifice is set up as follows: The concentric quartz tube (100 mm long, with 20 mm outer diameter and 18 mm inner diameter) has an orifice with a 70 μm diameter at its center and a stainless steel exit tube; IIF is fixed to the flange of a gate valve with an O-ring system. The sample gas is carried to the Li$^+$ ion reaction chamber of the mass spectrometer through the gate valve in such a way that the outlet of the exit tube is located 15 mm from the Li$^+$ ion emitter. This arrangement ensures that the analyte flows in an unperturbed stream from the IIF into the IAMS. The sample cell is made of alumina (5 mm diameter) and is placed in the platinum holder of the IIF.

The performance and applicability of the IIF-IAMS system is illustrated by polytetrafluoroethylene (PTFE) samples. The potential of the system for the analysis of oxidative pyrolysis has also been considered. The temperature-programmed decomposition of PTFE gave constant-slope plots of temperature versus signal intensity in a defined region and provided an apparent activation energy of 28.8 kcal mol^{-1} for the PFTE decomposition product $(CF_2)_3$. The IIF-IAMS experiments have been conducted to elucidate the effects of copper(II) chloride on the formation of aromatic compounds during the pyrolysis of polyethylene. Under the time-resolved pyrolysis conditions of IIF-IAMS, the effects of $CuCl_2$ were measured to compare the variation in ratios of alkenes and aromatics during polyethylene thermal degradation when pyrolysis temperature was increased [45]. Thermogravimetry experiments and X-ray powder diffraction analysis conducted under similar conditions to those used for IIF-IAMS enabled us to characterize the oxidation states of copper prevailing during the thermal degradation process of polyethylene in the presence of $CuCl_2$.

Bisphenol A The accumulation of plastics in the environment is a matter of great concern [46]. One of the available solutions, waste recycling also known as incineration, can be used to transform waste plastics into energy. However, the combustion of plastics can produce many gaseous products, the nature of which depends mainly on external conditions, such as temperature and oxygen availability. Accordingly, a series of pyrolysis experiments were conducted with the IIF–IAMS system to mimic the reductive environment and conditions within incinerators [47]. Li^+ IAMS has considerable advantages for product monitoring in the gas phase in comparison with traditional electron-impact MS. Unlike traditional MS that involves ionization by high-energy electrons, IAMS preserves the profiles of the product molecules much better, allowing their detection as adduct ions without any fragmentation. Among the pyrolysis products of polycarbonates, Bisphenol A (BPA) may be of most concern [48, 49]. BPA is an endocrine disruptor; although its acute toxicity is low, there is concern that long-term exposure to low doses of BPA may induce chronic toxicity in humans. Therefore, it is important to lower BPA emissions if possible [46]. Experiments were conducted with polycarbonate pyrolysis in nitrogen and air atmospheres [50]. In air, a BPA emission peak is observed at lower temperature than that observed in the nitrogen atmosphere. This clearly shows that there are two different mechanisms of BPA production, but the details are not yet known [51]. In further experiments, a mixture of polycarbonate and $CuCl_2$ in a ratio of 1:3 was pyrolyzed [47]. The amount of evolved BPA was almost 1/100 of that when no $CuCl_2$ was present. Thus, one can expect that BPA emission can be successfully reduced not only by the addition of $CuCl_2$ but also by other metal salts.

Cisplatin IIF-IAMS was evaluated as an analytic methodology for the measurement of the thermally labile, nonvolatile, and insoluble compound cisplatin, which is used as an anticancer agent in the treatment of testicular and ovarian cancers [52]. Takahashi et al. [53, 54] aimed to develop an improved method for the mass spectrometric determination of cisplatin, particularly in its molecular ion form. Full-scan spectra were obtained with 10 μg samples. This system provided cisplatin molecular ions as Li^+ ion adducts; this was the first reported instance of cisplatin Li^+ ion adducts. It is confirmed that cisplatin vaporized partially in molecular form. IIF combined with IAMS was also used to study the temperature-programmed decomposition of this drug [53]. The slope of the plot of signal intensity versus temperature for cisplatin decomposition from 225 to 249 °C was used to determine the apparent activation energy of 38.0 kcal/mol for the decomposition of cisplatin. This value of decomposition parameter is useful for predicting drug stability (shelf life). These studies demonstrate that IAMS can be a valuable technique for the direct mass spectral analysis and kinetic study of d-metal complex platinum anticancer agents.

Japanese Lacquer Films Lacquers are used ubiquitously as surface-coating materials for wood, porcelain, and metal. The main component of Japanese lacquer ("urushi" in Japanese) is urushiol, a brown liquid (boiling point, 200–210 °C) consisting of a mixture of several catechols, each substituted with a saturated or unsaturated alkyl chain of 15 or 17 carbon atoms [55]. The liquid that causes an allergic skin reaction in most people, is obtained from the sap of the Japanese lacquer tree (Rhus vernicifera) and can be polymerized to form lacquer films. Lacquer films have been used in Asian countries for thousands of years as durable, tough, functional, and beautiful coatings. Therefore, analytical methods for the chemical characterization of lacquers are important, especially for studies of the conservation and restoration of lacquered objects [56, 57].

Lacquers have been chemically analyzed by means of various techniques including chromatography/mass spectrometry(GC/MS), solid-state ^{13}C nuclear magnetic resonance spectroscopy, Fourier transform infrared spectroscopy, and X-ray photoelectron spectroscopy. However, of these conventional methods, only mass spectrometry provides information about the chemical components of lacquers [58]. Furthermore, most of these methods are time-consuming, and they demand large amounts of sample and complex pretreatment procedures. The products of pyrolysis of lacquer polymers have been studied by means of thermogravimetry/differential thermal analysis, infrared spectroscopy, and mass spectrometry by Miyakoshi et al. [59–61]. These investigators reported that two-stage pyrolysis-gas chromatography/electron impact mass spectrometry is effective for the rapid analysis of lacquer films with only small amounts of sample and no sample preparation. Many articles on the use of this method for the examination of lacquer art objects and archaeological lacquer materials have been published, including a useful review [62]. However, interpretation of electron impact mass spectra is complicated by molecular fragmentation. The electron impact mass spectra of urushiol, even in its unpolymerized state, show many different fragmentation ions. Therefore, the pyrolysis mass spectra of polymerized lacquer films can be expected to be even more complex.

IAMS with a temperature-programmed direct probe offers the opportunity to study directly the thermal degradation processes occurring in complex natural materials, as well as the chance to determine the identity of a lacquer source by identification of different types of lacquer monomer [63, 64]. A typical mass spectrum of the thermal products of a Japanese lacquer film was obtained by rapid heating up to 500 °C in N_2 gas environment (Fig. 6.7). Many peaks were found from m/z 315–345 (Fig. 6.7 inset). These peak clusters were identified as pure urushiol monomer molecules. Urushiol is a mixture of a catechol substituted with an alkyl chain of 15 or 17 carbon atoms; urushiol is a mixture of saturated and unsaturated molecules. Studies have revealed that lacquer can be roughly classified into three types: Japanese (or Chinese) lacquer (based on the urushiol monomer), Vietnamese lacquer (based on the laccol monomer), and Burmese lacquer (based on the thitsiol monomer). The mass spectral analysis of the urushiol monomer is interesting because identification of the type of lacquer monomer (urushiol, laccol, or thitsiol), can allow researchers to identify the source of ancient archeological lacquerware. Such identification is essential for conservation and restoration studies.

6.4.4 Radical Species in Pyrolysis Processes

Airborne Free Radicals The widespread occurrence of free radical intermediates in gas-phase reactions has frequently been the subject of kinetics studies (refer to

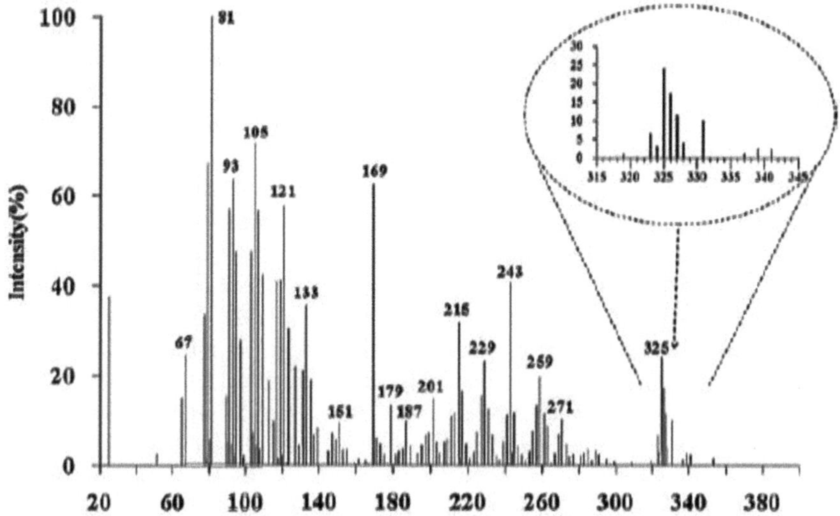

Fig. 6.7 A typical Li^+ ion attachment mass spectrum of the products from the pyrolysis of a Japanese lacquer film recorded as the furnace temperature increased from 50–500 °C. Samples were placed in the EGA furnace and heated linearly in a N_2 atmosphere at a programmed rate of 128 °C/min. The relative intensity is given as a percentage of the intensity of the carboxylic acid peak at m/z 81 ($C_2H_5COOHLi^+$). Inset: a partial mass spectrum from m/z 315 – 345. These peak clusters correspond to the urushiol monomer. (Reprinted with permission from Ref. [64]. ©2012, Elsevier)

Sect. 5.2.4). Nevertheless, considerable experimental difficulties involved in detecting, identifying, and measuring the concentrations of intermediate species exist [65, 66]. Westerberg et al. detected the emission of airborne free radicals in air samples during the injection molding, extruding, seam welding, and wire cutting of polyethylene and polystyrene plastics [67]. On the basis of electron spin resonance spectroscopy on the resulting spin adducts, those investigators suggested a degradation mechanism based on free radical reactions. Motivated by Westerberg's detection of various airborne free radicals in polyethylene processing fumes, the Fujii group investigated and identified the principal radical species produced in the pyrolysis of polyethylene by coupling an IIF with IAMS instrumentation [68]. All possible hydrocarbon products identified in the mass spectrum of polyethylene pyrolysis at 450 °C were classified by their formulas. Identification was made under the assumption that the only products produced were hydrocarbons. Spectra of the thermal decomposition products, detected by means of IAMS, clearly showed that radicals, such as C_nH_{2n+1} (n = 5–12) and C_nH_{2n-1}(n = 5–11), were the predominant species produced. Many of these thermally-decomposed species were identified for the first time by MS. These results provided direct evidence that free radicals are formed in the pyrolysis environment and are detectable with mass spectrometric techniques.

d-Metal Complex Radicals The observation of coordinatively unsaturated d-metal radical intermediate species in a reaction provides a motivation to study the intrinsic role of 17-electron organometallic free radicals [69–71]. Because radical intermediates have short lifetimes and their steady-state concentrations in a reacting system are low, detecting, and identifying such intermediates are always difficult tasks. EGA–IAMS [72, 73] was used to qualitatively analyze the d-metal radical products of the pyrolysis of $Mn_2(CO)_{10}$. The use of an atmospheric-pressure sampling inlet device to introduce the analytes, including radical intermediates, into the ion attachment mass spectrometer permitted real-time, continuous monitoring of the pyrolysis products. The results indicated that pyrolysis of $Mn_2(CO)_{10}$ produced two d-metal radicals as well as stable molecules, including $HMn(CO)_5$. The Li^+ adduct mass spectra of the pyrolysis products contained peaks at m/z 202 and 369 for $Mn(CO)_5Li^+$ and $Mn_2(CO)_9Li^+$, respectively, providing direct evidence that d-metal complex radicals were formed in the furnace. These results are significant in two main respects: To positively identify the free radicals by MS and the results provide evidence that the reactions in the pyrolysis process involved neutral radicals. The radicals may have reacted with each other, with other pyrolysis products, or with the residual gases present in the IIF to form $HMn(CO)_5$ and other substituted products.

Bisphenol A Biradical An unknown species has been detected in the analysis of the products in a pyrolysis of polycarbonate using Li^+ ion-attachment mass spectrometry (IAMS) [50]. The mass spectra exhibited a Li^+ adduct peak at m/z 233 that was tentatively assigned to bisphenol A (BPA) biradical [51]. Experimentally, this assignment was supported by the observation that the production rate increased under an inert nitrogen atmosphere. To further confirm the assignment, the stability of the BPA biradical to intramolecular rearrangement reactions as well as unimolecular decomposition has been analyzed via density functional theory calculations [B3LYP/6-311*G(3df,2p)]. The results show that the bisphenol A biradical is an open-shell biradical singlet that is stable to unimolecular decomposition. Although some of the proposed intramolecular rearrangement products have lower energies than those of the BPA diradical, these pathways have large reaction barriers and the kinetic lifetime of the radical is expected to be of the order of hours under the conditions of the experiment. The calculations also reveal that the bisphenol A diradical has large Li^+ affinities supporting the fact that these Li^+ complexes could be detected in the Li^+ ion attachment mass spectrometry. On the basis of these results the Li^+ adduct peak at m/z 233 detected in the pyrolysis of polycarbonate is assigned to the bisphenol A biradical.

Organic Peroxide Polyethylene glycol (PEG) is used ubiquitously as an industrial material in surfactants, cosmetics, ointments, packaging bags, dyes, detergents, and high-energy-density batteries. Therefore, suitable analytical methods are needed for PEG quality assurance [74], especially for medicinal and pharmaceutical uses. The thermal decomposition of polyethylene glycol was investigated [75]. Unstable products could be detected; for instance, many highly reactive organic peroxides, such as CH_3OOH and $HOCH_2OOH$, were found in this study. Classification analysis revealed 10 major compositional formulas among the product species:

$C_nH_{2n+2}O$, $C_nH_{2n+2}O_2$, $C_nH_{2n+2}O_3$, $C_nH_{2n+2}O_4$, $C_nH_{2n+2}O_5$, $C_nH_{2n+2}O_6$, $C_nH_{2n+2}O_7$, $C_nH_{2n}O$, $C_nH_{2n}O_2$, and $HO(CH_2CH_2O)_nH$ ethylene glycol oligomers. The Li^+ ion adduct mass spectra showed a characteristic profile in terms of both the appearance of unique components and the distribution of pyrolysis products. Among the products of the thermal decomposition of PEG, formaldehyde (HCHO) and organic peroxides were particularly interesting [76–78]. Formaldehyde, one of the 10 most abundant products, is a known human carcinogen. The activation energy of HCHO decomposition in a N_2 atmosphere to be 155.72 kJ mol^{-1}. This information is helpful in understanding HCHO emissions during PEG incineration. The detection of peroxides suggests that they may form during the incineration of PEG, which may have important environmental significances, such as formation, mechanism and contribution to aerosols. The existence of peroxide products may have implications for chemical evolution in incinerator systems [78].

6.5 ICR and Quadruple Ion Trap

Thermionic Source Inside an ICR Cell McMahon and Beauchamp developed [79] experimental methods which permit operation of the standard ICR cell in a trapped ion mode. Appropriate configurations of applied electrostatic fields permit trapping of ions in the source region of the ICR cell. Detection is effected after a suitable delay by drifting the ions from the source through the analyzer region. The minor modifications required do not inhibit normal operation of the cell, thus allowing for the full range of conventional ICR experiments with the additional capability of examining variation of ion abundance with time. The latter mode of operation greatly simplifies elucidation of reaction kinetics. The ion molecule reactions of alkali atoms with halogenated hydrocarbons have been investigated [80] using this technique. To investigate such reactions a thermionic source of alkali ions was mounted inside the source region of an ICR cell. Accepted values of reaction rate constants are reproduced. They also demonstrated that halide ion transfer reactions between carbonium ions can be employed to determine R^+-X^- heterolytic bond energies.

Thermionic Ion Source External to ICR Cell Although much has been made of the ability of FT-ICR MS to perform a variety of experimental events separated by time rather than space, many useful ionization sources are optimally operated at much higher pressure and/or do not function well in the presence of a strong magnetic field. Therefore, there has been increasing interest in ionization sources in which ions are formed external to the ICR ion trap, with subsequent ion extraction and injection into the ICR ion trap [81–83]. McIver et al. recognized that physically separated ionization and detection regions of the mass spectrometer and differential pumping between these regions would allow for the use of higher pressure ionization sources with FT-ICR MS [84]. Conceptually, an external ion injection device that requires no external acceleration lenses between the ionization source and the

ICR ion trap, no deceleration grids or lenses before the ion trap, and which allows for ion trapping in either a continuous or gated mode would be significantly simpler to use and easier to control the relative abundances of trapped ions.

Marshall et al. [85] demonstrate a method for transmitting externally formed ions into an ICR Ion trap. In an electrostatic ion guide, a potential difference is applied between a conductive cylinder and a rigid wire suspended along the central axis of the cylinder. The cylinder is then positioned between an ion source located outside the bore of a superconducting solenoidal magnet and an ion trap located at or near the center of the solenoid. An ion optics simulation program (for instance, SIMIONTM) predicts that low-energy ions entering the ion guide will spiral around the central wire and pass through the fringe field of the magnet to reach the ICR Ion trap. The theoretical predictions are borne out by experiments in which Na^+ and K^+ ions from a thermionic emitter are transmitted with high efficiency through the fringe field of the magnet to the ICR ion trap. A thermionic emitter was constructed by modifying the filament with the electron ionization source provided with the conventional FTMS mass spectrometer (Extrel 2000 FT-ICR). Solutions of KOH or KOH/NaOH in methanol were prepared and evaporated onto a rhenium filament. The lifetime of the filament m the high vacuum during the course of these experiments was x 4 h. The filament was replaced when the ion current dropped to an undetectable level as measured on the collector. The filament control was performed as usual in pulsed electron ionization experiments except that the signs of the potentials applied to the filament, grid, and collector were changed to allow for the production of a positive ion beam. The potential difference between the filament and the grid imparted the initial translational energy of the ions. This difference was typically 2–5 V. In their initial experiments, a high-resolution FT-ICR magnitude-mode mass spectrum of K^+ with resolving power of $m/\Delta m = 285{,}000$ was obtained. Mass resolving power in these preliminary experiments is limited both by pressure and by the initially large magnetron radius of the ion packet. In any case, it is important to note that ions may be successfully transmitted through the fringe field of the magnet by use of relatively small voltages on the ion guide.

Quadrupole Ion Trap with an External Thermionic Source By the use of mass-selective instability experiments, the ion trap mass spectrometer serves as both a reactor in which ion/molecule reactions occur, and a mass analyzer for the products of these reactions. The analytical use of ion trap mass spectrometers relies upon the method of ramping the radio frequency (RF) drive potential. The capabilities of ion traps to perform attachment reactions with alkali cations using classical scanning sequences have been exploited. Kinetic studies have shown that, the attachment efficiency is very high, near-collision efficiency, and illustrate how the present method is particularly well suited for ion trap mass spectrometers. Control of the attachment process may be readily performed by the use of a classical ionization sequence in some aspects similar to a tandem MS/MS scanning sequence.

Sablier et al. performed all experiments using a Varian Saturn III GC/MS research ion trap mass spectrometer with a quadrupole ion trap mass spectrometer equipped with an external thermionic ion source to generate the reagent Na^+ ions [86]. Samples were admitted into the ion trap either through a variable leak valve

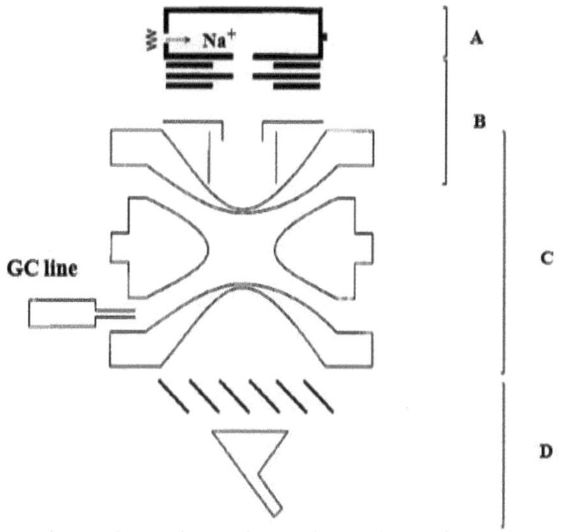

Fig. 6.8 Schematic diagram of the quadrupole ion trap setup, with the external ion source A, the interfacing lens system B, the ion trap manifold C and the detection system D. (Reprinted with permission from Ref. [86])

or via the gas chromatography (GC) line. The kinetic studies were performed by continuous sample introduction with a reaction time period of 1–800 ms. Figure 6.8 shows, in a schematic form, the instrumental set-up used for the experiments. The reagent sodium cations were generated in an external ionization source consisting of a modified EI/CI ionization source. The ionization source and the interface were constructed to inject externally generated ions through the aperture normally used to admit electrons along the axial axis of the trap (traditionally referred as the z-axis) in the upper-end-cap electrode. The emission current of alkali cations was expected to lie in the 10^{-7} to 10^{-6} A range [4, 87]. The emitter was simply mounted in place of the off-axis electron filament in the modified external ion source.

Applications of sodium ion attachment reactions are illustrated by the detection and characterization of explosives and some of their correlated pyrolytic degradation products. Detection limits for phthalate compounds are shown to reach the low ng range of injected samples, without any noticeable difficulties in the full scan mode of acquiring mass spectra. The most important aspect of this study is that it is the first direct demonstration of the applicability of alkali ion attachment reactions using a sodium cation emitter as a novel and sensitive technique of ionization for quadrupole ion trap mass spectrometry. The combination of alkali attachment with an ion trap may well represent a noticeable improvement in the selectivity and sensitivity of available ion trap technology, and it may enlarge the scope of use of this type of mass spectrometer. Application of sodium ion attachment reactions to ion trap as a method of ionization for the detection of explosives offers advantages for the direct determination of molecular weights, with a very low level of fragmentation in the resulting mass spectra. This simplifies the interpretation of the mass spectra and reasonably offers the possibility to distinguish between preionization decomposition and ion fragmentation. Additionally, sodium ion attachment permits characterization of degradation products resulting from pyrolytic processes, in contrast to difficulties in this regard with classical ionization methods. One of the main advantage of sodium attachment reactions for such analysis resides in the fact that conditions of ionization are not pressure dependent compared to conventional CI ionization processes.

6.6 Ion Attachment-TOF System

The time-of-flight mass spectrometry (TOFMS)/ion attachment reaction chamber setup seems to be interesting, especially for the analysis of nonvolatile and thermally labile compounds. Recently, Saito et al. made a unique development [88] in this area. They developed a new type of mass spectrometry based on a time-of-flight mass spectrometer combined with an ion attachment ionization technique (IA-TOF). Although high mass resolution generally requires a bulky TOFMS apparatus, they have attained high mass resolution with a relatively compact tabletop IA-TOF system. In contrast to electron ionization mass spectra, IA-TOF mass spectra are not complicated by peaks due to fragmentation of the molecular ion; the adduct ion formed in IA does not fragment.

The IA-TOF system mainly consists of an IA ion source, a differential pumping system, and a TOF mass spectrometer (Fig. 6.9). The main body of the IA-TOF sys-

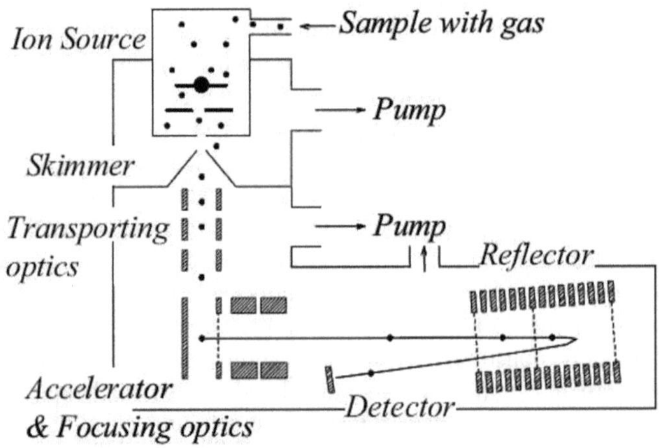

Fig. 6.9 Schematic view of the IA-TOF system. Gas-phase sample molecules are ionized by alkali metal ion attachment and become adduct ions in the IA ion source. The adduct ions are extracted through the aperture into the differential pumping region, pass through the skimmer, and are introduced into the TOF mass spectrometer. Because 100 Pa of N_2 gas is introduced into the IA ion source, a differential pumping system is used to evacuate the gas since the TOF mass spectrometer must be kept in a high vacuum condition. The orthogonal acceleration reflector TOF mass spectrometer realizes mass spectrometry with a high mass resolution over a wide mass range. (Reprinted with permission from Ref. [88]. ©2007, John Wiley and Sons)

tem is a tabletop size (50 cm x 80 cm). The Li^+ ions were produced by electrically heating the Li emitter and were cooled in an ambient third-body gas (N_2,100 Pa). Gas-phase sample molecules M introduced into the IA ion source were ionized by the Li^+ ion attachment and formed $[M + Li]^+$ adduct ions. The Li ions and the adduct ions were extracted with the N_2 gas through a 0.8-mm-diameter aperture and introduced into a differential pumping region. The step in production of the adduct ions is brought about by ion-dipole or ion-induced dipole attraction between the Li^+ ions and a sample molecule. The cross section of the interaction between an ion and a sample molecule varies inversely as the square root of the ion energy [89]. Because Li^+ ions have low energy, the cross section is a few ten times larger than that between a neutral atom and a sample molecule. The residence time of the Li^+ ions in the ion source was about 3.7×10^{-5} s because the mobility of Li^+ ions in N_2 is reduced to about 4 cm^2/V. Using the Wannier equation, the mean kinetic energy and the effective temperature of the Li^+ ions are estimated to be 4.6×10^{-2} eV and 530 K, respectively [90]. The emission current of the Li^+ ions was about 1×10^{-6} A [4, 87]. Therefore, abundant Li^+ ions remained in the ion source long enough to attach to the sample molecules. Preliminary simulations showed that a Li^+ ion collides with N_2 molecules about 1000 times in the ion source. Therefore, a Li^+ ion collides with a sample molecule approximately once in the ion source when the concentration of the sample molecule is 0.1 %. This estimation shows that the emission current of Li^+ ions determines the upper limit for the production of adduct ions in the IA source.

A TOF mass spectrometer mainly consists of a single-stage accelerator, ion-focusing optics, a two-stage reflector, and an ion detector. Ions are accelerated by a pulsed electric field in the accelerator, then pass through a field-free TOF drift region, are reflected by the reflector, and finally reach the detector. The dominant factors that limit the mass resolution of the present IA-TOF system are (a) the spread of the initial positions of the ions in the acceleration region, (b) the spread of the initial velocities (energies) of the ions in the acceleration region, (c) the temporal fluctuation of the electric fields of the ion-focusing optics and the reflector,(d) the fidelity and jitter of the pulsed acceleration electric field, and (e) the temporal response of the detection system. Factors (a) and (b) cause a spread in the kinetic energy after acceleration. In addition, factor (a) also introduces a spread in the flight length in the accelerator. A reflector scheme can effectively compensate for the spread of the TOF due to the spread of the kinetic energy. To further compensate for the effect due to factor (a), a two-stage reflector was used, compensating for both first- and second-order effects of initial position variation. To reduce the effect of factor (b),an orthogonal acceleration configuration with a high acceleration potential was adopted [91]. The orthogonal acceleration configuration eliminates the z component of the initial velocity parallel to the acceleration direction. The reflector further reduces the effect of the residual broadened z component of the initial velocity to a negligible level. In addition, a high acceleration potential U_a of up to 20 kV suppresses turn-around effects in the accelerator. The y component of the initial velocity is also eliminated by the orthogonal acceleration configuration, and the effect of the residual broadened y component is projected as a broadened profile on the ion detector plane. The x component of the initial ion velocity is projected as a shifted

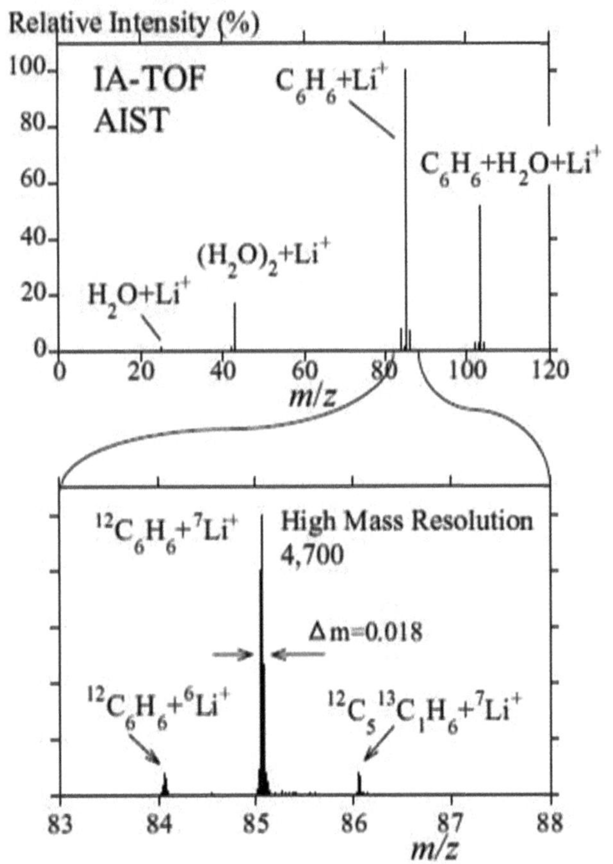

Fig. 6.10 Typical IA-TOF mass spectrum of a C_6H_6 sample diluted by N_2 gas. In IA, an alkali metal ion A^+ attaches to a sample molecule M, and an adduct ion $[M+A]^+$ is produced. The spectrum is fragment-free and provides a high mass resolution of 4700. The high mass resolution yields masses far more accurate than nominal masses, aiding identification of the empirical formula of the sample molecule. (Reprinted with permission from [88]. ©2007, John Wiley and Sons)

and broadened profile on the ion detector plane. However, optimum selection of the operating conditions of the ion-focusing optics and careful alignment of the detector minimize the broadening of the flight time profiles.

To evaluate the performance of this IA-TOF system in analyzing gas-phase organic compounds in ambient air, a benzene (C_6H_6) sample diluted with N_2 gas at atmospheric pressure was analyzed. A C_6H_6 sample at a partial pressure of 0.002% was prepared in a gas bottle and introduced into the IA ion source through a mass flow controller. The gas flow rate was 4 sccm (standard cubic centimeters per minute), and the pressure in the IA ion source was kept at 100 Pa. Figure 6.10 shows a typical IA-TOF mass spectrum of a C_6H_6sample. In IA, a lithium ion Li^+ attaches to a sample molecule M, forming an adduct ion $[M + Li]^+$. No fragments are produced during the IA process. Mass peaks around m/z 85 are derived from $[C_6H_6 + Li]^+$. Mass peaks around m/z 25 and 43 are due to trace water contaminants: $[H_2O + Li]^+$ and $[(H_2O)_2 + Li]^+$, and mass peaks around m/z 103 are due to $[H_2O + C_6H_6 + Li]^+$ cluster ions. Because IA-TOF system yields fragment-free spectra, mixtures of compounds can be directly mass-analyzed without a preseparation step. The low mass cut-off filter due to the pulsed acceleration successfully cut off the signals of the Li^+ ions. Samples originally in the liquid and solid phases were also analyzed by using a direct inlet probe with a heater. Using the Ultramark 1621 mass marker [92], it is demonstrated that the IA-TOF system has a high mass accuracy, $\Delta m/z < 0.02\,\%$, over a wide mass range. The mass accuracy was dominated by the mass resolution.

The newly developed IA-TOF system is capable of measuring mass spectra over a wide mass range in real time, and its wide dynamic range and high signal-to-noise ratio allow measurement of not only dominant components, but also minor components. These results show that the IA-TOF system is a versatile scheme for the real-time analysis of molecules over a wide mass range with a high-mass resolution and a high-mass accuracy.

6.7 Potable IAMS

Ion attachment mass spectrometry (IAMS) has proven to be a unique method that complements electron-impact ionization mass spectrometry for the determination of components in chemical processes and environments, such as microwave discharge plasma. The identification of intermediate free radicals and other species in chemical reactions is particularly challenging. One of the greatest advantages of IAMS is that it can be used to directly analyze gaseous compounds. The features that allow this system to detect the intermediate free radicals (refer to Sect. 5.2.4) and novel molecular species produced in various plasmas have been extensively explored (Sect. 5.2.7), and IAMS techniques can be used to identify and quantify compounds and mixtures under plasma and pyrolysis (Sect. 6.4) conditions. To extend the ion attachment technique to ion trap mass spectrometry (Sect. 6.5) or time-of-flight mass spectrometry (Sect. 6.6) has been realized.

However, the commercial IAMS apparatus is too cumbersome for use in field analysis of air samples. Unfortunately, the commercial apparatus is not practical for the study of atmospheric environments. Therefore, to develop a new, compact and fieldable IAMS system by eliminating the differential pumping stage is desirable. The primary objective for the development was to design a system that can be easily transported to the field and that can detect any chemical species at atmospheric pressure on a real-time basis. A single turbomolecular pump is employed as the vacuum system to fill the basic requirements for vacuum conditions; this is simple and cost effective. The optimal operation parameters for its use, the details and operation of the capillary leak inlet, and the system's analytical power in some preliminary applications are described also in this section.

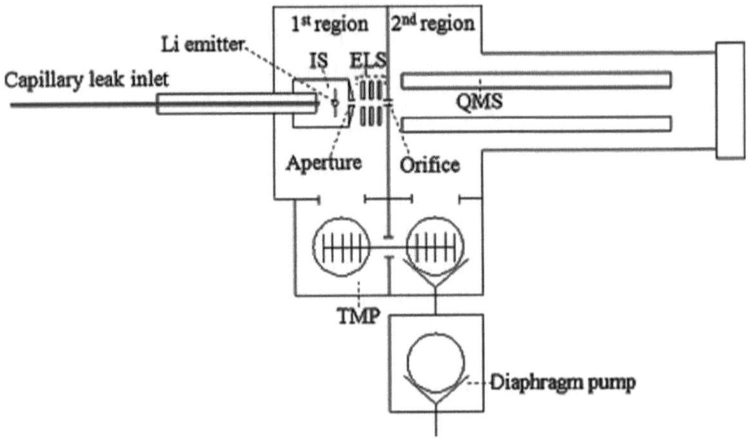

Fig. 6.11 A schematic drawing of new ion attachment mass spectrometer with the capillary sample inlet and a vacuum envelope with a wall separating ELS chamber from mass analyzer chamber. This capillary inlet is fixed on the front flange of the vacuum envelope. IC ionization chamber, ELS electrostatic lens system, QMS quadrupole mass spectrometer, TMP turbomolecular pump, RP rotary pump. The vacuum envelope is pumped by a single 230 L sec^{-1} turbomolecular pump with two ISO-100 inlet flanges (Pfeiffer-Vacuum TMH 261-250-010) plus a 250 mL min^{-1} rotary pump. Ionization chamber is closed-type, with a 1 mm ϕ aperture through which ionic species are passed. The typical operating conditions are: IS pressure with nitrogen gas used as a buffer gas, 40 Pa; pressure of ELS chamber, 8×10^{-2} Pa, pressure of the QMS chamber, 8×10^{-4} Pa. (Reprinted with permission from Ref. [93]. ©2012, Springer)

A compact ion attachment mass spectrometer was designed that is simple and small and fulfills all the basic requirements for IAMS: the system can be used to obtain only molecular ions by detecting any chemical species in real time [93]. This custom-made apparatus (Fig. 6.11) consists of a Li^+ ion attachment ion source into which a stream of gas from a capillary leak inlet is directed, an electrostatic lens system (ELS), and a quadrupole mass spectrometer and detector, all of which are installed in a vacuum-separated envelope. The system employs a single turbomolecular pump on the vacuum envelope instead of a differential pumping system.

Performance characteristics was investigated. A test sample of toluene was used to assess the system's linear response range and minimum detectable amount. Toluene was chosen because it is the most ubiquitous volatile organic compound in the urban atmosphere [94]. The signal response for toluene was linear over the chosen range of 7×10^{-11} g/s to 8×10^{-8} g/s, with a dynamic range greater than 10^4. A plot of four introduction rates of toluene against their corresponding peak heights produced an essentially straight line up to 8×10^{-8} g/s, where more than 5% of the total Li^+ reactant ions were used for the attachment. Taking the actual noise level (1.5×10^{-13} A) of the system as the ultimate limit on detection, the minimum detectable amount was calculated to be 1.3×10^{-12} g/s, on the assumption that the capability of the ion detection system in the electrometer is 4.5×10^{-13} A, at the sig-

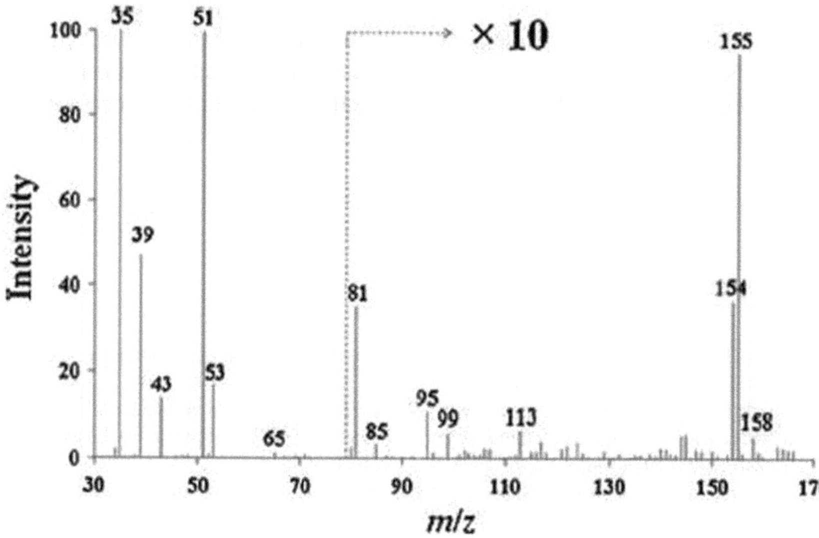

Fig. 6.12 Typical ion attachment mass spectrum of laboratory air. A stream of laboratory air was directed into the IAMS through a capillary leak inlet. (Reprinted with permission from Ref. [93]. ©2012, Springer)

nal-to-noise ratio of 3. Therefore, the minimum detectable amount of 1.3×10^{-12} g/s corresponds to a minimum detectable concentration of 2.1 ppb (vol/vol) if the reference compound enters the IAMS diluted by 30 mL/min of N_2 gas. For illustrative purposes, the system on laboratory air was tested. A typical mass spectrum of the air showed a mass peak at m/z 43 due to Li^+ ion adducts of water clusters, the intensity of which varies with relative humidity (Fig. 6.12). The other peaks, at m/z 35, 39, 51, 53,and 65, correspond to the major components of the air. The major components of the air, N_2, O_2, and CO_2, were identified by the presence of Li^+ ion adduct mass peaks at m/z 35, 39, and 51, respectively. The mass peak at m/z 53 was assigned to $N_2Li^+H_2O$. The mass peak at m/z 65 is due to Li^+ ion attached to acetone, which can be attributed to the presence of acetone (used as a solvent) in the room, and the mass peaks at m/z 85, 99, and 113 may be due to $C_6H_6Li^+$, $C_7H_8Li^+$, and $C_8H_{10}Li^+$, respectively. The remaining persistent peaks at m/z 81, 95, 154, 155, and 158 have not yet been assigned.

The newly developed instrument has several important features. It accepts high capacity direct introduction of samples, operates with the atmospheric sampling device, and allows easy coupling of various sample introduction sources to the mass spectrometer. It permits real-time detection of chemical species, possibly including radical intermediates and easy identification of compounds via the generation of ions that do not fragment. In addition, the capillary leak inlet system is simple and adaptable. In the near future, a large increase is expected in the use of IAMS for environmental and food chemistry research, especially in cases where direct air sample introduction is preferable. Potential applications include emission control and monitoring of volatile organic compounds in urban and rural environments, emissions from construction material and furniture, and emissions from industrial facilities and industrial fermentation and food production processes. Further uses also include the monitoring of catalytic processes, cigarette smoke, and breath.

References

1. Hodges RV, Beauchamp JL. Application of alkali Ions in chemical ionization mass spectrometry. Anal Chem. 1976;48:825–8.
2. Fujii T. Ion attachment mass spectrometry. In Gross M. editor. Encyclopedia of mass spectrometry. Vol. 6. Ionization methods. America Society for Mass Spectrometry. Elsevier: Amsterdam; 2007. pp. 327–34.
3. Sablier M, Fujii T. Mass spectrometry of free radicals: a methodological overview. In: Webb G, editor. Progress in chemistry, Sect. C (Phys Chem). New York: Royal Society of Chemistry; 2005. pp 53–99.
4. Fujii T. Quadrupole mass spectrometry in combination with lithium ion attachment for sampling at atmospheric pressure: Possible coupling to a superfluid critical chromatography. Anal Chem. 1992;64:775–8.
5. Sinha MP. Laser-induced volatilization and ionization of microparticles. Rev Sci Instrum. 1984;55:886–891.
6. McKeown PJ, Johnston MV, Murphy DM. Online single-particle analysis by laser desorption mass spectrometry. Anal Chem. 1991;63:2069–73.
7. Carson PG, Johnston MV, Wexler AS. Laser desorption/ionization of ultrafine aerosol particles. Rapid Commun Mass Spectrom. 1997;11:993–6.
8. Canagaratna MR, Jayne JT, Jimenez JL, Allan JD, Alfarra MR, Zhang Q, Onasch TB, Drewnick F, Coe H, Middlebrook A, Delia A, Williams LR, Trimborn AM, Northway MJ, DeCarlo PF, Kolb CE, Davidovits P, Worsnop DR. Chemical and microphysical characterization of ambient aerosols with the aerodyne aerosol mass spectrometer. Mass Spectrom Rev. 2007;26:185–222.
9. Jayne JT, Leard DC, Zhang X, Davidovits P, Smith KA, Kolb CE, Worsnop DR. Development of an aerosol mass spectrometer for size and composition analysis of submicron particles. Aerosol Sci Technol. 2000;33:49–70.
10. Drewnick F, Hings SS, DeCarlo PF, Jayne JT, Gonin M, Fuhrer K, Weimer S, Jimenez JL, Demerjian KL, Borrmann S, Worsnop DR. A new Time-of-Flight Aerosol Mass Spectrometer (ToF-AMS)–instrument description and first field deployment. Aerosol Sci Technol. 2005;39:637–58.
11. DeCarlo PF, Kimmel JR, Trimborn A, Northway MJ, Jayne JT, Aiken AC, Gonin M, Fuhrer K, Horvath T, Docherty KS, Worsnop DR, Jimenez JL. Field-deployable, high-resolution, time-of-flight aerosol mass spectrometer. Anal Chem. 2006;78:8281–9.
12. Northway MJ, Jayne JT, Toohey DW, Canagaratna MR, Trimborn A, Akiyama K, Shimono A, Jimenez JL, DeCarlo PF, Wilson KR, Worsnop DR. Demonstration of a VUV lamp photoionization source for improved organic speciation in an aerosol mass spectrometer. Aerosol Sci Technol. 2007;41:828–39.
13. Bombick D, Pinkston JD, Allison J. Potassium ion chemical ionization and other uses of an alkali thermionic emitter in mass spectrometry. Anal Chem. 1984;56:396–402.
14. Fujii T, Ogura M, Jimba J. Chemical ionization mass spectrometry with use of alkali ion attachment to molecule. Anal Chem. 1989;61:1026–9.
15. Eiceman GA, Karpas Z. Ion mobility spectrometry. 2nd ed. Boca Raton: CRC Press; 2005. pp. 370.

16. Wilkins CL, Trimpin S. Ion mobility spectrometry—mass spectrometry: theory and applications. Boca Raton: CRC Press; 2010. pp. 374.
17. Haq FU. Construction of thermionic alkali-ion sources. J Phys E: Sci Instrum. 1986;19:275–6.
18. Tan TL, Ong PP, Fong TM, Soo KA. Studies of rubidium aluminosilicates Rb ions as thermionic emitters. Int J Mass Spectrom Ion Process. 1994;34:221–8.
19. Tabrizchi M. Thermal ionization ion mobility spectrometry. Anal Chem. 2003, 75, 3101–6.
20. Tabrizchi M, Hosseini ZS. An alkali ion source based on graphite intercalation compounds for ion mobility spectrometry. Meas Sci Technol. 2008;19(075603):6.
21. Brown M E. Introduction to thermal analysis: techniques and applications (hot topics in thermal analysis and calorimetry). 2nd ed. Dordrecht: Springer; 2013. p. 280.
22. Haines PJ. Thermal methods of analysis: principles, applications and problems. Dordrecht: Springer; 1995. p. 300.
23. Tsuge S, Ohtani H, Watanabe C. Pyrolysis—GC/MS data book of synthetic polymers: pyrograms, thermograms and ms of pyrolyzates. Oxford: Elsevie; 2011. p. 420.
24. Materazzi S, Gentili A, Curini R. Applications of evolved gas analysis. Part 2: EGA by mass spectrometry. Talanta. 2006;69:781–94.
25. Materazzia S, Vecchiob S. Evolved gas analysis by mass spectrometry. Appl Spectrosc Rev. 2011;46:261–340.
26. Takahashi S, Tsukagoshi M, Kitahara Y, Juhasz M, Fujii T. Design and performance of an evolved gas analysis ion attachment mass spectrometer. Rapid Commun Mass Spectrom. 2010;24:2625–30.
27. Kitahara Y, Takahashi S, Kuramoto N, Sala M, Tsugoshi T, Sablier M, Fujii T. Ion attachment mass spectrometry combined with infrared image furnace for thermal analysis: evolved gas analysis studies. Anal Chem. 2009;81:3155–8.
28. Fujii T. A new method for thermal analysis: ion-attachment mass spectrometry (IAMS). J Therm Anal Calorim. 2012;110:17–25.
29. Tsugoshi T, Nagaoka T, Nakamura M, Shiokawa Y, Watari K. Application of ion attachment mass spectrometry to evolved gas analysis for in situ monitoring of porous ceramic processing. Anal Chem. 2006;78:2366–9
30. Juhasz M, Kitahara Y, Fujii T. Thermal decomposition of vitamin C: an evolved gas analysis-ion attachment mass spectrometry study. Food Chem. 2011;129:546–50.
31. Smith EA, Oehme FW. Acrylamide and polyacrylamide: a review of production, use, environmental fate and neurotoxicity. Rev Environ Health. 1991;9:215–28.
32. Tareke E, Rydberg P, Karlsson P, Eriksson S, Toernqvist M. Analysis of acrylamide, a carcinogen formed in heated foodstuffs. J Agric Food Chem. 2002;50:4998–5006.
33. Tutas M, Saglam M, Yuksel M. Pyrolysis product of polyacrylamide by pyrolysis–gas chromatography. J Anal Appl Pyrolysis. 1991;22:129–37.
34. Smith EA, Prues SL, Oehme FW. Environmental degradation of polyacrylamides. I. Effects of artificial environmental conditions: Temperature, Light, and pH. Ecotoxicol. Environ. Saf. 1996;35:121–35.
35. Tutas M, Saglam M, Yuksel M, Guler C. Investigation of the thermal decomposition kinetics of polyacrylamide using a dynamic TG technique. Thermochim Acta. 1987;111:121–6.
36. Kitahara Y, Okuyama K, Ozawa K, Suga T, Takahashi S, Fujii T. Thermal decomposition of acrylamide from polyacrylamide: time-resolved pyrolysis with ion-attachment mass spectrometry. J Therm Anal Calorim. 2012;110:423–9.
37. Knumann R, Bockhorn H. Investigation of the kinetics of pyrolysisof PVC by TG-MS-analysis. Combust Sci Technol. 1994;101:285–99.
38. Vernin G, Chakib S, Rogacheva SM, Obretenov TD, Parkanyi C. Thermal decomposition of ascorbic acid. Carbohyd Res. 1998;305:1–15.
39. Juhász M, Takahashi S, Kitahara Y, Fujii T. Thermal decomposition of pyridoxine: an evolved gas analysis-ion attachment mass spectrometry study. Rapid Commun Mass Spectrom. 2012;26:759–64.
40. Gregory III JF, Hiner ME. Thermal stability of vitamin B_6 compounds in liquid model food systems. J Food Sci. 1983;48:1323–7.

41. Lešková E, Kubíková J, Kováčiková E, Košická M, Porubská J, Holčíková K. Vitamin losses: retention during heat treatment and continual changes expressed by mathematical models. J Food Comp Anal. 2006;19: 252–76.
42. Takahashi S, Suga T, Kitahara Y, Fujii T. Evolved gas analysis of $Ti(C_5H_5)_2Cl_2$ by means of Li^+ Ion attachment mass spectrometry. J Phys Chem A 2012;116:865–9.
43. Guo M, Sun H, McArdle HJ, Gambling L, Sadler PJ. Ti(IV) uptake and release by human serum transferrin and recognition of Ti(IV)-transferrin by cancer cells: understanding the mechanism of action of the anticancer drug titanocene dichloride. Biochemistry 2000;39:10023–33.
44. Tel'noi VI, Rabinovich IB. Thermochemistry of organic compounds of transition metals. Russ Chem Rev. 1977;46: 689–705.
45. Sablier M, Sala M, Kitahara Y, Takahashi S, Fujii T. Influence of copper chloride for the formation of aromatic compounds during polyethylene pyrolysis. J Anal Appl Pyr. 2010;89:178–82.
46. Lin CH, Lin HY, Liao WZ, Dai SH. A. novel chemical recycling of polycarbonate (PC) waste into bis-hydroxyalkyl ethers of bisphenol A for use as PU raw materials. Green Chem. 2007;9:38–43.
47. Sala M, Kitahara Y, Takahashi S, Fujii T. Effect of atmosphere and catalyst on reducing bisphenol A (BPA) Emission during thermal degradation of polycarbonate. Chemosphere. 2010;78:42–5.
48. vomSaal FS, Hughes C. An extensive new literature concerning low dose effects of bisphenol A shows the need for a new risk assessment, Environ Health Perspect. 2005;113:926–33.
49. Vandenberg LN, Maffini MV, Sonnenschein C, Rubin BS, Soto AM. Bisphenol-A and the great divide: a review of controversies in the field of endocrine disruption. Endocr. Rev. 2009;30:75–95.
50. Kitahara Y, Takahashi S, Tsukagoshi M, Fujii T. Formation of bisphenol A by thermal degradation of poly(bisphenol A carbonate). Chemosphere. 2010;80:1281–4.
51. Arulmozhiraja S, Coote ML, Kitahara Y, Juhász M, Fujii T. Is Bisphenol A biradical formed in the pyrolysis of polycarbonate?. J Phys Chem A. 2011;115:4874–81.
52. Lippert B. Cisplatin: chemistry and biochemistry of a leading anticancer drug. Weinheim: Wiley-VCH; 1999. p. 567.
53. Takahashi S, Kitahara Y, Nakamura M, Shiokawa Y, Fujii T. Temperature-resolved thermal analysis of cisplatin by means of Li^+ ion attachment mass spectrometry. Phys Chem Chem Phys. 2010;12:3910–3.
54. Juhász M, Takahashi S, Fujii T. Temperature-resolved thermal analysis of cisplatin by evolved gas analysis–mass spectrometry. J Anal Appl Pyr. 2011;91:114–8.
55. Qin M, Mitchell JD, Vogl O. Oriental lacquer. 10. The South East Asian lacquer. J Macromol Sci A: Pure Appl Chem. 1996;33:1791–1803.
56. Webb M. Lacquer: technology and conservation (conservation and museology series). London: Butterworth-Heinemann; 2000; p. 200.
57. Snyder DM. An overview of oriental lacquer: art and chemistry of the original high-tech coating. J Chem Educ. 1989;66:977–80.
58. Occolowitz JL. Mass spectrometry of naturally occurring alkenyl phenols and their derivatives. Anal Chem. 1964;36:2177–81.
59. Niimura N, Miyakoshi T, Onodera J, Higuchi T. Identification of ancient lacquer film using two-stage pyrolysis-gas chromatography/mass spectrometry. Archaeometry. 1999; 41:137–49.
60. Honda T, Lu R, Kitano N, Kamiya Y, Miyakoshi T. Applied analysis and identification of ancient lacquer based on pyrolysis-gas chromatoraphy/mass spectrometry. J. Appl. Polym. Sci. 2010;118:897–901.
61. Lu R, Ma X-M, Kamiya Y, Honda T, Kamiya Y, Okamoto A, Miyakoshi T. Identification of Ryukyu lacquerware by pyrolysis-gas chromatography/mass spectrometry. J Anal Appl Pyrolysis. 2007;80:101–10.

62. Kumanotani J. Enzyme catalyzed durable and authentic oriental lacquer: a natural micro-gel-printable coating by polysaccharide–glycoprotein–phenolic lipid complexes. J Prog Org Coat. 1998;34:135–46.
63. Tsukagoshi M, Kitahara Y, Takahashi S, Tsugoshi T, Fujii T. Characterization of Japanese lacquer liquid and films by means of evolved gas analysis-ion attachment mass spectrometry. Anal Methods. 2011;3:1943–7.
64. Tsukagoshi M, Kitahara Y, Takahashi S, Fujii T. Pyrolysis analysis of Japanese lacquer films: direct probe–Li^+ ion attachment mass spectrometry versus pyrolysis/gas chromatography/mass spectrometry. J Anal Appl Pyr. 2012;95:156–63.
65. Halonen L. 30th International Symposium on free radicals. J Phys Chem A. 2010;114:4697–5034.
66. Perkins MJ. Radical chemistry: the fundamentals. New York: Oxford University Press; 2001. p. 92.
67. Westerberg LM, Pfaffli P, Sundholm F. Detection of free radicals during processing of polyethylene and polystyrene plastics. Am Ind Hygiene Assoc J. 1982;43:544–6.
68. Kitahara Y, Takahashi S, Tsukagoshi M, Fujii T. Free radicals produced from thermally-irradiated polyethylene polymers: an ion attachment mass spectrometric study. Chem Phys Lett. 2011;507:226–8.
69. Crabtree RH. The organometallic chemistry of the transition metals. 6th ed. New York: John Wiley, 2014. p. 520.
70. Rastrelli F, Bagno A. Predicting the NMR spectra of paramagnetic molecules by DFT: application to organic free radicals and transition-metal complexes. Chem-Eur. J. 2009;15:7990–8004.
71. Mach K, Novakova J, Raynor JB. Electron spin resonance spectroscopy of pentacarbonylmanganese ($Mn(CO)_5$) radicals generated in the gas phase thermolysis of decacarbonyldimanganese ($Mn_2(CO)_{10}$). J Organomet Chem. 1992;439:341–5.
72. Kitahara Y, Fujii T. Evolved gas analysis—ion attachment mass spectrometric observation of $Mn(CO)_5$ and $Mn_2(CO)_9$ radicals produced by $Mn_2(CO)_{10}$ pyrolysis. Res Chem Intermed. 2012;38:233–9.
73. Kitahara Y, Fujii T. Evolved gas analysis–ion attachment mass spectrometric analysis of decacarbonyldimanganese pyrolysis. J Therm Anal Calorim. 2012;11:431–5.
74. Lanigan RS. Final report on the safety assessment of PEG-2, -3, -5, -10, -15, and -20 Cocamine. Int J Toxicol. 1999;18:43–50.
75. Kitahara Y, Takahashi S, Fujii T. Thermal analysis of polyethylene glycol: evolved gas analysis with ion attachment mass spectrometry. Chemosphere. 2012;88:663–9.
76. Frey MM, Hutterli MA. Contrasting atmospheric boundary layer chemistry of methylhydroperoxide (CH_3OOH) and hydrogen peroxide (H_2O_2) above polar snow. Atmos. Chem Phys. 2009;9:3261–76.
77. Sauer F, Limbach S, Moortgat GK. Measurements of hydrogen peroxide and individual organic peroxides in the marine troposphere. Atmos Environ. 1997;31:1173–84.
78. Hua W, Chen ZM, et al. Atmospheric hydrogen peroxide and organic hydroperoxides during PRIDE-PRD'06, China: their concentration, formation mechanism and contribution to secondary aerosols. Atmos. Chem Phys. 2008;8:6755–73.
79. McMahon TB, Beauchamp JL. A versatile trapped ion cell for ion cyclotron resonance spectroscopy. Rev Sci Instrum. 1972;43:509–12.
80. Wieting RD, Staley RH, Beauchamp JL. Reactions of Alkali ions with organic molecules in the gas phase. Low energy pathways for carbonium ion formation and novel methods for generating alkali ion complexes with π- and n-donor bases. J Am Chem Soc. 1975;97:924–6.
81. Henry KD, Quinn JP, McLafferty FW. High-resolution ESI mass spectra of large molecules. J Am Chem Soc. 1991;113:5447–49.
82. Watson CH, Kruppa G, Wronka J, Lauklen FH. Continuous-flow fast-atom bombardment on an external-ion-source fourier transform ion cyclotron resonance mass spectrometer. Rapid Commun Mass Spectrom. 1991;5: 249–51.

83. Barshick CM, Eyler JR. A glow discharge ion source with Fourier transform ion cyclotron mass spectrometric detection. J Am Soc Mass Spectrom. 1992;3:122–7.
84. McIver TT Jr, Hunter RL, Bowers WD. Coupling a quadrupole mass-spectrometer and a fourier-transform mass-spectrometer. Int J Mass Spectrom Ion Proc. 1985;64:67–77.
85. Limbach PA, Marshall AG. An electrostatic ion guide for efficient transmission of low energy externally formed ions into a Fourier transform ion cyclotron resonance mass spectrometer. Int J Mass Spectrom Ion Proc. 1993;125:135–43.
86. Faye T, Brunot A, Sablier M, Tabet JC, Fujii T. Sodium ion attachment reactions in an ion trap mass spectrometer, Rapid Commun Mass Spectrom. 2000;14:1066–73.
87. Dzidic I, Kebarle P. Hydration of the alkali ions in the gas phase. Enthalpies and entropies of reactions $M^+(H_2O)_{n-1}+H_2O=M^+(H_2O)_n$. J Phys Chem. 1970;74:1466–74.
88. Saito N, Nanjyo J, Taneda Y, Shiokawa Y, Tanimoto M. Development of a tabletop time-of-flight mass spectrometer with an ion attachment ionization technique. Rapid Commun Mass Spectrom. 2007;21:2654–62.
89. Gioumousis G, Stevenson DP. Reactions of gaseous molecule ions with gaseous molecules. V. Theory, J Chem Phys. 1958;29:294–9.
90. McFarland M, Albritton DL, Fehsenfeld FC, Ferguson EE, Schmeltekopf AL. Flow-drift technique for ion mobility and ion-molecule reaction rate constant measurements. I. Apparatus and mobility measurements. J Chem Phys. 1973;59:6610–19.
91. Guilhaus, M.; Selby, D; Mlynski, V. Orthogonal acceleration Time-of-Flight Mass Spectrometry. Mass Spectrom Rev. 2000, 19, 65–107.
92. Moini M. Ultramark 1621 as a calibration/reference compound for mass spectrometry. Rapid Commun Mass Spectrom. 1994;8:711–4.
93. Takahashi S, Nakamura M, Fujii T. Design and performance of a compact Li^+ ion attachment mass spectrometry system with an atmospheric sampling device. J Am Soc Mass Spectrom. 2012;23:547–52.
94. Yamamoto N, Okayasu H, Murayama S, Mori S, Hunahashi K, Suzuki K. Measurement of volatile organic compounds in the urban atmosphere of Yokohama, Japan, by an automated gas chromatographic system. Atmos Environ. 2000;34:4441–6.

Chapter 7
Cationization Mass Spectrometry for Condensed-Phase Samples

W. M. A. Niessen

7.1 Introduction

Electron ionization (EI) is the gold standard in analyte ionization by mass spectrometry (MS). As an alternative to EI, a wide variety of alternative ionization techniques have been discovered and developed for use in MS. A characteristic feature of many of these ionization techniques is that often even-electron ions are generated as a result of ion-attachment processes, whereas in EI odd-electron ions are generated due to expelling of an electron. In contrast to EI, in these ion-attachment processes generally only very little energy transfer and, therefore, very little internal energy effects are involved. As a result, ions related to the intact analyte molecule are generated with little fragment ions, again in contrast to EI, where several electron volts (eVs) of energy is transferred to the analyte ion during the ionization process, resulting in extensive in-source fragmentation. The term "soft ionization techniques" is frequently used. In positive-ion mode, ion attachment mostly involves the generation of protonated molecules ($[M+H]^+$) by attachment of a proton at appropriate sites in the molecule with (relatively) high proton affinity. However, attachment of other cations, e.g., Alkali$^+$-ions as well as other metal ions, may also be observed. In the context of this book, the discussion primarily focusses at Alkali$^+$-cationization, although some attention is also paid to cationization by other metal ions as well as to anion attachment.

Soft ionization can be achieved in the gas phase, from the liquid phase, or from the solid phase. Gas-phase cationization techniques have been discussed in previous chapters, especially in Chaps. 5 and 6. Typical examples of liquid-phase ionization techniques are thermospray ionization (TSI) and electrospray ionization (ESI), which are especially important in combining liquid chromatography and mass spectrometry (LC–MS). A wide variety of solid-phase ionization or desorption/ionization techniques have been developed over the years, including field

W. M. A. Niessen (✉)
hyphen MassSpec, de Wetstraat 8, 2332 XT Leiden, The Netherlands
e-mail: mail@hyphenms.nl

T. Fujii (ed.), *Ion/Molecule Attachment Reactions: Mass Spectrometry*,
DOI 10.1007/978-1-4899-7588-1_7

desorption (FDI), fast-atom bombardment (FAB), and matrix-assisted laser desorption ionization (MALDI). Principles and application of these condensed-phase ionization techniques are discussed in this chapter, obviously with special focus to $Alkali^+$-cationization processes.

For most of the soft ionization techniques, there is a free choice in the type of mass spectrometer used, thus quadrupole, ion-trap, sector, time-of-flight (TOF), Orbitrap, and Fourier-transform ion-cyclotron mass spectrometers (FT-ICR-MS) have been applied (see Sects. 4.3 and 4.4). For structure elucidation, the use of a soft ionization technique in most cases implies the application of tandem mass spectrometry (MS–MS or MS^n, see Sects. 4.3 and 4.4). A wide variety of MS–MS instruments have been developed, including tandem-quadrupole (TQ) instruments, TOF–TOF-MS instruments, and ion-trap instruments, but also a variety of hybrid instruments, e.g., the quadrupole–time-of-flight (Q–TOF) instrument. Although several ion dissociation techniques have been developed, collision-induced dissociation (CID) by far is the most widely applied technique (Sect. 4.3.2). Most MS–MS instruments only provide low-energy CID (< 100 eV) [1], whereas sector and TOF–TOF instruments also provide high-energy CID (> 1 keV) [2]. Some differences observed with either low-energy or high-energy CID are outlined in discussion of applications of cationization mass spectrometry in Sect. 7.5.

7.2 Condensed-Phase Ionization Techniques

A wide variety of soft ionization methods is available, some of which are extensively used such ESI and MALDI, while some others are only applied for specific applications such as FDI and secondary ion mass spectrometry (SIMS). Some of the older techniques can be considered to be outdated and obsolete such as TSI, ^{252}Cf plasma desorption ionization (PDI), and FAB, because more powerful and/or more user-friendly alternatives have been developed. The various condensed-phase ionization techniques are briefly discussed in this section. The use of metal-ion cationization in chemical and desorption ionization was already reviewed in 1992 [3].

7.2.1 Field Desorption Ionization

Field (desorption) ionization (FDI) was first described by Beckey in 1969 [4]. In FDI, the sample solution is deposited on a 10-μm-diameter FDI emitter, which is activated to provide for whiskers or microneedles on the surface [5, 6]. The emitter is kept at a high potential (> 5 kV) in the high-vacuum ion source, mostly of a sector instrument (see Fig. 7.1). By passing through a current, slow heating of the emitter is achieved. As heating of the emitter continues, nonvolatile analytes can be desorbed and ionized by various mechanisms. High local electrical fields at the

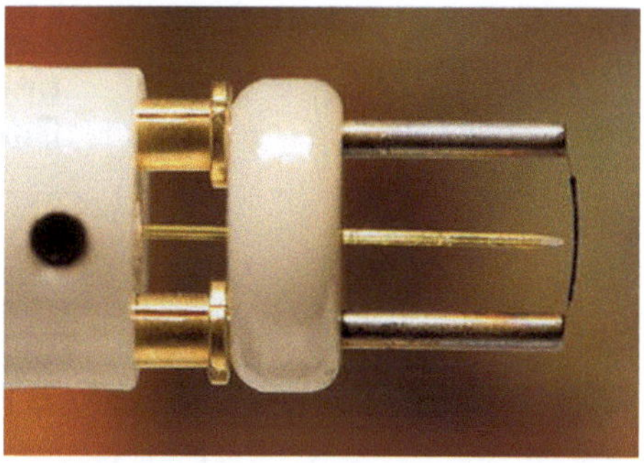

Fig. 7.1 Picture of a liquid injection field desorption ionization probe tip (LIFDI-MS). (Reprinted from [9] with kind permission from Springer Science and Business Media)

tip of the microneedle enable electron tunnelling from the sample molecules into the emitter (field ionization). Desorption of ions, i.e., $M^{+\bullet}$, is also possible under these conditions. In addition, preformed ions due proton or alkali-metal-ion attachment may be extracted from the condensed phase as a result of the high local field strength. As a result, $M^{+\bullet}$ and cationized molecules are observed in positive-ion FDI spectra with little fragmentation. If fragment ions are observed, they are often thermally induced as a result of the increasing emitter current (and temperature) during the experiment.

Because of the experimental difficulty of the technique and because more user-friendly and to some extent more powerful alternatives have become available, FDI is not frequently applied anymore, except for some specific applications. In this respect, an important development is liquid injection field desorption ionization (LIFDI), which enables sample application to the emitter without breaking the vacuum (see Fig. 7.1) [7, 8]. The specific applications where FDI and LIFDI are still applied comprise the analysis of some organometallic compounds [9, 10], ionic liquids [11], and compound classes, such as (cyclo)paraffins, aromatic hydrocarbons, and nonpolar sulfur compounds (thiophenes) [7, 12–14], not readily amenable to ESI or MALDI. For such nonpolar analytes, mainly molecular ions $M^{+\bullet}$ are observed, whereas for some more polar compounds, $[M+H]^+$ and/or sodiated molecules ($[M+Na]^+$) may be observed, e.g., for glycosides (Sect. 7.5.2), lipids (Sect. 7.5.4), and peptides (Sect. 7.5.5). A detailed overview on technology and applications of FDI-MS was provided by Schulten et al. [15, 16].

7.2.2 *Fast-Atom Bombardment Ionization*

FAB was introduced in 1981 by Barber et al. [17, 18]. In FAB [17–20], the analyte of interest is dissolved in an appropriate matrix solvent, such as glycerol, diethanolamine and other rather polar solvents with low vapor pressure. The solution is applied as a thin film onto a metal target, which subsequently is brought in a beam of high-energy particles (see Fig. 7.2). FAB differed from the already existing SIMS method [21] in two experimental aspects, i.e., the use of a beam of atoms instead of a beam of ions as primary particles, and the use of a liquid matrix for dissolving the

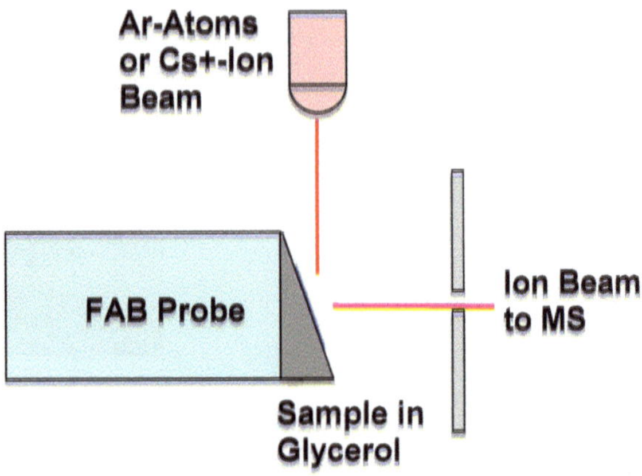

Fig. 7.2 Schematic diagram of an ion source for fast-atom bombardment (FAB-MS)

analyte molecules. Subsequently, it was shown that the use of a liquid matrix is the major experimental difference. Identical spectra can be produced for a given analyte dissolved in a particular matrix solvent by bombardment with either fast atoms (FAB) or fast ions (liquid secondary ion mass spectrometry, LSIMS). An advantage of LSIMS is the ability to focus the primary beam, e.g., from a Cs^+ ion gun.

Three ionization mechanisms appear to be important in FAB, i.e., desorption of preformed ions by energy transfer upon particle impact, desolvation of preformed ions in the splash droplets resulting from disruption of the liquid layer upon particle impact, and gas-phase ion-molecule reactions in the interface layer of the liquid and the vacuum (the so-called selvedge). The efficiency of analyte ionization by FAB is clearly related to the liquid matrix. The intensity of the secondary ion beam, its duration in time, and the ability to ionize highly labile, polar, ionic, and high molecular-mass analytes can all to a large extent be attributed to the matrix. It determines the success or failure of the FAB ionization. The empirical observations of matrix influence on secondary ion currents can be rationalized by studying the physical properties of the matrix solvents such as viscosity, solubility of the analyte, protolytic properties, surface activity, and dielectric constants [22]. Widely used matrix solvents are glycerol, thioglycerol, tetraglyme, and ethanolamine. Thioglycerol is more volatile than glycerol, but shows better protolytic properties in the production of protonated molecules; diethanolamine is a suitable proton acceptor in negative-ion FAB. Mixtures of various solvents may be used to balance various properties within the matrix. Adjustment of the pH, addition of alkali-metal ions, and/or addition of surfactants may have favorable effects on the ionization efficiency. The matrix is assumed to play a major role in the solution chemistry leading to preformed ions, the renewal of the surface layer upon desorption of the preformed ions, and the reduction of the energy required for desorption by solvating and separating the analytes on the surface. Relatively hydrophobic analytes are more readily desorbed than hydrophilic ones; obviously, pH, co-solvents and sample interferences are important in this respect because hydrophobicity and solubility must be balanced.

In FAB spectra, mostly even-electron species are detected formed by protonation or cationization with alkali-metal ions in positive-ion mode and by deprotonation in negative-ion mode. The (few) fragment ions observed can be explained by elimination of neutral molecules from the initial even-electron species. Next to analyte-related peaks, (abundant) background peaks are observed, which can be attributed to protonated matrix clusters, e.g., $[(\text{Glycerol})_n+\text{H}]^+$ with $n=1$–10. A high chemical noise, also called peak-at-every-mass or incoherent fragmentation, is a characteristic feature of FAB, the origin of which is not well understood.

Until the introduction of ESI-MS in the mid-1990s, FAB-MS has been widely applied in the analysis of a wide variety of biomolecules. Early data involving MS–MS of even-electron molecules, both $[\text{M}+\text{H}]^+$ and $[\text{M}+\text{Alkali}]^+$, greatly contributed to the structure elucidation strategies that are still applied today, although other ionization techniques, i.e., ESI-MS and MALDI-MS, are used to generated the primary precursor ions. This becomes clear when reviewing the importance of fragmenting FAB-generated $[\text{M}+\text{Alkali}]^+$-ions for compound classes as antibiotics (Sect. 7.5.1), glycosides (Sect. 7.5.2), oligosaccharides and glycans (Sect. 7.5.3), lipids (Sect. 7.5.4), and peptides (Sect. 7.5.5).

7.2.3 Matrix-Assisted Laser Desorption Ionization

Matrix-assisted laser desorption ionization (MALDI) was simultaneously introduced in 1988 by two research groups [23, 24]. In 2002, Koichi Tanaka, together with John Fenn (see Sect. 7.2.6) and the NMR spectroscopist Kurt Wüthrich, received the Nobel Prize for Chemistry for his ground-breaking work in the identification and structure elucidation of biological macromolecules.

In a typical MALDI experiment (see Fig. 7.3), e.g., for peptide or protein analysis, 0.3–1 µl of an aqueous analyte solution is mixed with 0.5–1 µl of a ~5 mM solution of an appropriate matrix, e.g., 2, 5-dihydroxybenzoic acid (DHB), sinapinic acid, or α-cyano-4-hydroxycinnamic acid (CHCA), in 50 % aqueous acetonitrile containing 0.1 % trifluoroacetic acid, and then deposited onto a metal target. Other

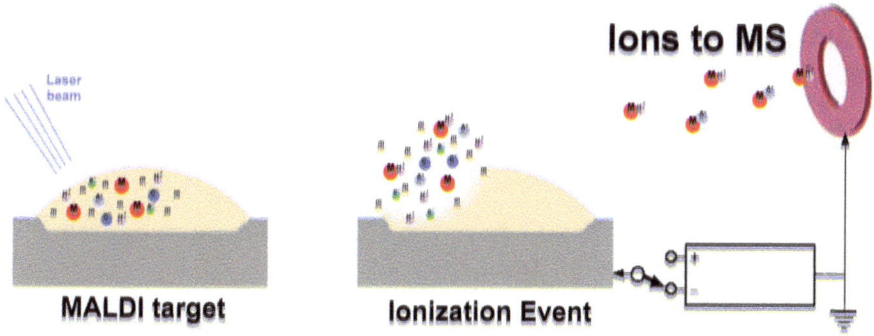

Fig. 7.3 Schematic diagram of an ion source for matrix-assisted laser desorption ionization (MALDI-MS)

matrices may be needed for other compound classes (see below). Upon drying, co-crystallization of matrix and analyte molecules takes place. When these crystals are laser bombarded with photons which fit to the absorption maximum of the matrix, e.g., with 337 nm from a N_2 laser for the matrices mentioned, gas-phase analyte ions are generated in the selvedge which can be mass analyzed by a time-of-flight mass spectrometer (TOF-MS) [25–28]. The analyte ionization results from efficient electronic excitation of the matrix and subsequent transfer of the energy to the dissolved analyte molecules, which are desorbed and analyzed as protonated or cationized molecules. Unlike ESI, where ion envelopes of multiple-charge ions are generated, mostly single-charge $[M+H]^+$ are generated in MALDI-MS, together with less abundant double-charge $[M+2H]^{2+}$ and proton-bound dimeric $[2M+H]^+$ ions.

The ionization process is not fully understood [29, 30]. Generally, a two-step process is assumed to take place. The laser energy is absorbed by the matrix molecules, which are desorbed and ionized by protonation. In the hot plume generated in this ablation step, proton transfer between matrix ions and analyte molecules leads to protonated analytes. Instead of by protonation, cationization by for instance $Alkali^+$-ions present in the sample preparation may take place.

High molecular-mass compounds, e.g., in excess of 200 kDa, can be analyzed using the MALDI-MS. Nowadays, MALDI-MS plays an important role in the characterization of various classes of biomolecules. In various ways, it is important in the analysis of peptides and proteins, being an essential tool in the current proteomics research (Sect. 7.5.5). Furthermore, MALDI-MS is important in the analysis of lipids (Sect. 7.5.4), oligosaccharides and glycans (Sect. 7.5.3), oligonucleotides (Sect. 7.5.6), synthetic polymers (Sect. 7.5.7), and even of small molecules like drugs and antibiotics (Sect. 7.5.1). Moreover, MALDI-MS plays an important role in two emerging application areas of MS, being imaging mass spectrometry [31] and identification of bacteria and microbial fingerprinting [32].

Surface-enhanced laser desorption ionization (SELDI) is a variation of MALDI. A modified target surface is applied to achieve biochemical affinity with the analyte molecules [33]. In SELDI-MS, the protein mixture is spotted onto a surface with a specific (bio)chemical functionality, e.g., cation- or anion-exchange materials, hydrophobic materials, or materials with immobilized metal affinity, lectin, or even protein or antibody affinity. Specific proteins in the mixture bind to the surface, while others can be removed by washing. Thus, on-target biomolecular interactions are used as part of the measurement strategy. The specific binding to the SELDI target acts as a sample pre-treatment and/or analyte isolation step. After washing, a matrix is applied and the experiments proceeds like in MALDI-MS. SELDI-MS is currently extensively used for clinical diagnostics and in clinical biomarker discovery studies [33].

7.2.4 Other Desorption Ionization Techniques

Several other desorption ionization methods have been described, including laser desorption ionization and SIMS [21, 34]. In ^{252}Cf plasma desorption ionization

(PDI), the sample is dissolved in a volatile solvent and deposited on a nitrocellulose target material, which is subsequently bombarded with the ^{252}Cf fission fragments [33–37]. Many mass spectra of PDI-MS show the occurrence of $[M+Na]^+$ next to $[M+H]^+$ [37]. PDI-MS has been extensively used for the analysis of biological macromolecules, but has been superseded by MALDI. Atmospheric-pressure desorption ionization methods are briefly discussed in Sect. 7.2.7.

7.2.5 Thermospray Ionization

The development of the thermospray ionization interface (TSI) started in the mid-1970s in the laboratories of Vestal at the University of Houston. The long-term research project aimed at the development of an LC–MS interface which is compatible with 1 ml/min of an aqueous mobile phase and capable to provide both EI and solvent-independent chemical ionization (CI) [38]. The initial interface was a highly complex system, involving vaporizer heating using a CO_2 laser [38] or a set of hydrogen flames [39]. Subsequently, the system was greatly simplified with respect to vaporizer design and vacuum system. Finally, direct electrically heated vaporizers were applied and a 0.3-m^3/s mechanical pump was connected directly to the outlet side of the ion source [40]. In a TSI interface, a jet of vapor and small droplets is produced by means of a heated vaporizer tube into a low-pressure region. Nebulization is due to the disruption of the liquid by the expanding vapor formed by the partial and rapid evaporation of the liquid. A considerable amount of heat is transferred to the solvent in the nebulization process, which assists in the subsequent desolvation of the droplets in the low-pressure region. Efficient pumping at the ion source enables the introduction of up to 2 ml/min of aqueous solvents into the MS vacuum system [41, 42]. TSI has been the most widely applied LC–MS interface in the 1980s and early 1990s. In the mid-1990s, it rapidly started to lose territory in favor of interfaces based on atmospheric-pressure ionization (API), i.e., ESI and atmospheric-pressure chemical ionization (APCI) (Sect. 7.2.6). As an LC–MS interface and an ionization technique, TSI is obsolete and no longer used, where it is still used as a nebulization technique in flame-furnace atomic-absorption spectrometry [43].

A typical TSI-MS system consists of a gas-tight cylindrical tube with the vaporizer probe at one end and the pump-out line at the other (see Fig. 7.4). The latter is connected to a rotary pump, equipped with a (liquid-nitrogen) cold trap to avoid pump-oil contamination by solvent vapors. The ion source contains a sampling cone acting as the entrance slit to the mass analyzer, and a repeller or a retarding electrode opposite or slightly downstream to the sampling cone. A filament behind an electron entrance slit and a discharge electrode may be positioned upstream. The source block is heated by cartridge heaters, and a temperature sensor is placed further downstream to monitor the vapor jet temperature.

With the development of TSI, Vestal et al. [39, 44–46] also introduced a new ionization technique. They demonstrated that collision of the vapor-droplet beam from

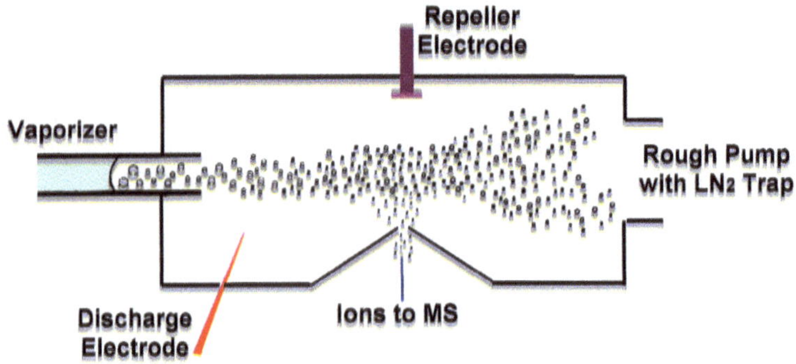

Fig. 7.4 Schematic diagram of a thermospray ion source (TSI-MS). No filament is present in this source

the nebulizer with a nickel-plated copper plate leads to soft ionization of analytes. Subsequently, it was found that the collision was not a vital step in the process [45]. The presence of a volatile buffer or acid in the mobile phase appeared more important in TSI, i.e., in charging the droplets produced and in generating preformed ions in solution. The ionization phenomena were explained in terms of the ion-evaporation (IEV) model (see Sect. 7.3) [46].

In practice, four modes of ionization can be distinguished in TSI, i.e., two liquid-based ionization modes, ion-evaporation and thermospray buffer ionization (applied in ~60% of the applications), and two electron-initiated ionization modes, filament-on ionization and discharge-on ionization (applied in ~40% of the applications). With ionic analytes and preformed ions in solution, ion evaporation seems to be most important. In this mode, the concentration of the volatile electrolytes in the mobile phase should be carefully optimized. With neutral analytes, TSI buffer ionization is predominant: Ionization takes place by either gas-phase ion-molecule reactions or rapid proton-transfer reactions upon transition from the liquid to the gas phase due to the desolvation of the droplet, i.e., at the interface of the liquid droplet and the gas phase (the selvedge). For this ionization mode, the addition of ammonium acetate or any other volatile buffer to the LC effluent is obligatory. In absence of a buffer, with nonaqueous mobile phases, or with mobile phases that contain over 50% organic modifier, either the filament-on or the discharge-on mode must be used. In the filament-on mode, high-energy electrons (0.4–1.0 keV) emitted from a heated filament are accelerated into the ion source. In the discharge-on mode, a continuous gas discharge is used to generate electrons. The electrons initiate medium-pressure CI, in which solvent vapors are applied as reagent gas (solvent-mediated CI). Thus, the filament-on and discharge-on modes show greater versatility in terms of applicability range when the buffer is left out, although enhanced performance may be observed with buffer present. For most compounds, the positive-ion mode is more sensitive than the negative-ion mode. For a proper operation of TSI, the careful optimization of a variety of mostly interrelated experimental parameters is required, i.e., the source block temperature, the vaporizer temperature, the repeller potential, the solvent flow-rate, and the mobile-phase composition.

Under these conditions, mainly protonated molecules $[M+H]^+$ are observed in positive-ion mode. In ion-evaporation and thermospray buffer ionization modes, $Alkali^+$-cationization may be observed as well, but generally not as frequently as in ESI operation. In general, little attention was paid to such effects in TSI-MS.

7.2.6 Electrospray Ionization

Electrospray ionization (ESI) as we know it today can be attributed to John Fenn, who in 2002 received the Nobel Prize of Chemistry for his contribution to the identification and structure elucidation of biological macromolecules. However, Fenn was not the first to use electrospray nebulization for sample introduction into a mass spectrometer. In the mid-1960s, Dole et al. [47, 48] investigated the possibility to transfer proteins from the liquid phase to the gas phase by electrospraying dilute solutions in a nitrogen bath gas. The vacuum interface needed for an efficient coupling of an atmospheric-pressure ESI ion source to a mass spectrometer has been developed by Yamashita and Fenn [49, 50]. In the same time frame, several other research groups investigated similar approaches. Evans et al. [51, 52] developed electrohydrodynamic ionization and Zolotai et al. [53, 54] "field evaporation of ions from solution." In both approaches, analytes are electrosprayed under high-vacuum conditions from nonvolatile solvents like glycerol. Addition of sodium iodide, necessary for conductivity reasons, stimulates the formation of $[M+Na]^+$ and $[M+(glycerol)_n+Na]^+$ for compounds like saccharides, nucleosides, and small peptides [55]. More from a fundamental point-of-view, Iribarne and Thomson [56–59] investigated the direct emission of ions from liquid droplets, generated by pneumatic nebulization of a liquid solution in an atmospheric-pressure chamber and charged by random statistical charging using an induction electrode positioned close to the nebulizer. Solvated single-charge ions are formed in the evaporating spray. Their theoretical description of the process is later adapted to explain the ionization mechanisms of both TSI [44–46] and ESI [60–63].

In ESI, an analyte solution, e.g., the mobile phase from an LC column, is nebulized into an API source as a result of a strong electric field, eventually assisted by N_2 as a nebulizing gas and heating. Small, highly charged droplets (1–10 μm) are generated (Fig. 7.5). Gas-phase ions are generated in the process of droplet evaporation and field-induced electrohydrodynamic disintegration of the droplets. The gas-vapor mixture (N_2 and mobile-phase solvents) with analyte ions is sampled by the ion-sampling orifice into the vacuum interface and toward the mass spectrometer. Desolvation and collisional cooling of the ions occur when they move through the vacuum interface toward the high-vacuum mass analyzer. In most cases, either $[M+H]^+$ or $[M-H]^-$ is generated, depending on the operating polarity, but other ions like $[M+Na]^+$, $[M+K]^+$, or $[M+CH_3COO]^-$ may be generated as well (or instead). Alkali-metal cationization in ESI seems to primarily depend on the (residual) concentration of alkali-metal ions in the sample and on the properties on the analytes. It seems the presence of vicinal hydroxy or other oxygen groups greatly promoted the formation of alkali-metal cationization.

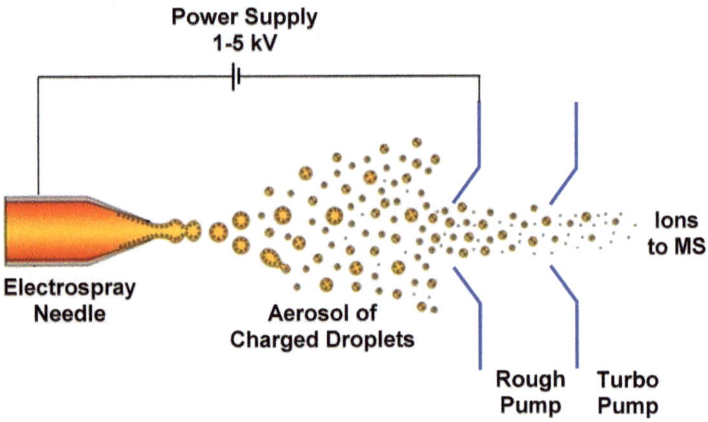

Fig. 7.5 Schematic diagram of an electrospray ion source (ESI-MS)

When studying the ionization behavior of a particular compound or compound class in ESI-MS, often direct infusion of the analyte solution into the ESI source is performed. Although this provides a first impression on whether analyte ionization is feasible, it has frequently been shown that the appearance of the ESI-MS spectrum may drastically change between direct infusion and on-line LC–MS operation, e.g., [64], and even between LC–MS analysis of standard solution and extracts from biological samples. Such changes especially comprise changes in relative abundance between $[M+H]^+$ and ions due to $Alkali^+$-cationization and $H^{+/}Alkali^+$-exchange. Compounds with strong affinity to $Alkali^+$-ions still show $[M+Alkali]^+$ in LC–MS, despite the fact that the $Alkali^+$-ion are separated from the analyte by LC, i.e., the $Alkali^+$-salts elute unretained in reversed-phase liquid chromatography (RPLC). This finding also implies that, at least part of, the $Alkali^+$-cationization is a liquid-phase process.

As ESI has become the ionization method of choice in both the MS analysis of biological macromolecules and the on-line LC–MS coupling, significant research efforts and instrumental developments and improvements have been reported and/or implemented in ESI devices. The orthogonal rather than axial positioning of the electrospray needle in the ESI ion source is important in reducing contamination of the ion sampling orifice. A wide variety of ion-sampling orifices have been developed, including glass capillaries, heated stainless-steel capillaries, sampling cones, and flat sampling orifices, often protected against contamination by counter-current N_2 flows. Ion transmission in the vacuum interface has been improved by the use of continuously more advanced RF-only ion focusing and transport devices, that is: next to RF-only quadrupoles, hexapoles, and octapoles also the implementation of ion funnels [65] and travelling-wave stacked-ring ion guides [66]. Whereas in most analytical applications of ESI-MS low pressure is pursued in the vacuum interface, the preservation of protein complexes in native-MS is best achieved at somewhat higher pressures in this region [67]; valves have been implemented to allow pressure adjustments in native-MS experiments.

Another instrumental development is based on the fact that the generation of smaller droplets is more favorable in terms of droplet evaporation during ESI, of sensitivity and the ability to preserve non-covalent molecular associates. Thus, nanoelectrospray ionization (nESI) has been developed [68], where the analyte is sprayed from a gold-coated fused-silica capillary with a tip diameter of 1–5 μm rather than from capillaries with a 100–150-μm tips that are used in conventional (pneumatically assisted) ESI. In nESI, flow-rates as low as 20 nl/min can be nebulized. Thus, gentler operating conditions (temperature, gas flows, needle voltage) can be achieved. In order to more readily implement nESI in LC–MS operation, integrated chip-based nano-LC–nESI devices have also been developed [69].

The ionization mechanism of ESI is not fully understood [60–63]. The two prevailing models are the *ion-evaporation model* of Iribarne and Thomson [56–59] and the *charge-residue model* of Dole [47, 48]. Both models assume that analyte molecules are present in solution as preformed ions in solution, which can be achieved either by choosing an appropriate pH of the solution or the mobile phase or, which is relevant in the context of the chapter, by providing alkali-metal ions to stimulate alkali-metal cationization. According to the *charge-residue model,* the sequence of solvent evaporation and electrohydrodynamic droplet disintegration proceeds until only one preformed analyte ion is present in the microdroplet. By evaporation of the solvent, the preformed analyte ion is released to the gas phase. According to the *ion-evaporation model,* gas-phase ions are generated from the highly charged microdroplets, because the local field strength is high enough for preformed ions to be emitted into the gas phase (see Sect. 7.3). Although the two models are to some extent complementary, the relative importance of either mechanism in the actual ion production of a particular analyte is difficult to decide. Especially under high flow-rate conditions during LC–MS analysis of small molecules, gas-phase ion-molecule reactions, e.g., proton-transfer reactions, at the interface of droplets and the gas phase seems to contribute to the analyte ionization as well.

ESI enables soft ionization of highly labile and nonvolatile compounds such as (oligo)nucleotides, (oligo)saccharides, peptides, and proteins without significant fragmentation. In the analysis of biomacromolecules, an ion envelope of multiple-charge ions, $[M+nH]^{n+}$ or $[M–nH]^{n-}$, is generated, from which the molecular weight of the molecule can be accurately calculated (better than 0.01 %) using software procedures. Next to protonated molecules, other cationization products involving NH_4^+, Na^+ and K^+ may be observed in the positive-ion mode. In negative-ion mode, $[M+HCOO]^-$-, $[M+CH_3COO]^-$-, or $[M+CF_3COO]^-$-anionized species may be observed next to the deprotonated molecule $[M–H]^-$, depending on the mobile-phase additives and the physicochemical properties of the analyte. Spectral features are discussed in more detail in Sect. 7.4.

ESI-MS is very widely applied, for instance in all application areas discussed in Sect. 7.5. This is partly due to the fact that until today, ESI-MS is the most convenient and most widely applied ionization technique for LC–MS. ESI-MS helped to open new application areas for MS.

In ESI, as with most other ionization techniques discussed here, the addition of alkali-metal ions (other types of metal ions) can be used to extend the applicability range. In a technique called "coordination ionspray" (CIS-MS), cationization agents, like especially Ag^+, are applied in the ESI-MS analysis of nonpolar analytes to induce coordination and thereby cationization. CIS-MS was introduced by Bayer et al. [70]. As "ionspray" is a trade name of AB-Sciex, the term coordination electrospray ionization (CIS) is used throughout this text.

Another ionization technique, which seems closely related to TSI and ESI, is atmospheric-pressure chemical ionization (APCI). In APCI, the solvent stream, e.g., the effluent from an LC column, is pneumatically nebulized into a heated vaporizer zone, where (almost) complete evaporation of the aerosol droplets is achieved [71–73]. Analyte ionization is initiated by electrons from a downstream corona discharge needle. The electrons act as primary source of ionization of the solvent or mobile-phase constituents, which in turn by gas-phase ion-molecule reactions in the API source ionize the analyte molecules, mostly by proton-transfer reactions, i.e., formation of $[M+H]^+$ in positive-ion and $[M–H]^-$ in negative-ion mode. There are also some results, indicating the Na^+-cationization can take place under APCI conditions. Atmospheric-pressure photoionization (APPI) is an ionization technique closely related to APCI. In APPI, the analyte ionization is initiated by light from a vacuum-ultraviolet lamp, e.g., a Kr-lamp, instead of by means of a corona discharge. Next to direct photoionization of the analytes, gas-phase ion-molecular reactions greatly contribute to the ionization in APPI [74, 75].

7.2.7 Atmospheric-Pressure Desorption Ionization Techniques

Desorption electrospray ionization (DESI) may serve as an example of the many atmospheric-pressure surface ionization technique that has recently been introduced [63, 76]. In DESI, the high-velocity spray of charged microdroplets from a (pneumatically assisted) electrospray needle is directed at a surface, which is mounted in front of the ion-sampling orifice of an API source (see Fig. 7.6). Surface constituents are released from the surface and ionized. These gas-phase ions can be introduced to and observed by MS [77]. In this way, DESI-MS enables for instance the analysis of drugs in tablets or natural products in plant parts without extensive sample pre-treatment or prior separation. In addition, DESI-MS and some of its related surface ionization techniques enable chemical imaging of surfaces such as thin-layer chromatography (TLC) plates and tissue sections [78].

Somewhat earlier than DESI, atmospheric-pressure matrix-assisted laser desorption ionization (AP-MALDI) was introduced [79–81]. Initially, an angled probe tip was bombarded with photons from a laser. Desorbed ions were transferred by a stream of nitrogen toward the ion sampling orifice of an API-MS instrument. Although initially a TOF-MS was used, the potential of AP-MALDI with ion-trap instruments was also explored [82]. An attractive feature of the latter setup is the ease at which multistage MS–MS (MS^n) can be achieved. Subsequently, several ion-source geometries have been developed and applied [80]. AP-MALDI-MS also plays a role in the emerging mass spectrometry imaging techniques.

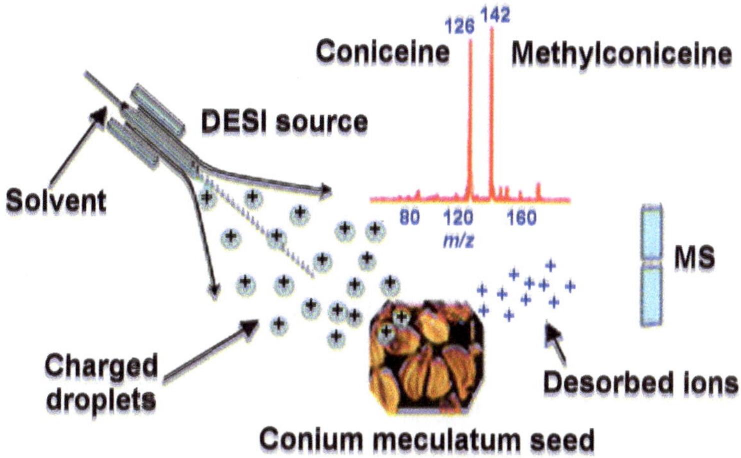

Fig. 7.6 Schematic diagram of the setup and ion source for desorption electrospray ionization (DESI-MS). (Adapted from N. Talaty et al. Analyst, 130, 2005, 1624–1633 with permission of the PCCP Owner Societies)

In addition to DESI and AP-MALDI, a large variety of other, sometimes closely related, atmospheric-pressure desorption ionization techniques have been introduced in the past decade, connected to a huge number of acronyms. Van Berkel et al. [76] tried to classify these emerging techniques into four categories, i.e., (1) thermal desorption ionization, (2) laser desorption/ablation ionization, (3) liquid-jet and gas-jet desorption ionization, and (4) liquid extraction surface sampling probe ionization.

In the first category, the actual analyte ionization is mostly based on APCI. Commercially available techniques like direct analysis in real time (DART) [83] and atmospheric-pressure solids analysis probe (ASAP) [84] are examples of this category. In DART, the ionization process involves interaction of analyte molecules on a surface, e.g., a tablet, a leaf or a glass rod dipped in or rubbed with a liquid sample, with electronically or vibronically excited atoms or molecules in a helium carrier stream. The analyte ions formed are introduced into an API-MS [83]. In ASAP, the sample is introduced into the ion source by means of a glass rod, from which analyte molecules are vaporized and ionized at atmospheric pressure [84]. The second group involves techniques were desorption/ablation is achieved by means of a laser, whereas the actual ionization is mostly done with ESI or APCI. The AP-MALDI technique [79–81], discussed earlier, is classified in this category as well. In the third group, DESI and related techniques, for instance, based on sonic spray ionization [85, 86], such as desorption sonic spray ionization (DSSI) [87], are classified. The last group contains techniques based on analyte extraction using a confined liquid stream with either a liquid microjunction or a sealed surface contact. Ionization is achieved by ESI or APCI. One may question whether the latter techniques should be considered as desorption ionization techniques.

7.3 Ionization Mechanisms

In two interesting review papers, Arpino and Guiochon [88] and Vestal [89] independently concluded that the various soft ionization methods have some common features. In some way or another, a matrix is involved in which the analyte molecules are dissolved. This matrix may be a specific compound mixed with the analyte to achieve analyte ionization, i.e., sinapinic acid, DHB, or CHCA matrix used in MALDI, nitrocellulose in PDI, or glycerol in FAB. Alternatively, the matrix is the liquid phase from which droplets are generated, i.e., in TSI and ESI. Two general processes appear to be important in the ionization mechanism, i.e., (1) the formation of analyte ions in the sample matrix prior to evaporation or desorption and (2) rapid evaporation prior to ionization, which can be effected by very rapid heating or by sputtering by high-energy photons or particles. In this respect, these soft ionization techniques are considered to be "energy-sudden" techniques [89, 90]. In general, it is assumed that the energy deposited on the sample surface can cause (gas-phase) ionization reactions to occur near the interface of the solid or liquid and the vacuum (the so-called selvedge) or provide preformed ions in the condensed phase with sufficient kinetic energy to leave their environment. Thus, desorption of preformed ions from the matrix appears to be the common mechanism. The energy needed in the desorption of analyte ions can be applied in a number of ways, as indicated by the variety of methods discussed in the previous section. The mass spectra obtained are characterized by the occurrence of various cationized molecules, such as $[M+H]^+$, $[M+Na]^+$, and $[M+K]^+$, and possibly some fragments from unimolecular dissociations. Brief discussions on ionization mechanisms were included in the descriptions of the various techniques in the previous section.

Nebulization ionization is the process involved in the analyte ionization in EHI [55], TSI [44–46], and ESI [60–63], where no primary source of ionization, i.e., a filament or a discharge electrode, is applied. The mechanism of nebulization ionization is not fully understood. A number of competing processes appear to take place. The general understanding can be summarized as follows: Upon nebulization, charged droplets of a few μm ID are generated. The fate of these droplets is determined by a number of competing processes. Charge-preserving solvent evaporation results in smaller droplets with a higher charge (surface) density. When the repulsive forces due to the surface charges exceed the forces due to surface tension, the droplets explode or disintegrate as a result of field-induced Rayleigh or electrohydrodynamic instabilities. Multiple smaller droplets, often called microdroplets, are generated. Due to a sequence of repetitive electrohydrodynamic droplet disintegration and continuous solvent evaporation, microdroplets containing only one charged molecule may be generated. By soft desolvation of this droplet, the ions will be free in the gas phase and amenable to mass analysis. This step is assumed to take place in the *charge-residue model,* proposed by Dole et al. [47, 48]. However, it is also argued and demonstrated that at a certain droplet-size/charge ratio, field-induced ion evaporation of preformed ions from the solution may take place, as proposed in the *ion-evaporation model* by Iribarne et al. [56–59]. The resulting evaporated ions in the gas phase are amenable to mass analysis. In addition to this, gas-phase electrolyte ions, e.g., NH_4^+, may be formed during these processes as well. These ions may act as reagent gas ions in gas-phase ion-molecule reactions, which actually are to be considered as chemical ionization processes [91]. The gas-phase ions generated by either mechanism will most likely be solvated again, as they are generated in a humid high-pressure region with multiple interactions between ions and their surrounding neutrals from mobile-phase constituents. In their way toward the ion-sampling orifice, in the region between the orifice and the skimmer, as well as in the RF-only multipole device between the skimmer and the baffle to the mass analyzer region, desolvation of these solvated cluster ions will take place. Furthermore, the ions and their solvated clusters generated may be collisionally activated due to the many ion-molecule collisions in the high-pressure ion source. This may occasionally result in the formation of fragment ions. It is impossible to distinguish the importance of these various processes in the ionization of a particular molecule. Irrespective of the exact ionization mechanism, the formation of preformed ions in solution seems to be an essential step in most of these mechanisms. This may be achieved by adjusting the mobile-phase pH, by addition of cationizing or anionizing additives, by inducing (permanent) charge via analyte derivatization, and/or by keeping the ionic strength of the solution low.

7.4 Spectral Features in Cationization Mass Spectrometry

Cationization processes in MS clearly demonstrate, that the appearance of the mass spectrum of a particular analyte strongly depends on the experimental conditions, e.g., on the presence or absence of certain additives in the sample or, when an LC mobile phase is concerned, in the solvents.

An important feature of cationization MS is that the type of ions generated is distinctly different from the one generated in EI. The primary ions generated in EI are molecular ions, $M^{+\bullet}$, due to expelling an electron from the neutral species. By definition, molecular ions are odd-electron ions ($OE^{+\bullet}$). A substantial amount of internal energy is put into the molecular ion, resulting in rapid in-source compound-specific fragmentation. The mixture of the intact molecular ion and the fragment ions is subsequently mass analyzed. In contrast, most soft-ionization techniques like ESI and MALDI generate either positively or negatively charged even-electron ions (EE^+), often $[M+H]^+$ or $[M–H]^-$. No or hardly any in-source fragmentation occurs. If fragmentation is desirable, it has to be induced by increasing the internal energy of the ion, for instance by CID in an MS–MS experiment. The fragmentation of $OE^{+\bullet}$ and EE^+ is distinctly different, especially because radical-induced reactions opens up a number of fragmentation pathways in $OE^{+\bullet}$ that remain unavailable for EE^+.

For an ion with a known elemental composition, its $OE^{+\bullet}$ and EE^+ character can be readily derived using the nitrogen rule, which states that a molecule, $M^{+\bullet}$ or any other $OE^{+\bullet}$ with an even mass or mass-to-charge ratio (*m/z*) should contain zero or

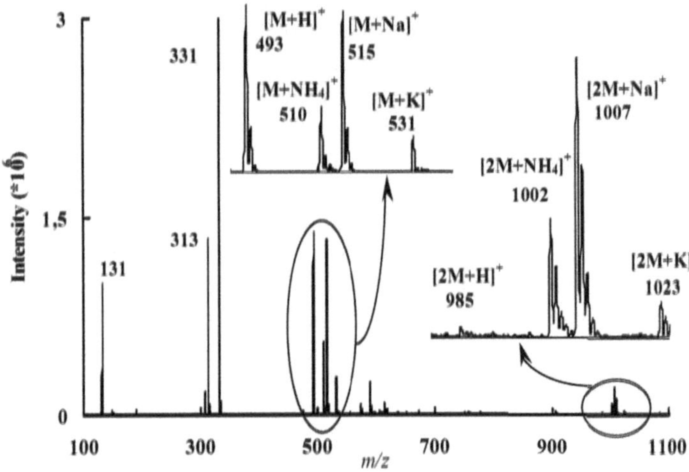

Fig. 7.7 ESI-MS mass spectrum of the iridoid glycoside globularin, showing characteristic ions discussed in the text, i.e., $[M+H]^+$ with *m/z* 493, $[M+NH_4]^+$ with *m/z* 510, $[M+Na]^+$ with *m/z* 515, and $[M+K]^+$ with *m/z* 531 as well as the adduct-bound dimers, i.e., [2M+NH4]$^+$ with *m/z* 1002, $[2M+Na]^+$ with *m/z* 1007, and $[2M+K]^+$ with *m/z* 1023. The ions with *m/z* 131, 313, and 331 are fragment ions. (Reprinted from [163] with permission of Wiley; copyright 2007, John Wiley and Sons, Ltd)

an even number of *N*-atoms, whereas an EE^+ with an even *m/z* should contain an odd number of *N*-atoms. Thus, an $OE^{+\bullet}$ with odd *m/z* contains an odd number of *N*-atoms, and an EE^+ with an odd *m/z* zero or an even number of *N*-atoms.

The most important ionization process with the soft ionization techniques in positive-ion mode is protonation, i.e., formation of $[M+H]^+$. Next to protonation, ion-attachment or cationization processes with sample-related Na^+- or K^+-ions may contribute to the ionization of polar analytes in these ionization techniques. $[M+Na]^+$ and $[M+K]^+$ have been reported for virtually all soft ionization techniques discussed in Sect. 7.2. The ESI-MS mass spectrum of the iridoid glycoside globularin may serve as an example of this behavior, showing $[M+H]^+$, $[M+NH_4]^+$, $[M+Na]^+$, and $[M+K]^+$ (see Fig. 7.7, see also Sect. 7.5.2). It should be emphasized that the terms "protonated molecule," "sodiated molecule," and "potassiated molecule" should be applied rather than the frequently, but erroneous term "molecular ion." The term "molecular ion" is reserved for radical ions! Also the terms "quasimolecular ion" or "pseudomolecular ion" should not be used [92].

In mass spectra, where both $[M+H]^+$ and $[M+Na]^+$ and/or $[M+K]^+$ are present, the cationized species are readily recognized by a 22-Da (21.982 Da) or 38 Da (37.956 Da) mass shift, respectively, relative to the $[M+H]^+$. Sodium is monoisotopic, whereas potassium, being an "A+2"-element, shows an M+2 isotopic contribution of 7.2 %. Often, the occurrence of $[M+Na]^+$ and $[M+K]^+$ can actually be attributed to the presence of Na^+- and/or K^+-ions in the (often) biological matrices analyzed, but also residual alkali-metal ions (typically <1 mM) in the solvents used in sample preparation (or in LC mobile phases) may play a role, e.g., [93]. $[M+Na]^+$ and $[M+K]^+$, often indicated as so-called adduct ions, are especially observed for compounds containing multiple (vicinal) hydroxyl and/or carboxyl groups. Theoretical calculation with Na^+-cationization of disaccharides have shown that stable structures arise from tridentate interaction of Na^+ with the ring oxygen and two hydroxy oxygen atoms [94].

One should be aware that (apparent) $[M+Na]^+$-ions may also be observed if compounds with acidic functional groups, e.g., carboxylic acids, sulfonates, sulfates, or phosphates, are analyzed in positive-ion mode. One may question whether actual Na^+-cationization takes place with these types of molecules. The observed 22-Da (or 38-Da for the apparent $[M+K]^+$) can also be attributed to an exchange of the acidic hydrogen for sodium, thus resulting in the formation of protonated sodium salts, e.g., $[(M–H+Na)+H]^+$, $[(M–2H+2Na)+H]^+$, or sodiated sodium salts, e.g., $[(M–H+Na)+Na]^+$. From their *m/z* values, these H^+/Na^+-exchange products cannot be discriminated from Na^+-adduct ions $[M+Na]^+$. This behavior is frequently observed for peptides and oligonucleotides, carrying multiple carboxylic acid and phosphoric acid groups, respectively. Cationization processes, both adduct-ion formation and $H^+/Alkali^+$-exchange, occur in ESI-MS and MALDI-MS as well as in other condensed-phase ionization techniques, which are often based on the formation of preformed ions in the condensed phase. In negative-ion mode, especially in LC–MS with ESI-MS, ion-attachment processes and/or adduct-ion formation with solvent constituents may occur as well, resulting in ions like $[M+HCOO]^-$, $[M+CH_3COO]^-$, or $[M+Cl]^-$, depending on the additives used.

There is some debate on the origin of the adduct ions. It may be argued that some of the adducts, especially the ones involving $H^+/Alkali^+$-exchange, are already generated in the condensed phase and transferred to the gas-phase as such. On the other hand, for ESI-MS, for instance, gas-phase processes leading to Na^+-cationization and even H^+/Na^+-exchange have been demonstrated as well [95–97]. For Li^+-cationization of peptides, occurring at the amide bond, theoretical calculations were made, demonstrating that sufficient affinity between the Li^+-ion and the amide bond exists [95]. This would point to other cationization sites than what would be expected based on $H^+/Alkali^+$-exchange at carboxylic acid groups. Gas-phase production of $[M+Na]^+$-ions of peptides using a dual ESI source, i.e., one to introduce the protonated peptide and one to introduce a sodium diethyldithiocarbamate anion, via ion-ion reactions has also been demonstrated [96, 97]. Several examples are available, where addition of Li^+ to the solvent effectively suppresses Na^+-cationization in favor of Li^+-cationization, point toward higher affinity of many analytes for Li^+-ions than for Na^+-ions.

For MALDI-MS, van Kampen et al. [98], and their references therein] showed the importance of (1) the affinity of the analyte toward the $Alkali^+$, which seems to be primarily governed by the charge density of the $Alkali^+$; (2) the gas-phase availability of $Alkali^+$, which seems to be related to the lattice energies in the alkali-cation salt, and thus to the counter anion used; and (3) the matrix applied, which in turn may affect ion separation within the salt. For polyethylene glycol (PEG) with an alkali-halide salt, the most abundant $[M+Alkali]^+$-ion observed depends on the matrix, which decides between two opposing effects, i.e., the gas-phase availability, which is highest for Li^+, and the affinity of PEG to $Alkali^+$, which is highest for Cs^+. In DHB, $[M+K]^+$ is observed as a compromise, whereas in meso-tetrakis(pentafluorophenyl)porphyrin (F20TPP), $[M+Li]^+$ is observed instead. The latter is somewhat explained from the high affinity of Li^+ to F20TPP, resulting in gas-phase $[F20TPP+Li]^+$, that subsequently may act as Li^+ donor in the selvedge. F20TPP was found to be a useful matrix for small-molecule MALDI-MS analysis [98]. In another study, involving MALDI-MS of neutral oligosaccharides using a 3:1 mixture of mixture of 2,5-DHB and 1-hydroxyisoquinoline, doped with equimolar amounts of alkali salts, it was found that the affinity of oligosaccharides for alkali-metal ions was found to increase in the order $Li^+ < Na^+ < K^+ < Cs^+$ [99].

The exact nature of the $Alkali^+$-cationization products, generated under various conditions using ESI-MS and MALDI-MS, can certainly be studied in considerable more detail.

In both ESI and MALDI mass spectra, ions due "adduct-bound" dimers may be observed, e.g., $[2M+H]^+$ or $[2M+Na]^+$ in positive-ion mode and $[2M–H]^-$ in negative-ion mode, e.g., [100]. Adduct-bound dimers, i.e., $[2M+NH4]^+$, $[2M+Na]^+$, and $[2M+K]^+$, were observed in the ESI-MS mass spectrum of the iridoid glycoside globularin (see Fig. 7.7). The adduct-bound dimers generally occur only at higher concentrations and their relative intensity is a function of the analyte concentration. For an unknown compound, the occurrence of adduct-bound dimers may lead to ambiguity in the determination of the molecular mass. Addition of alkylamines to the mobile phase has been proposed in order to prevent adduct-bound dimer formation [101]. In this way, formation of $[M+Na]^+$-ion may be reduced as well in favor of $[M+alkylamine+H]^+$-ions [64].

ESI is a rather unique ionization technique among the soft ionization techniques, as it enables the easy formation of multiple-charge ions, e.g., $[M+nH]^{n+}$. Double-charge ions of proteins are sometimes observed in MALDI mass spectra. With proteins, the multiple charging in ESI leads to an ion envelope of multiple-charge ions, from which with great accuracy the molecular weight of the protein can be derived [102, 103]. Advanced software tools, based on maximum entropy algorithms, have been developed for this purpose [104–106]. Similar series of multiple-charge ions can be observed in the negative-ion mode for proteins [107, 108] and especially for oligonucleotides [103, 109, 110]. Proteins, peptides, and especially oligonucleotides may show apparent sodium adducts, which may also be considered as H^+/Na^+-exchange products, i.e., $[(M–mH+mNa)+nH]^{n+}$ with peptides and proteins in positive-ion mode and $[(M–mH+mNa)–nH]^{n-}$ with oligonucleotides in negative-ion mode. Larger oligosaccharides are also prone to the formation of multiple-charge ions. Given the high affinity of saccharides to sodium, this often involves $[M+nNa]^{n+}$ in positive-ion mode [111]. Adduct formation with other $Alkali^+$-ions, especially Li^+, has frequently been used as well, especially for structure elucidation purposes.

In some studies, Na^+- or K^+-salts are actually added to the sample to stimulate and/or direct the cationization process. Obviously, such addition does not have to be limited to Na^+ and/or K^+. Addition of Li^+ or Ag^+ has been frequently applied as well. Both elements show characteristic isotopic contributions, i.e., lithium shows an 8.0% isotope peak at the monoisotopic *m/z* of $[M+Li-1]^+$, whereas silver is a so-called A+2-element, with an isotope peak of ~93% at the monoisotopic *m/z* of $[M+Ag+2]^+$. One of the objectives in the addition of lithium salts is that Li^+-cationized molecules more readily fragment in collision-cell CID (Sect. 4.3.3) than $[M+Na]^+$ and $[M+K]^+$ [112]. Under such conditions, it is observed that the fragmentation of $[M+Li]^+$ often shows somewhat different characteristics than the fragmentation of $[M+H]^+$, thus potentially providing an additional tool in structure elucidation. Similarly, the addition of silver salts may enable the ionization of compounds that are not readily amenable to the soft ionization techniques, for instance because they lack a protonation site. This behavior is exploited in, for instance, CIS-MS [70]. Again, $[M+Ag]^+$-ions are found to exhibit to some extent different fragmentation behavior, as is demonstrated with some examples in subsequent sections. In this perspective, it is not surprising that researchers have explored fragmentation characteristics of cationized molecules involving other cations. Again, some examples are given below.

7.5 Applications

In this section, applications of cationization mass spectrometry are discussed for several compound classes. Attention is paid to mass spectral features of these compound classes with the soft ionization techniques discussed in Sect. 7.2. Cationization with alkali-metal ions as well as with other metal ions may also have a distinct influence on the fragmentation in MS–MS and/or MS^n. Considerable attention is paid to this aspect as well. The discussion is not meant to be comprehensive, as the body of relevant literature is too big to be reviewed completely. Therefore, it is tried to highlight important features and typical applications.

7.5.1 Small Molecules

Cationization of small organic molecules has been applied as an alternative to protonation for a wide variety of compounds not readily amenable to protonation. Grade and Cooks [34] reported Ag^+-, Pt^+- and Li^+-cationization of compounds like anthracene, 1,4-dicyanobenzene, *p*-aminobenzoic acid in SIMS. $[M+Na]^+$- and $[M+K]^+$-ions have been reported to frequently occur with many small molecules in FDI-MS, FAB-MS, ESI-MS and MALDI-MS. Perhaps, $[M+Na]^+$-ions are most frequently reported for ESI-MS, but this may also be due to the fact that ESI-MS, also in LC–MS, has a very broad application range. In fact, ESI-MS has greatly extended the use of MS and created routine analytical procedures in many new application areas. Na^+-cationization or adduct formation to $[M+Na]^+$ is widely observed in the positive-ion LC–MS analysis of small molecules using ESI-MS, including pesticides, drugs and their metabolites, antibiotics, steroids, natural products, and other compound classes. As indicated before, apparent $[M+Na]^+$-ions may also be observed for compounds with acidic functions, as H^+/Na^+-exchange may occur, e.g., [93].

In ESI-MS, and also in APCI-MS, the appearance of the mass spectrum depends on the mobile-phase composition. In addition, the appearance of mass spectra may change by applying instruments from different manufacturers and/or with different ion source geometries. Some results with the cholesterol reducing agent simvastatin may serve as an example. In acetonitrile/water or acetonitrile/aqueous ammonium acetate (1 mmol/L, adjusted to pH 4.0), simvastatin shows $[M+Na]^+$ (Sciex ionspray interface and API-III) [113]. Under similar mobile-phase conditions but on a different instrument (Sciex turboionspray and API-365), $[M+H]^+$ and $[M+NH_4]^+$ were most abundant. With 2 mmol/L ammonium acetate in the mobile phase, the $[M+NH_4]^+$ actually is the most abundant ion, next to $[M+H]^+$, $[M+Na]^+$ and $[M+K]^+$ being present [114, 115]. Attempts to force the ionization toward the generation of $[M+Na]^+$ by the addition of sodium acetate in the case of simvastatin compromised the performance of the method.

Small-molecule analysis by LC–MS is often directed at quantitative analysis based on precursor-ion–product-ion transitions in selected-reaction monitoring (SRM). In this, the occurrence of $[M+Na]^+$ is unfavorable, because (1) it may result to spreading the charge over a variety of ionic species, the relative abundance of which may change with the salt concentration of the (biological) samples analyzed, and (2) $[M+Na]^+$ of many compounds are not efficiently fragmented in collision-cell CID in a TQ instrument, which is the instrument of choice in quantitative bioanalysis. The addition of ammonium salts or formic acid to the mobile phase may suppress the formation of alkali-metal cationization, but this may compromise the chromatography. An on-line ion-suppressor, based on a cation-exchange resin, to eliminate alkali-metal ions prior to ESI-MS was reported as well [116]. For coupling of high-performance anion-exchange chromatography (HPAEC) to MS, in-line cation-exchange membrane desalters have been described [117, 118].

If reduction of sodium adduct and H^+/Na^+-exchange turns out not to be successful, e.g., because the analysis does not have adequate proton accepting functionality or a high affinity to $Alkali^+$-ions, the addition of 0.1–1 mmol/L sodium acetate to the mobile phase may actually direct the ionization to the $[M+Na]^+$, which can then be analyzed in LC–MS with ESI-MS and by using selected-ion monitoring (SIM). This has for instance been the strategy chosen for the quantitative analysis of a thiol compound, a drug in development, in dog plasma [93]. In this type of strategies, the sodium acetate can be added to the mobile phase or introduced by post-column infusion at a T-piece between LC column and ESI ion source.

The effect of the mobile-phase composition on the ESI-MS mass spectra of the chemotherapeutic agent paclitaxel has been investigated [64]. The use of different mobile-phase additives was evaluated. Paclitaxel is prone to formation of $[M+Na]^+$- and $[M+K]^+$-ions and for quantitative bioanalysis of this drug, control over the adduct formation is important to obtain reliable results. The use of either docedylamine ($C_{12}N$)/acetic acid, providing predominantly $[M+C_{12}N+H]^+$-ions, or acetic acid/sodium acetate, providing predominantly $[M+Na]^+$-ions, were proposed [64].

With small molecules showing various adduct ions in ESI-MS, it seems useful to use the sum of the adduct signals in quantitative analysis using SIM. This has been done in an application were concentrations of ginkgolides and bilobalide were determined in 26 commercial gingko products from different sources [119]. Bilobalide showed both $[M+H]^+$, $[M+NH_4]^+$, $[M+Na]^+$ and $[2M+Na]^+$. Higher fragmentor voltages to promote in-source CID resulted in a decrease of the $[M+NH_4]^+$ response, but an increase of the other responses. The spectra of the gingkolides were dominated by $[M+NH_4]^+$-ions, whereas at higher fragmentor voltages, the total response and that of $[M+H]^+$ and $[M+Na]^+$ increased somewhat. Increase of the $[M+Na]^+$-response at higher fragmentor voltage may partly be due to the fact that the $[M+Na]^+$-ion is not fragmented, whereas $[M+H]^+$ and $[M+NH_4]^+$ are. As for quantification of bilobalide based on $[M+NH_4]^+$, an RSD as high as 17% was obtained for standard solutions, it was decided to monitor both $[M+H]^+$, $[M+NH_4]^+$, $[M+Na]^+$ in SIM. This resulted in an improved RSD of 6% [119].

Another interesting example of dealing with adduct formation in LC–MS based on ESI-MS is the bioanalysis of decitabine, used in the treatment of patients with myelodysplastic syndromes with promising activity in the treatment of acute leukemia [120]. A quantitative bioanalytical method, based on hydrophilic interaction chromatography (HILIC) and ESI-MS–MS was developed. During initial method development, abundant $[M+Na]^+$- and $[M+K]^+$-ions were observed, whereas $[M+H]^+$ showed relatively low abundance. Significant analyte/internal standard co-suppression and inter-batch matrix effects were observed when using $[M+Na]^+$ as the precursor ion for quantification. Addition of 2 mmol/L lithium acetate to the mobile phase effectively suppressed the formation of sodium and potassium adducts in favor of the formation of $[M+Li]^+$. The CID mass spectrum of the $[M+Li]^+$ as precursor shows two fragments, the most abundant one being the lithiated triazine base with *m/z* 119, next to the ion with *m/z* 123 which is the lithiated anhydrosugar. A validated method was developed using an SRM transition based on fragmentation of $[M+Li]^+$ [120].

Some compounds can be analyzed in both positive-ion and negative-ion mode. Because of the occurrence of intense alkali-metal cationization, it is sometimes preferred to perform the analysis in negative-ion mode. This was done for the analysis of the amphoteric or zwitterionic surfactant cocamidopropylbetaine (CAPB), which in the negative-ion mode predominantly shows $[M–H]^-$, whereas in the positive-ion mode a range of ions is observed, including monomeric, dimeric, and trimeric protonated, sodiated and potassiated molecules, mixed double-charge ions, and a fragment ion [121].

Some compound classes have been widely analyzed using ESI-MS, enabling statements on the likeliness of Na^+-cationization. Pesticide classes that are prone to the formation of $[M+Na]^+$, often next to $[M+H]^+$, are carbamate and organophosphorus pesticides, and various classes of herbicides like phenylureas, sulfonylureas, and chloracetanilides. Antibiotics that readily show $[M+Na]^+$ next to $[M+H]^+$ are sulfonamides, betalactams, tetracyclines, aminoglycosides, and macrolides. The influence of mobile-phase additives on the ESI-MS mass spectra of tetracyclines has been systematically investigated in more detail [122]. In FAB-MS, macrolide antibiotics show either $[M+H]^+$ or $[M+Alkali]^+$, after addition of Li^+-, Na^+-, or K^+-ions to the FAB matrix [123].

The condensed-phase ionization of polyether ionophore coccidiostats and antibiotics mostly results in $[M+Na]^+$-ions. This has been reported for FAB-MS of lasalocid, septamycin and monensin with desorption directly from TLC plates [124] and for maduramicins and other polyether antibiotics [125]. In ESI-MS, they are also mostly analyzed as $[M+Na]^+$. The presence of multiple cyclic ether units, a free carboxylic acid group at one end, and a terminal alcohol group at the other enable them to form stable complexes with various metal cations, e.g., alkali cations, Ca^{2+}, Mg^{2+}, and Cu^{2+}, via intramolecular head-to-tail hydrogen bonding [126]. In MALDI-MS using DHB as matrix, such coccidiostats show $[M+Na]^+$ as well as some $[M+K]^+$ and $[(M-H+Na)+Na]^+$ [127].

The anthelmintic avermectins are 16-membered macrocyclic lactone derivatives, which lack protonation sites. They have been analyzed as $[M+NH_4]^+$ in positive-ion ESI-MS or APCI-MS, as $[M+Na]^+$ in positive-ion ESI-MS, or as $[M-H]^-$ in negative-ion APCI-MS. Different ionization approaches were tested for the residue analysis of abamectin. Best sensitivity was achieved in positive-ion ESI-MS using $[M+NH_4]^+$. Similar sensitivity was achieved with $[M+Li]^+$, but this provided only one fragment ion, which is not favorable regarding EU regulatory rules concerning confirmation of identity in residue studies [128].

In FAB-MS, macrolide antibiotics show either $[M+H]^+$ or $[M+Alkali]^+$, after addition of Li^+-, Na^+-, or K^+-ions to the FAB matrix [123]. If $Alkali^+$-cationization of the analyte takes place, the charge localization is often different from the protonation site. For macrolide antibiotics like erythromycin A, for instance, it was demonstrated that protonation takes place at the dimethylamino group in the amino-sugar substituent, whereas the $Alkali^+$-ion is localized at the highly oxygenated lactone aglycone [123]. With $[M+H]^+$ as precursor ion, fragment ions consistent with loss of the sugar substituents or with loss of the lactone aglycone are observed. With $[M+Alkali]^+$ as precursor ion, a series of high-*m/z* fragment ions are observed, consistent with small molecule neutral losses from the sugar substituents, cross-ring fragmentation of the sugar substituent, and finally loss of the complete sugar units, with charge retention on the lactone aglycone. This difference in fragmentation behavior is illustrated for erythromycin A in Fig. 7.8 [123]. If, however, the charge localization site is not very different, no differences may be observed in fragmentation. This is, for instance, the case for tetracyclines, where Li^+- and Na^+-ions are bound at the tricarbonylmethyl group, whereas protonation is at the dimethylamino substituent of the same ring [129].

Given the difference in properties and possibly the charge location between the proton and the Na^+-ion, one would expect differences in fragmentation behavior between $[M+H]^+$ and $[M+Na]^+$ as precursor ion. Obviously, it is not easy to systematically investigate this for a wide variety of compound classes. In order to facilitate the dereplication of natural products, Fredenhagen et al. [130] compiled a natural-products mass-spectral library, containing the mass spectra of 1020 com-

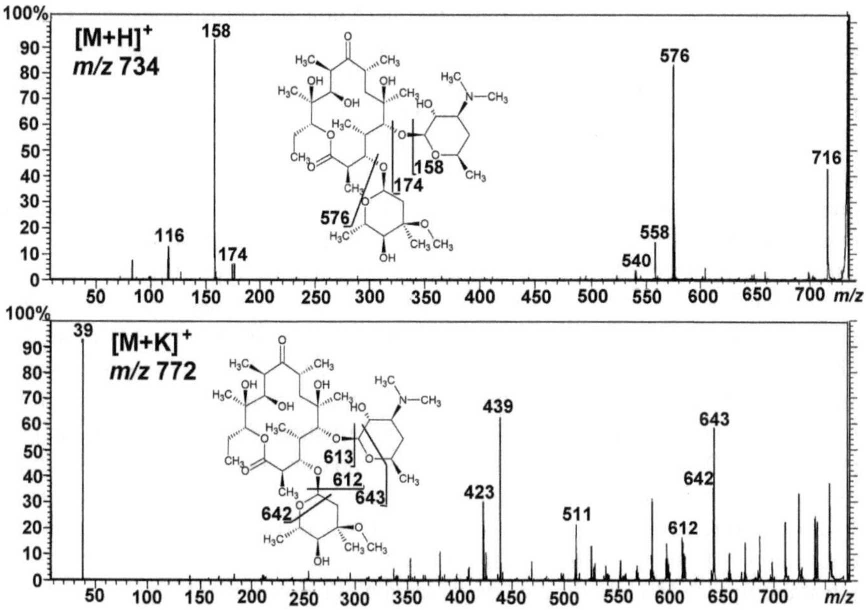

Fig. 7.8 High-energy CID mass spectra of erythromycin A with either the FAB-generated $[M+H]^+$-ion (*top spectrum*) or $[M+K]^+$-ion (*bottom spectrum*) as precursor ions. (Reprinted from [123] with kind permission from Springer Science and Business Media)

mercially available natural products, based on ion-trap MS^2 mass spectra. For 72 % of the compounds, $[M+H]^+$ was used as precursor ion, for 6 % $[M+Na]^+$, whereas, other precursor ions, e.g., $[M+2H]^{2+}$, $[M+H–H_2O]^+$, or $[M–H]^-$, were used for the other compounds. Some special attention was paid to the fragmentation behavior for compounds that showed both $[M+H]^+$- and $[M+Na]^+$-ions. For 30 % of these, the fragmentation patterns were basically the same, although differences in the relative intensity of the product ions may be observed. The latter may be explained by different affinities for the proton or the Na^+-ion. For the other compounds, some common fragmentation behavior was observed for $[M+Na]^+$-ions. In carboxylic acid, the loss of CO_2 is favored. In compounds with a cyclic ketal, such as mocimycin, ring opening into a β-dicarbonyl moiety results in more stable Na^+-chelates. A subsequent retro-aldol condensation generates a characteristic fragment with *m/z* 651, which is not observed with $[M+H]^+$ as precursor ion. For phenolic compounds, tropylium-like ions are observed with $[M+Na]^+$ and not with $[M+H]^+$. The position of the Na^+ seems to determine the fragmentation, rather than the bond stability [130]. Different fragmentation characteristics in CID mass spectra with $[M+H]^+$ as precursor or with $Alkali^+$-cationization products (or with other metal ions) as precursor ion are outlined with more examples below.

Carotenoids are tetraterpenoid compounds with 40 carbon atoms, consisting of eight isoprene molecules. The carotenes are unsaturated hydrocarbons with either a linear structure like lycophene or a partly cyclic structure like α- and β-carotene.

The xanthophylls like astaxanthin are also unsaturated hydrocarbons that contain additional oxygen atoms, e.g., in the form of alcohol, epoxide or ketone functions. These may be conjugated with glycosides or fatty acids. The MS analysis of carotenoids has recently been reviewed [131]. Interestingly, somewhat against expectation, carotenes are readily ionized using ESI-MS, APCI-MS or MALDI-MS [132–135]. However, they show unusually ionization characteristics. In positive-ion mode, carotenes show a competition between formation of $M^{+\bullet}$ and $[M+H]^+$, the relative abundance being somewhat depending on the solvent conditions [133]. For xanthophylls and their fatty acid esters, next to $M^{+\bullet}$ and $[M+H]^+$, also $[M+Na]^+$ may be observed, at least in ESI-MS and MALDI-MS [135]. In negative-ion APCI, $M^{-\bullet}$ as well as $[M–H]^-$ may be formed for carotenes, and mainly $[M–H]^-$ for the xanthophylls [132–134]. Characteristic fragmentation, involving losses of butene, toluene, and combinations thereof, is also observed, together with quite some low-*m/z* fragment ions, which often are not very structure informative. By addition of sodium acetate to MALDI sample preparations of astaxanthin ester mixtures, the low-*m/z* fragments were greatly reduced, and much simpler spectra with structure-informative fragments were obtained [135]. This simplified the profiling of complex carotenoid ester samples from microalgae. It should also be noted, that one of the earliest applications of CIS-MS was the Ag^+-cationization of carotenoids [136].

Steroids comprise another compound class with poor ionization characteristics in ESI-MS, as they lack basic or acidic moieties to direct (de) protonation [137, 138]. Adduct formation of anabolic steroids in ESI-MS was investigated by Pozo et al. [138]. Different mobile-phase additives were applied, i.e., formic acid, Na^+ and NH_4^+. Different behavior was also observed for either methanol or acetonitrile as organic solvent. In the current discussion, focus is given to Na^+-cationization issues. Na^+-cationization is important for steroids with a conjugated keto-group, which form both $[M+H]^+$ and $[M+Na]^+$. Addition of 0.1 mmol/L Na^+ results in $[M+Na+CH_3OH]^+$ when methanol was present and in $[M+H]^+$ when acetonitrile was present. Adduct formation can be reduced by addition of ammonium formate. Steroids with an unconjugated keto group show $[M+Na+CH_3OH]^+$, with methanol present, and only a poor response of $[M+H+CH_3CN]^+$, when acetonitrile is present [138]. More recently, Ag^+-cationization in CIS-MS has been proposed for the residue analysis of anabolic androgenic steroids [139].

In order to facilitate the MALDI-MS analysis of hydrocarbons and wax esters, the use of lithium 2,5-dihydroxybenzoate (Li-DHB) as matrix has been proposed [140]. This matrix proved successful in the analysis of C_{24}- to C_{40}-hydrocarbons, di- and triglycerides, wax esters, and PEGs; $[M+Li]^+$-ions are observed for these analytes. Subsequently, it has been shown that salts of other organic aromatic acids (lithium benzoate, lithium salicylate, lithium vanillate, lithium 2,5-dimethoxybenzoate, lithium 2,5-dihydroxyterephthalate, lithium α-cyano-4-hydroxycinnamate, and lithium sinapate) as MALDI matrices as alternative to Li-DHB provided up to 9×-improved signals [141].

Ag^+-cationization in CIS-MS was, among other, applied to determine the molecular-weight distributions of non-boiling petroleum fractions [142]. The fractions were analyzed using a toluene/methanol/cyclohexane (60:28:12 %) solvent system.

The mass spectra show bimodal distributions ranging from *m/z* 300–3000 and from *m/z* 3000 to 20,000. The results obtained by CIS-MS for the saturated, aromatic, and polar fractions in a bitumen are in qualitative agreement with published data obtained with gel permeation chromatography and/or FDI-MS [142].

Interesting fragmentation is observed upon CID of Ag^+-cationized amines, diamines, aminocarboxylic acids, and alkyl benzyl ethers [143–145]. Both Ag^+-containing and non-Ag^+-containing fragment ions are observed. The latter can be considered as immonium ions, $[(Alkyl)_2CH{=}NH_2]^+$, generated by the loss of AgH due to a 1,2-elimination. If CH_3- and C_6H_5-substituents are present at the α-*C* relative to the amino-*N*, the loss of $AgCH_3$ and AgC_6H_5 may be observed [143, 144]. This hypothesis of 1, 2-elimination has been tested by D-labelling studies [145]. The losses of AgH or $AgCH_3$ have also been observed in the characterization of Ag^+-cationized ferrocenyl catalyst complexes [146].

7.5.2 Glycosides

A glycoside is a natural product comprising of an aglycone with *O*-linked glycans attached to it by means of glycosidic bonds. The aglycone can be, for instance, a steroid, a flavonoid, or an iridoid.

Early accounts on the formation of cationized molecules were related to the analysis of glycosides by means of FDI-MS [147, 148]. The FDI mass spectrum of chlorogenic acid, for instance, showed ions with *m/z* 354 due to $M^{+\bullet}$, *m/z* 355 due to $[M+H]^+$, *m/z* 377 due to $[M+Na]^+$, *m/z* 393 due to $[M+K]^+$, as well as two fragment ions with *m/z* 336 due to $[M–H_2O]^{+\bullet}$ and *m/z* 180 to $M^{+\bullet}$ of caffeic acid (loss of the glycan part). These are typical characteristics of glycosides in FDI-MS. The occurrence of $[M+Na]^+$ and $[M+K]^+$ was attributed to contamination in the solvent used in sample preparation [147]. Similarly, Na^+-cationization was observed in the characterization of ginsenosides after LC fractionation/purification and FDI-MS [16].

Flavonoids, flavonoid glycosides and other flavonoid conjugates are extensively studied using MS techniques. The topic is extensively reviewed, e.g., [149–153]. Data for FDI-MS, FAB-MS, and LSIMS-MS show that most flavonoid glycosides show both $[M+H]^+$-, $[M+Na]^+$-, and $[M+K]^+$-ions, often in combination with some structure-informative fragmentation. Given the fact that flavonoids and their glycosides are present in complex mixtures in their natural sources, LC–MS techniques involving ESI-MS or APCI-MS have been used more recently. Similar ionization characteristics are observed.

For structure elucidation of flavonoids and their glycosides, mostly $[M+H]^+$ in positive-ion mode or $[M–H]^-$ in negative-ion mode is selected as precursor ion. Under these conditions, characteristic fragmentation involves the loss of the glycosidic groups, often step by step, thus enabling some sequence determination, unless isomeric sugar groups are attached. In this respect, one should note that in some cases for diglycosides, e.g., for rutin, fragment ions corresponding to $[M+H–162]^+$ are observed, which are actually due to a rearrangement and the loss of the inner glucose residue [152–154]. This internal residue loss may be observed when fragmenting $[M+H]^+$, but is not observed for $[M+Na]^+$ [155, 156].

CID mass spectra of kaempferol 3-O-glycosides, based on $[M+Na]^+$ as precursor ion, have been studied using ESI-MS^3 in an ion-trap instrument in order to get more information on the glycosidic part [157]. In the MS^2 spectrum, the major fragment ion is due to the neutral loss of the aglycone, thus resulting in $[M+Na]^+$ of the glycosidic part. Subsequent fragmentation of the latter ions in MS^3 showed different mass spectra between four isomeric kaempferol 3-O-rhamnosylhexosides and four isomeric kaempferol 3-O-glucosylhexosides, thus enabling to obtain linkage information based on cross-ring cleavages [157]. The possibility to apply cross-ring cleavages to differentiate glycosidic linkages in disaccharides was previously demonstrated using $[M+Li]^+$ or $[M+Na]^+$ in FAB-MS–MS [158, 159] and ESI-MS–MS [160].

In positive-ion ESI-MS, most glycosides show both $[M+H]^+$, $[M+Na]^+$, $[M+K]^+$ and eventually, if an ammonium salt is present in the mobile phase, $[M+NH_4]^+$. This holds, for example, for flavonoid glycosides [149–153], saponins [161], steroid glycosides, iridoid glycosides [162, 163], and (tetrahydrofuran) lignan glycosides [164, 165]. Fragmentation of Alkali$^+$-cationized molecules as precursor ion in MS–MS provides specific information, often different from what is obtained with $[M+H]^+$ as precursor ion. As an example, the ESI-MS mass spectrum of globularin, an iridoid glycoside, as studied by El-Safi et al. [163], is shown in Fig. 7.7. CID mass spectra after fragmentation of either $[M+H]^+$ or $[M+Na]^+$ as precursor ion are also interestingly different (see Fig. 7.9) [163]. Both $[M+H]^+$ or $[M+Na]^+$ with *m/z* 493 and 515, respectively, show the loss of the glucose moiety, either as glucose (loss of 180 u) or as anhydroglucose (loss of 162 u), although with lower abundance for the $[M+Na]^+$. In the CID mass spectrum of $[M+H]^+$, this is followed by the loss of cinnamic acid (loss of 148 u). The protonated cinnamic acid as well as the cinnamic acylium ion is also observed (ions with *m/z* 149 and 131, respectively). In the CID mass spectrum of $[M+Na]^+$, the ions with *m/z* 185 and 203 correspond to the $[M+Na]^+$ of anhydroglucose and glucose, respectively. This indicates that the proton is preferentially kept by the aglycone, whereas the Na^+-ion is preferentially kept by the glucose part. Cross-ring fragmentation in $[M+Na]^+$ in the iridoid skeleton leads to the fragment ions with *m/z* 283, containing the cinnamic acid moiety, and *m/z* 255, containing the glucose moiety [163].

After FDI-MS, ginsenosides showed abundant $[M+Na]^+$-ions, next to $[M+2Na]^{2+}$-ions. In addition, a series of fragment ions were observed, consistent with losses of water, arabinose, glucose, or combination thereof [16]. FAB-generated $[M+Li]^+$ of iridoid glycosides were subjected to high-energy CID and provide a wider range of fragments than observed for the ESI-MS-generated $[M+Na]^+$ [162]. However, nowadays, high-energy CID is hardly applied anymore, although high-energy CID can still be obtained in MALDI-TOF–TOF instruments [2].

The differentiation of four isomeric saponins, analyzed as $[M+Li]^+$ in ESI-MS, was demonstrated as well. The saponins investigated had two sugar moieties, either 1-2-linked (sophorose) or 1-4-linked (cellobiose), which could be differentiated

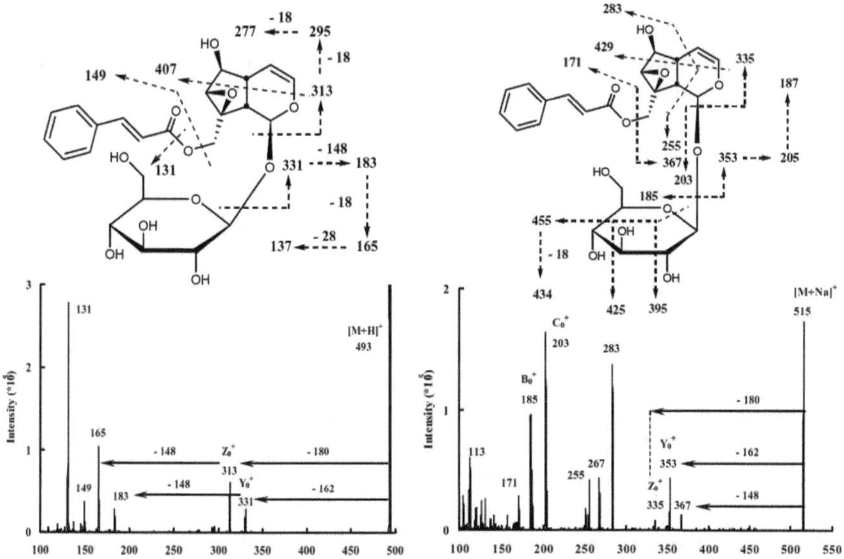

Fig. 7.9 Low-energy CID mass spectra of globularin with either the ESI-generated $[M+H]^+$-ion (*left-hand spectrum*) or $[M+Na]^+$-ion (*right-hand spectrum*) as precursor ions. See text for further explanation and interpretation. (Reprinted from [163] with permission of Wiley; copyright 2007, John Wiley and Sons, Ltd)

from cross-ring cleavages. Specific fragments were also found to discriminate between four possible methylation sites at the aglycone [161].

In the structural characterization of flavonoid glycosides, metal-ion cationization with other than alkali-metal ions has been extensively investigated by the group of Brodbelt [152, 166, 172] using ESI-MS. It was shown that up to 100-fold improved intensity, compared to $[M+H]^+$, could be obtained by metal complexation of the flavonoids and their glycosides with transition metal ions such as Cu^{2+} and Co^{2+} and 2,2′-bipyridine or 4,7-diphenyl-1,10-phenanthroline as an auxiliary neutral ligand. This resulted in [Flavonoid–H+Metal+Ligand]$^+$-ions [166, 167]. Such complexes can also be used to differentiate between 1-2- and 1-6-linked disaccharide isomers of flavonoid glycosides by MS–MS [167, 168]. Subsequently, other metal complexes were evaluated, including metal ions like Al^{3+} [169], Ag^+ [170], Mn^{2+} [171]. In the end, this results in a toolbox with enhanced structure elucidation strategies for flavonoids and their glycosides, based on ESI-MS, ion-trap MS^n, and tuneable transition metal-ligand complexation [172].

7.5.3 Sugars, Glycans, and Oligosaccharides

An oligosaccharide is a saccharide polymer containing a relatively small number of monosaccharides, *O*-linked by glycosidic bonds. The term glycan refers to the carbohydrate portion of a glycoconjugate, such as a glycoprotein, glycolipid, or the natural product glycoside discussed in Sect. 7.5.2.

Compared to the two other classes of biomacromolecules, i.e., proteins and oligonucleotides, the structure of oligosaccharides and glycans can be much more complex. This is due to the presence of up to four stereochemical centers in a hexose, which leads to 16 different hexose structures. In addition to "simple" neutral CHO-sugars, e.g., glucose (Glc), galactose (Gal), mannose (Man), there are also acidic sugars, e.g., glucuronic acid (GluA), sugars containing *N*-acetyl groups, e.g., *N*-acetyl-glucosamine (GlcNAc) and *N*-acetyl-galactosamine (GalNAc), or both, e.g., *N*-acetyl-neuraminic acid or sialic acid (Neu5Ac). Most monosaccharide units in mammals have a D-configuration, as determined by the stereochemistry of the C5-carbon, but some, fucose (Fuc) for instance, have an L-configuration. A hexose may be in solution in an acyclic form or in the form of a five-membered or six-membered ring. Based on the stereochemistry of the anomeric carbon, there are two types of glycosidic bonds: An α-glycosidic bond occurs between two carbons of the same stereochemistry, whereas in the β-glycosidic bond two carbons with different stereochemistry are involved. In addition, there is a range of possible glycosidic linkages, i.e., the 1-1-, 1-2-, 1-3-, 1-4-, and 1-6-linkage. Whereas proteins and oligonucleotides are linear copolymers, branching is possible in oligosaccharides and glycans, obviously via different glycosidic linkages. Thus, full characterization of an oligosaccharide is always quite challenging as it requires the identification of the (possibly isomeric) monomeric units, the determination of the sequence and/or the branching of these units, as well as their linkage type. A lot is already known about glycan structures, e.g., the *N*-linked glycan of a glycoprotein always has a common pentasaccharide core structure of a GlcNAc, which is β-1-4-linked to another GlcNac, which is β-1-4-linked to a Man, which is α-1-6- and α-1-3-linked to two other Man units. For *O*-linked glycans, however, at least six different core structures are known. From the core structure, more extensive glycans with different levels of complexity can be formed [111].

From the early days onward, FAB-MS has been applied to the analysis of oligosaccharides and glycan-containing biomolecules, such as the glycosides discussed in Sect. 7.5.2, glycolipids (Sect. 7.5.4), glycopeptides, and glycoproteins. Using FAB-MS, underivatized oligosaccharides could be analyzed up to ~4 kDa, whereas charge-introducing derivatization strategies extended this range up to ~6 kDa [173]. In addition, FAB-MS analysis revealed the presence of, for instance, methyl, acyl, phosphate, sulfate functional groups, which were more difficult to define by (in those days) conventional carbohydrate analysis methods, such as GC–MS of permethylated sugar monomers. Interestingly, the peracetylated or permethylated derivatives provided the best results in FAB-MS as well, i.e., a ~100-fold improvement relative to underivatized oligosaccharides. Under these conditions, $[M+H]^+$ is observed with sequence-informative fragment ions [173]. FAB-MS also played an important role in the initial characterization of the *N*-linked glycans in glycoproteins [174]. Given the multiple vicinal oxygen atoms, sugars are prone to cationization with $Alkali^+$-ions. It was found that $Alkali^+$-cationization in FAB-MS provides more intense signals for oligosaccharides and glycans. CID of FAB-generated

$[M+Na]^+$-ions enabled discrimination between different linkage positions [159]. Next to FAB-MS, PDI-MS was evaluated for the MS analysis of oligosaccharides as well [175].

MALDI-MS has turned out to be a very powerful tool in oligosaccharide characterization, as demonstrated in some early reviews by Harvey on the topic [176, 177] and the subsequent two-year updates [178–183]. In MALDI-MS, glycopeptides and glycoproteins provide $[M+H]^+$, as protonation can take place at the peptide backbone. Neutral oligosaccharides and glycans, after being released from the protein backbone, yield $[M+Na]^+$ with little fragmentation, whereas acidic glycans show fragments due to losses of the Neu5Ac units. Fragmentation can of course be induced when either post-source decay in a reflectron-TOF-MS instrument or CID in TOF–TOF-MS instrument is applied. Combining MALDI with ion-trap, Q–TOF, or FT-ICR-MS instruments also provides additional means for fragmentation of the MALDI-generated ions. A wide variety of matrices have been investigated for their applicability in oligosaccharide analysis. The most widely used ones are DHB, super-DHB, which is a mixture of 2,5-DHB and 2-hydroxy-5-methoxybenzoic acid, and a 3:1 mixture of mixture of 2,5-DHB and 1-hydroxyisoquinoline [99] for neutral oligosaccharides, and 2,4,6-trihydroxyacetophenone (THAP) for acidic oligosaccharides. For neutral oligosaccharides, predominantly $[M+Na]^+$-ions are generated. By doping with appropriate salts, other cationized species can be formed. The affinity of oligosaccharides for alkali-metal ions was found to increase in the order $Li^+ < Na^+ < K^+ < Cs^+$ [99]. The series of review papers on MALDI-MS of oligosaccharides and glycans [177–183] started to cover applications of MALDI-MS in the fields of, among others, plant-derived carbohydrates, *N*- and *O*-linked glycans from glycoproteins, but also glyco(sphingo)lipids and related compounds, and glycosides. Subsequently, new application areas were added as developments in sugar science and MS technology were proceeding, e.g., an increasing number of sugar derivatization strategies, the characterization of glycosylation profiles of therapeutic biopharmaceuticals like monoclonal antibodies, fragmentation of carbohydrates, studies on glycosyl-transferase and glycosidase enzymes, glycan arrays, MALDI imaging, and the use of ion mobility spectrometry. Many of these applications are presented in tabular form [178–183]. In the most recent review [183], Harvey concludes that MALDI-MS continues to be a major technique for carbohydrate analysis, despite the growing competition from ESI-MS. A major advantage of MALDI-MS over ESI-MS is the absence of multiple-charge ions, thus leading to an easier to interpret and cleaner glycan or oligosaccharide profile. Despite the spectacular progress in the past 15 years, there are still challenges, e.g., in identifying the correct structure and in the analysis of sialylated and sulfated glycans. In this respect, it is interesting to note that "old-fashion" approaches like permethylation are gaining popularity again [183].

Next to MALDI-MS, the use of ESI-MS is important in oligosaccharide and glycan characterization, especially because it can be combined with on-line LC separation of complex mixtures. The most important LC methods in glycan and oligosaccharide separation are HILIC [184], HPAEC, and, after derivatization by reductive amination with 2-aminobenzamide or 2-aminoacridone, RPLC on either C_{18} or porous graphitized carbon materials [185].

The MS analysis and characterization of oligosaccharides have been reviewed by Zaia [111]. The ionization of oligosaccharides in ESI-MS is limited by their hydrophilic character. $[M+H]^+$ is only observed for oligosaccharides and glycans that contain an aminosugar, e.g., GlcNAc, or when protonation sites have been introduced via derivatization. Otherwise, $[M+Na]^+$ or $[M+NH_4]^+$ are observed. By application of appropriate metal ion salts, cationization with Li^+, Ca^{2+}, Mg^{2+}, Co^{2+}, Cu^{2+}, Mn^{2+}, or Ag^+ may be achieved as well. Most oligosaccharides, and especially the acidic ones, can be analyzed as $[M–H]^-$ in negative-ion mode as well [111]. Different derivatization strategies have been used. Permethylation of oligosaccharides is applied for linkage analysis in GC–MS, but in ESI-MS or MALDI-MS analysis, permethylation can be useful as well. Alternatively, saccharide derivatization, mostly via reductive amination of the carbonyl group in the reducing sugar unit, can be performed to introduce a chromophore or fluorophore for UV or fluorescence detection after LC separation, to facilitate the RPLC separation of oligosaccharides, and/or to direct the fragmentation in MS–MS. Quite a variety of amines have been applied in this reductive amination reaction [111, 186].

Significant progress has also been made in understanding the fragmentation of oligosaccharides. A nomenclature system has been developed and is extensively used to annotate fragment ions in MS–MS spectra of oligosaccharides [187]. Computational studies provided us with insight in the stability of glycosidic bonds and led to general rules on the fragmentation of $[M+Na]^+$-ion of oligosaccharides [94]. The calculations revealed that cleavage of α-Glc, α-Gal, β-Man, α-Fuc, β-GlcNAc and β-GalNAc linkages are more easy than cleavages of β-Glc, β-Gal, and α-Man linkages, whereas the Neu5Ac bond is very labile. Stabilities of glycosidic bonds were calculated, and seem to be in good agreement with experimental findings [94].

The fragmentation of oligosaccharides, glycans and glycopeptides has been reviewed in detail by Zaia [111]. Attention is paid to characteristic features in the fragmentation of $[M+H]^+$, $[M–H]^-$, and various alkali and earth alkali cationized ions. Differences observed may be rationalized by the fact that protonation results in a localized charge, leading to charge-induced cleavages, whereas the alkali-metal ion undergoes coordination with several oxygen atoms, typically in a tridentate interaction [94], leading to more charge-remote fragmentation mechanisms. At an early stage, it was established that $[M+Na]^+$-ions in high-energy CID are more prone to cross-ring cleavages than $[M+H]^+$-ions [188]. Later on, it was demonstrated that such cross-ring cleavages can also be achieved in low-energy CID on a sector–quadrupole hybrid instrument [158], in ion-trap instruments [160], and in post-source decay in MALDI-TOF-MS [176]. This feature has effectively been applied in linkage analysis. Fragmentation yields seem to be inversely related to the size of the cation, i.e., $H^+ > Li^+ > Na^+ > K^+ > Rb^+ > Cs^+$ [177]. Three fragmentation pathways seem to be important in the fragmentation of alkali-metal ion cationized oligosaccharides: (1) cleavage of the glycosidic bond, (2) cross-ring cleavage, and (3) cleavage of the bond between the metal ion and the sugar. Detailed information is also provided on the fragmentation of permethylated and peracetylated oligosaccharides, which still may generated $[M+Na]^+$, eventually next to $[M+H]^+$, of reductively aminated carbohydrates, which mostly generated $[M+H]^+$, linkage analysis from MS–MS spectra of various types of ions using either high-energy or low-energy CID, and "internal-residue" losses (see Sect. 7.5.2) [111].

7.5.4 Lipids and Related Compounds

An eight-category lipid classification system has been proposed [189]: (1) fatty acyls including fatty acids, fatty alcohols, and fatty esters, (2) glycerolipids including mono-, di-, and triacylglycerols, (3) glycerophospholipids including phosphocholines (GPCho), phosphoglycerols (GPGro), phosphoethanolamines (GPEtn), phosphoserines (GPSer), phosphoinositols (GPIno), and phosphatidic acids (GPA), (4) sphingolipids including sphingoid bases, ceramides, phosphosphingolipids and glycosphingolipids, (5) sterol lipids including sterols, cholesterol and its derivatives, steroids, and bile acids, (6) prenol lipids including isoprenoids and polyprenols, (7) saccharolipids including acylaminosugars and their glycosides and acyltrehaloses and their glycosides, and (8) polyketides including macrolide and aromatic polyketides like aflatoxins [189]. The MS analysis and characterization of some of these lipid classes is briefly discussed in this section, with emphasis on $Alkali^+$-cationization.

Fatty acids are long-chain alkane carboxylic acids, which may contain double bonds, hydroxy, and/or carbonyl groups. Mammalian fatty acids have an even number of carbon atoms. Fatty acids have great biological significance, by itself and esterified with other biomolecules. Among these are fatty acids like eicosapentaenoic acids (C_{20}-ω_3-fatty acid), which are precursors of important compound classes like prostaglandins, leukotrienes, and thromboxanes, and docosahexaenoic acid (C_{22}-ω_3-fatty acids), which is the primary structural component of the human brain and skin. Fatty acids can form esters with a wide variety of biomolecules. Fatty acid mono-, di- and triacylglycerols form an important class of lipids, often called fats. Other component classes involve ester formation with compounds like bile acids, steroids, and sugars. Phospholipids consist of a glycerol esterified with one or two fatty acids and containing a polar head group. This leads to the six phospholipid subclasses indicated above (GPCho, GPGro, GPEtn, GPSer, GPIno, and GPA). Phosphocholines (with two fatty acid esters) and lysophosphocholines (with just one fatty acid ester) are the most abundant phospholipids in human plasma. Sphingolipids form a complicated class of lipid-related molecules. They consist of a sphingoid base, an *N*-acyl fatty acid, and a polar head group. In mammals, the major sphingoid base is sphingosine ($HO–CH_2–CH(–NH_2)–CH(–OH)–(CH_2)_{14}–CH_3$). Different polar head groups lead to sphingolipid subclasses, e.g., with OH for a ceramide, a carbohydrate for a glycosphingolipid, or a phosphorylcholine for a phosphonosphingolipid.

The MS analysis of lipids and related compounds has been frequently reviewed. An elaborate review of the MS and MS–MS analysis of lipids, bile acids, and steroids, covering a range of analyte ionization techniques, has been compiled by Griffiths [190]. The MS analysis of long-chain lipids, especially focusing on analyte ionization using ESI-MS and MALDI-MS, has been reviewed by Murphy and Axelsen [191]. Comprehensive lipid profiling using LC–MS with either ESI or APCI is reviewed by Cajka and Fiehn [192], whereas more general reviews on mass spectrometric methods in lipidomics are available as well [193, 194].

Using soft-ionization techniques, fatty acids, including prostaglandins and related acidic compounds, are analyzed mostly as deprotonated molecules $[M–H]^-$ in negative-ion mode, although the formation of lithiated lithium salts $[(M–H+Li)+Li]^+$, to be analyzed in positive-ion mode, can have certain advantages for structure elucidation (see below).

FDI-MS has been applied to the characterization of natural waxes and other types of lipid-related compounds [195, 196]. For components extracted from waxes, such as fatty alcohols and diols, fatty acids, and fatty acid esters, $M^{+\bullet}$ and $[M+H]^+$ were observed. The same holds for free fatty acids, cholesterol and its esters, and triglycerides extracted from human plasma, whereas $[M+H]^+$, $[M+Na]^+$, and/or $[M+K]^+$ were observed for GPCho from human plasma. FDI-MS proofs to be a versatile technique for lipid profiling [196]. Cs^+-cationization has been applied in the FDI-MS analysis of glycolipids such as β-hydroxyacyltrehaloses [197].

After its introduction in 1981, FAB-MS has also been widely applied in the characterization of lipids and related compounds. Both $[M+H]^+$ and $[M+Na]^+$ are generally observed in the positive-ion FAB-MS analysis of phospholipids [198, 199]. This was applied in, for instance, phospholipid profiling in extracts from the archaebacterium *Halobacterium cutivubrum* [199].

An important contribution to the field is the discovery of charge-remote fragmentation (CRF) in $[(M–H+Li)+Li]^+$-ions of fatty alcohols, fatty acids, and esters by Adams and Gross [200]. High-energy CID was performed on $[(M–H+Li)+Li]^+$ generated by FAB-MS. The mechanism of CRF involves C–C-bond cleavages along the alkyl chain and yields a neutral alkene, H_2, and an unsaturated fatty acid product ion. CRF was already known from long-chain fatty-acid anions in negative-ion mode [200]. In fact, CRF of $[(M–H+Li)+Li]^+$ in positive-ion mode provides more structural information than CRF of $[M–H]^-$ in negative-ion mode. For a saturated fatty acid, a series of ions differing 14 u is observed. The double bond of an unsaturated fatty acid chain interrupts this smooth pattern and results in a gap of 54 u. The gap is 54 u, because cleavage of both the double bond and the vinylic bond is unfavorable, whereas cleavage of the allylic bond is favorable. Figure 7.10 shows high-energy CID spectra of palmitic acid and 8,11-octadecadienoic acid with the $[(M–H+Li)+Li]^+$-ion as precursor ion [202]. CRF in high-energy CID has been extensively used for determination of double-bond positions in fatty acids [201–203]. The interesting features in fragmentation of lithiated fatty acids also apply to other functional groups, e.g., hydroxy and carbonyl groups as well as alkyl branching, as for instance observed in prostaglandins [204]. The CRFs observed in FAB-MS with high-energy CID in a sector instrument [205] are not observed in ESI-MS with low-energy CID in a triple-quadrupole instrument; although a wealth of structural information was still obtained, enabling discrimination between isomeric unsaturated fatty acids [206–208].

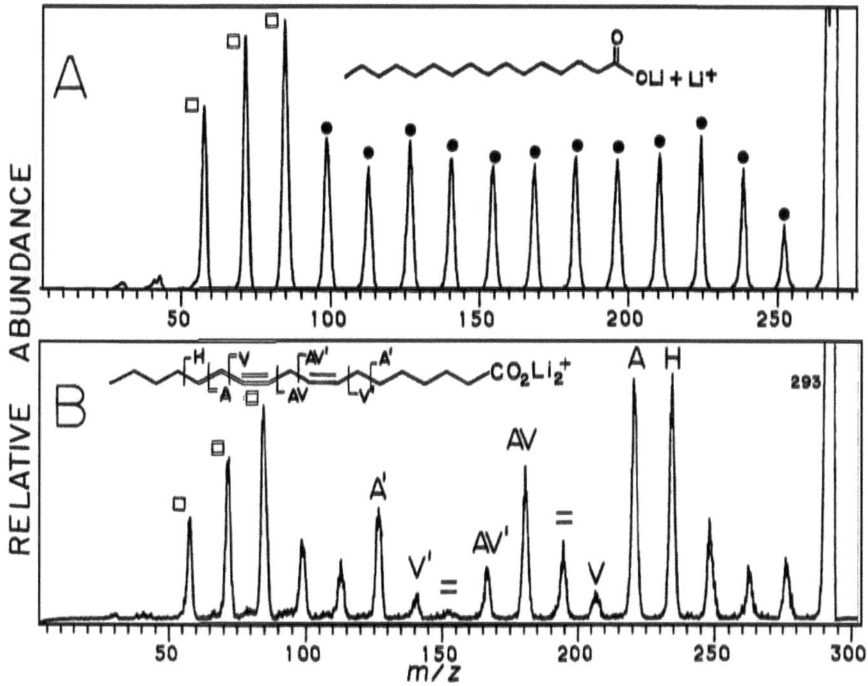

Fig. 7.10 High-energy CID mass spectra of FAB-generated [(M–H+Li)+Li]$^+$-ions of palmitic acid and 8,11-octadecadienoic acid, demonstrating charge-remote fragmentation. See text for further explanation and interpretation. (Reprinted from [202] with kind permission from Springer Science and Business Media)

Mono-, di-, and triacylglycerols are not readily ionized in ESI-MS, unless a lithium or ammonium salt is added to the mobile phase, thus resulting in $[M+Li]^+$ or $[M+NH_4]^+$ [192]. APCI-MS should be used instead to generate $[M+H]^+$ [209]. CID fragmentation of $[M+H]^+$ of triacylglycerols results in three ions in the middle-*m/z* range of the spectrum due to losses of the individual fatty acids from the glycerol backbone. Due to steric hindrance, the loss of the fatty acid in *sn-2* position leads to a less abundant peak. In this way, the order of the fatty acids on the glycerol backbone can be established. In addition, less abundant peaks due to protonated fatty acids and/or the fatty acid acylium ions (with difference of 18 u) may be observed in the low-*m/z* region of the spectrum. In the MS–MS spectra of $[M+Li]^+$ of triacylglycerols, some additional features are observed, as illustrated for a (16:0/18:1/18:0)-triacylglycerol (palmitic, oleic, and stearic acid; $[M+Li]^+$ with *m/z* 867) in Fig. 7.11 [210]. Instead of three, six peaks are now observed in the middle-*m/z* region of the spectrum. They occur in pairs differing 6 u. They can be attributed to the loss of either the fatty acid (R–COOH) or the lithium salt of the fatty acid (R–COOLi), i.e., the ions with *m/z* 605 and 611 are due to the loss of the *sn-1* palmitic acid (16:0), and the less abundant ions with *m/z* 579 and 585 to the loss of the *sn-2* oleic acid (18:1).

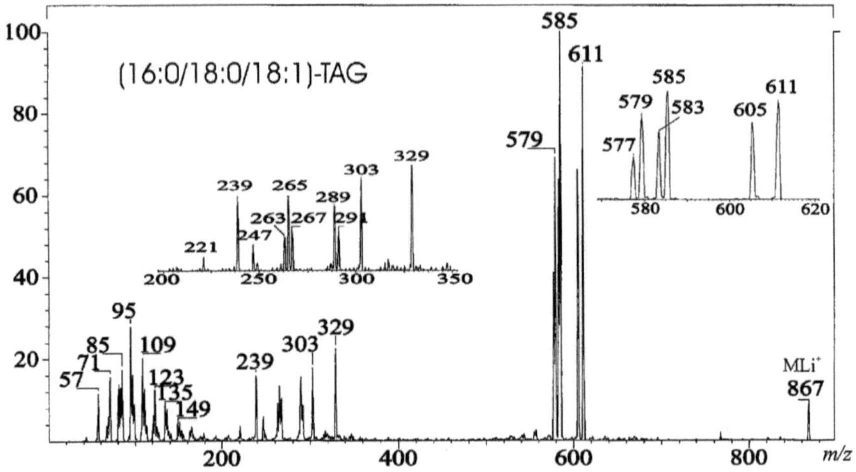

Fig. 7.11 Low-energy CID spectrum of ESI-generated $[M+Li]^+$-ions of (16:0/18:0/18:1)-triacylglycerol. See text for further explanation and interpretation. (Reprinted from [210] with kind permission from Springer Science and Business Media)

In the low-*m/z* region, lithiated fatty acids $[R–COOH+Li]^+$ and the fatty acid acylium ions $[R–C≡O]^+$ are observed, e.g., with *m/z* 291 and 267 for the *sn-3* stearic acid (18:0). Ions due to combined losses of either the *sn-1* or the *sn-3* fatty acid and the *sn-2* fatty acid may be observed as well. This elimination involves the loss of a free fatty acid and an α,β-unsaturated fatty acid, thus leading to ions with *m/z* 331 due to the loss of the *sn-1* palmitic acid and the *sn-2* oleic acid and with *m/z* 303 due to the loss of the *sn-3* stearic acid and the *sn-2* oleic acid [210]. At the very low-*m/z* range, indications are found, suggesting that CRF occurs under these conditions as well. In fact, infusion of triacylglycerols in a 2-mM lithium acetate-containing solution under in-source CID conditions generates $[(RCOOH–H+Li)+Li]^+$-fragment ions of the fatty acids present in the triacylglycerol in the source. This enables their selection as precursor ion in an MS–MS experiment. The low-energy CID spectrum shows the characteristic pattern of CRF and allows determination of the double-bond positions [210].

Unlike the fatty acids and the triacylglycerols, most other lipid classes, including most phospholipids and sphingolipids, can readily be ionized in positive-ion ESI-MS to generate $[M+H]^+$ and $[M+Na]^+$. With an ammonium salt in the mobile phase, $[M+NH_4]^+$ may also be observed for GPIno, GPGro, cholesteryl esters, and ceramides. ESI-MS in the negative-ion mode is also readily applicable to all classes of phospholipids, and to ceramides [192].

Phospholipids have been studied by means of many condensed-phase ionization techniques. In the FAB-MS mass spectra of GPCho, the three most abundant ions were $[M+H]^+$, $[M+Na]^+$, and a fragment ion $[M+Na–59]^+$ due to the loss of trimethylamine. CID mass spectra with $[M+Na]^+$ as precursor ion resulted in the loss of the fatty acids and the loss of choline phosphate, thus $[M+Na–183]^+$,

$[M+Na–184]^{+•}$, and $[M+Na–205]^{+}$. Fragmentation of the latter results in CRF of both fatty acids [211].

In an elaborate paper, Hsu and Turk [212] reviewed the fragmentation of various glycerophospholipid classes after ESI-MS and low-energy CID in both triple-quadrupole MS–MS and ion-trap MS^n. They compared fragmentation processes for the various phospholipid classes using $[M+H]^+$, $[M+Li]^+$, and/or $[(M–H+Li)+Li]^+$ as precursor ion in the positive-ion mode and using $[M–H]^-$ as precursor ion in the negative-ion mode. Different structural information can be derived from the mass spectra obtained in different modes of operation, when applicable [212].

Sphingolipids have been studied by soft-ionization methods, with FDI-MS, FAB-MS, ESI-MS, and MALDI-MS being the most widely used techniques. FDI-MS was applied for the characterization of glycosphingolipids with up to five saccharide units [213]. Both FDI-MS and LSIMS-MS were applied in the analysis of acidic glycosphingolipids, i.e., carrying asialo or sulfatide units [214]. In both cases, both $[M+H]^+$ and $[M+Na]^+$ was observed, whereas with the acidic compounds $[(M–H+Na)+Na]^+$-ion were observed as well. Similar observations have been made using other condensed-phase ionization methods [215, 216].

Based on $[M+H]^+$- or [M–H]-ions as precursor ions, CID mass spectra provide information on the polar head group, the sphingoid base, and *N*-acyl fatty acid [215, 216]. Ann and Adams [205, 217] showed that the fragmentation pathways can be influenced by generating Alkali$^+$-cationized molecules. They studied cationization in FAB-MS with Li^+, Na^+, K^+, Rb^+, and Cs^+. Information-rich CID mass spectra are obtained from $[M+Li]^+$- and $[M+Na]^+$-ions, providing the length of the sphingoid base and of the fatty acid, but also information on locations of double bonds and hydroxy groups in the fatty acid chain [217]. The latter involves CRF in the high-energy CID on a sector instrument. More recently, FAB-MS with high-energy CID was applied to the characterization of cerebrosides, which are glycosphingolipids. Fragmentation of $[M+Na]^+$ provided information on the compositions of the 2-hydroxy fatty acids, the long-chain sphingoid bases, and the sugar moiety at the polar head group. Location of the double bonds in the fatty acids and the methyl branching on the sphingoid base could also be determined [218]. Low-energy CID in combination with ESI-MS generally provides less information on sphingolipids, i.e., limited to the polar head group, the sphingoid base, and *N*-acyl fatty acid. CRF of the fatty acid chain is not achieved [206, 207]. Perhaps, a strategy involving a combination of in-source CID and MS–MS, as outlined above for triacylglycerols [210], may be applicable to sphingolipids as well.

MALDI-MS is generally performed at (complex) lipids mixtures, although post-LC-column fractionation systems have been developed as well. Additionally, MALDI-MS imaging of TLC plates is useful technology in lipid analysis [219]. The application of MALDI-MS in lipid analysis and lipidomics has been extensively reviewed [193, 194, 220, 221]. The most widely applied matrices in lipid analysis are DHB and 2,6-dihydroxyacetophenone (DHAP). Unlike in ESI-MS, MALDI-MS provides positive-ion response for all phospholipid classes. Individual components may be observed as $[M+H]^+$, $[M+Na]^+$, $[M+K]^+$, or adduct ions with additional H^+/Na^+- or H^+/K^+-exchange, thus significantly complicating the interpretation and (relative) quantification of individual components in

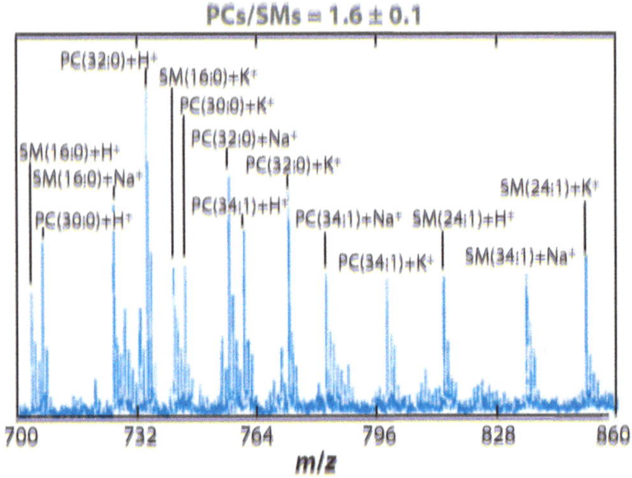

Fig. 7.12 Lipid profiling in bovine lens tissue by MALDI-MS. Glycerophosphocholines (*PC*) and phosphonosphingolipids (*SM*) profile of the outermost cortex region, fixed in 2.5 % formaldehyde buffered at pH 7.4. (Reprinted with permission from [222]; copyright 2004, American Chemical Society)

the profile. Besides, the most abundant ions are obtained for GPCho, carrying a permanent charge. As a result, ionization suppression of other phospholipid classes may occur. MALDI-MS plays an important role in lipidomics by tissue imaging [193, 221]. The features discussed here are nicely demonstrated in Fig. 7.12, showing a zoomed mass spectrum (*m/z* 700–860) of the outermost cortex region of a bovine lens tissue, fixed in 2.5 % formaldehyde sodium-phosphate-buffered solution (pH 7.4). $[M+H]^+$-, $[M+Na]^+$-, and $[M+K]^+$-ions are observed for a variety of GPCho and phosphonosphingolipids [222].

Charge localization after Li^+- or Na^+-cationization of lipids seems to be at the acidic and ester functions; the alkali-metal ions seem not to interact with the double bonds in an unsaturated fatty acid. The same holds for instance for Ba^{2+}-cationization of fatty acids in FAB-MS [202]. The effect of wide range of metal ions, i.e., Li^+, Na^+, K^+, Cu^+, Sr^{2+}, Ba^{2+}, Mn^{2+}, Fe^{2+}, Co^{2+}, Ni^{2+}, Cu^{2+}, Zn^{2+}, Sc^{3+}, V^{3+}, and Cr^{3+}, on the ionization and fragmentation characteristics of various classes of phospholipids, i.e., GPEtn, GPGro, and GPSer, has been investigated [223]. Double- and triple-charge metal cations do not form complexes with GPCho, probably due to the zwitterionic character of the analytes. Notable differences in fragmentation between the various cationized species were observed. For instance, Co^{2+}-GPEtn complexes upon CID showed preferential loss of the *sn-1* fatty acid over the *sn-2* fatty acid, where with Co^{2+}-GPGro complexes the reverse behavior is observed [223].

Ag^+-cationization of lipids, as performed under CIS-MS conditions (see Sect. 7.2.6), relies on the interaction of Ag^+ as a soft Lewis acid with the *C–C*-double bond as soft Lewis base. This is probably the background of the LC separation of unsaturated lipids and fatty acids using Ag^+ -loaded columns [224–226]. Ag^+

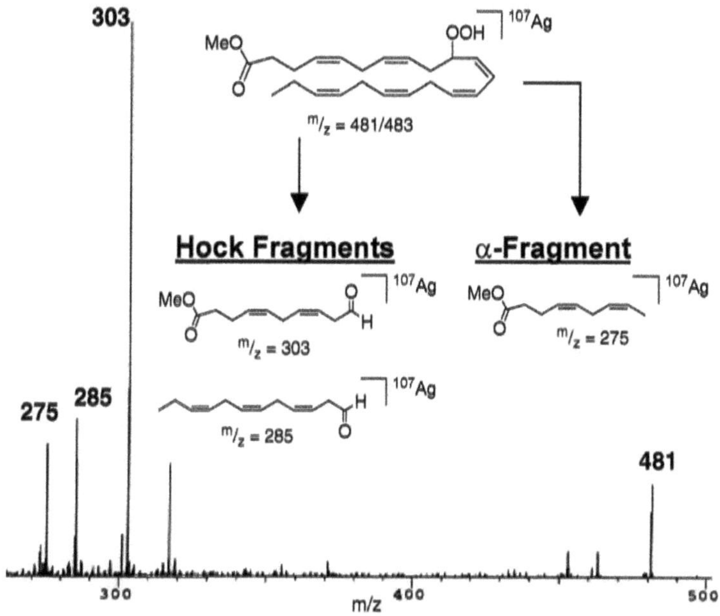

Fig. 7.13 Low-energy CID mass spectrum of an Ag^+-cationized component in a methyl docosahexaenoate oxidation mixture, separated by LC and mass analyzed using Ag^+-CIS-MS, showing characteristic fragments obtained under these conditions. See text for further explanation and interpretation. (Reprinted from [228] with kind permission from Springer Science and Business Media)

-cationization in CIS-MS has been applied to the characterization of triacylglycerols in vegetable oils [225, 226], of products of lipid peroxidation such as hydroperoxides and polyperoxides of cholesterol linoleate and cholesterol arachidonate [227], docosahexaenoate ester hydroperoxides [228], and ω_3-fatty acid eicosapentaenoic acids [229]. Structure informative fragmentation is achieved under these conditions. For instance, $[M+Ag]^+$ of a methyl docosahexaenoate oxidation product shows two so-called Hock fragments and an α-fragment, both involving a cleavage next to the hydroperoxide group (see Fig. 7.13) [228]. This enables determination of the position of the hydroperoxide group in the fatty acid chain [229].

7.5.5 Peptides and Proteins

Proteins and peptides in their primary structure are linear polyamide copolymers of L-α-amino acids. There are 20 DNA-encoded amino acids, and a wide variety of other (modified) amino acids. Proteins are peptides with more than 40–50 amino acids, which in fact is quite an arbitrary number. Before being biologically active, a protein, after translation from its gene, will (partly) fold into α-helices and β-sheets, which subsequently fold into a tertiary structure. For some proteins, a quaternary structure exists, consisting of multiple protein subunits and eventually cofactors. An important aspect in their biological functions is the decoration of the protein by post-translational modifications (PTMs), including phosphorylation, glycosylation and a wide range of other modifications. Initially, MS analysis of proteins was primarily directed at elucidating the primary structure, that is: in determining the amino acid sequence. More recently, especially due to powerful condensed-phase ionization techniques such as MALDI-MS and ESI-MS, MS has extended its potential to characterization of PTMs, relative and absolute protein quantification to establish protein expression levels in cell, and even to non-covalent complexes, e.g., protein-protein complexes, but also complexes of proteins with other type of molecules including DNA, sugars, lipids, and drugs.

There is probably no single compound class where more MS-based research and applications have been performed for than for peptides and proteins. Both EI-MS and basically all condensed-phase ionization techniques have been applied in the characterization of peptides and proteins. Illustrative review papers have been published on the analysis of peptide using PDI-MS [230], FAB-MS [231], and FDI-MS [232]. Routine intact protein analysis by MS became available with the introduction of ESI-MS [233, 234], showing an ion envelope of multiple-charge ions (see Sect. 7.4) [102, 103], and MALDI-MS, showing primarily single- and double-charge ions (see Sect. 7.4). In peptides and proteins, protonation predominantly takes place at the *N*-terminal amino group and at the basic amino acids Lys, His, and Arg [233, 234]. Alkali$^+$-cationization may be due to H$^{+/}$Alkali$^+$-exchange phenomena, involving the *C*-terminal carboxylic acid and the carboxylic acid group in Glu and Asp side chains, or due to chelation of the Alkali$^+$-ion either at the carbonyl oxygen at the *C*-terminal amino acid [235] or at amide bonds (see Sect. 7.4) [95]. If the *C*-terminal carboxylic acid group is converted in a methyl ester, cationization can only take place at the amide backbone units [236]. Alkali$^+$-cationization may adversely affect spectrum deconvolution procedures in ESI-MS. However, Alkali$^+$-cationization of peptides has also been reported for other condensed-phase ionization techniques. K$^+$-cationization of penta- and hexapeptides derived from substance P were observed using FDI-MS [237].

Microcystins are hepatotoxic cyclic heptapeptides, that are prone to $[M+H]^+$ formation and H$^+$/Na$^+$-exchange, as two carboxylic side groups are present. This may lead to a series of ions, i.e., the single-charge ions $[M+H]^+$, $[(M–H+Na)+H]^+$ or $[M+Na]^+$, $[(M–2H+2Na)+H]^+$ or $[(M–H+Na)+Na]^+$, and $[(M–2H+2Na)+Na]^+$, and the double-charge ions $[M+2H]^{2+}$, $[(M–H+Na)+2H]^{2+}$ or $[M+H+Na]^{2+}$, $[(M–2H+2Na)+2H]^{2+}$ or $[(M–H+Na)+H+Na]^{2+}$, and $[(M–2H+2Na)+H+Na]^{2+}$ [238]. This obviously impairs the detection of microcystins by LC–MS and compromises accuracy and precision in quantitative analysis. In screening for microcystins, based on precursor-ion analysis of double-charge ions and the common product ion with *m/z* 135, $[C_6H_5–CH_2–CH=O–CH_3]^+$, a fragment of the side chain of the unique β-amino acid 3-amino-9-methoxy-2,6,8-trimethyl-10-phenyldeca-4,6-dienoic acid (ADDA), a wide variety of precursor ions is found, as the *m/z*-135 ion is generated by any of the double-charge ions indicated. Addition of ammonium formate, either to the mobile phase or in a pulsed post-column way, greatly reduces the formation of sodiated species in favor of $[M+H]^+$ an $[M+NH_4]^+$ [238].

Fragmentation of peptides and proteins by CID and other ion dissociation techniques in MS–MS has been extensively investigated. Fragmentation mechanisms have been elucidated in detail [239, 240]. Interpretation strategies and nomenclature systems have been developed [231, 241–243]. High-throughput protein identification strategies in complete proteomes have been developed, relying not so much on interpretation of MS–MS spectra, but rather on bioinformatics tools and database searching strategies [244–250]. Proteomics strategies and workflows [250] have been implemented in various research fields, directed at protein biomarker discovery, drug discovery, understanding cell functioning and cellular signalling cascades, and development and characterization of biopharmaceuticals. MALDI-MS strategies involving peptide mapping strongly contribute to current proteomics research [250–253], e.g., in clinical diagnostics and (cancer) biomarkers discovery and analysis [254].

Low-energy CID of protonated peptides mainly involves cleavage at the peptide bond, resulting in two complementary ions, an acylium ion (or protonated oxazolone derivative) and a protonated peptide, which according to the nomenclature rules [231, 241] are called b- and y-ions, respectively. High-energy CID opens a wider range of peptide fragmentation, including other backbone cleavages and internal cleavages within amino acids, whereas electron-capture and electron-transfer dissociation (ECD and ETD) provide fragmentation at other bonds, mainly leading to c- and z-ions [255].

Fragmentation of metal ion cationization products of peptides $[M+Metal]^+$ has also been investigated, initially using FAB-MS on a sector instrument [235, 256]. The fragmentation of $[M+Metal]^+$-ion of peptides (with Metal is Li^+, Na^+, Ag^+, Cu^+) results in the (subsequent) loss of the *C*-terminal amino acid residue (most likely as CO and the amino-acid imine, $[HN=CHR_{AA}+H]^+$). Thus, $[b_{n-1}+Metal+OH]^+$- and $[a_{n-1}+Metal–H]^+$-ions are formed (with an *m/z* difference of 46 u). This enables *C*-terminal sequencing. This characteristic fragmentation is not observed for $[M+K]^+$- and $[M+Rb]^+$-ions. In fact, the $Alkali^+$-metal ions are chelated by the carbonyl oxygens at the *C*-terminal of the peptide. This catalyzes the hydrolysis of the amide bond, due to a nucleophilic attack of the *C*-terminal carboxylate anion, which leads to expelling the *C*-terminal amino acid residue. In effect, rearrangement of an OH of the *C*-terminal amino acids takes place [235, 256]. Using MS^n, part of the *C*-terminal sequence of a peptide can be established along these lines, as demonstrated in the composite MS^n mass spectrum of the sodiated GVYVHPV-peptide, shown in Fig. 7.14 [257]. There has been some debate on the exact mechanism involved in this *C*-terminal residue loss [235, 258, 259]. The same type of fragmentation can be achieved after $Alkali^+$-ion production by MALDI or ESI [258, 259].

Sabareesh and Balaram [236] compared the fragmentation of $[M+H]^+$ and $[M+Na]^+$ of modified peptides of which the *N*-terminal was blocked by *t*-butyloxycarbonyl (Boc) group (residue mass 100 Da) and the *C*-terminal was converted into a methyl ester. With $[M+H]^+$ as precursor ion, the loss of CH_3OH, which formally

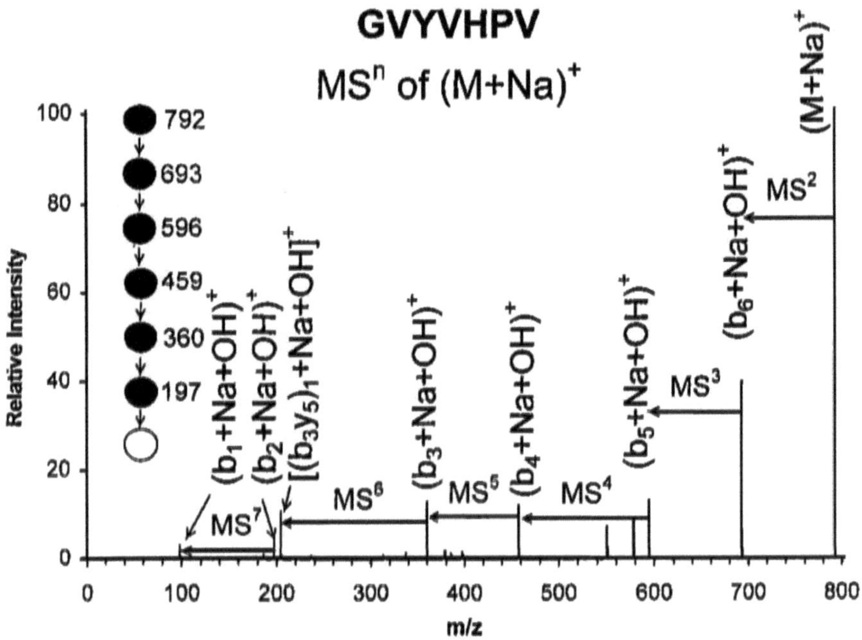

Fig. 7.14 Composite MS^n mass spectrum of $[M+Na]^+$ as precursor ion of a peptide with a sequence GVYVHPV. Due to sequential *C*-terminal losses in subsequent MS–MS steps, almost the complete sequence of the peptide can be derived. (Reprinted with permission from [257]; copyright 1998, American Chemical Society)

results in a b-ion, resulted in a more abundant ion than the loss of Boc (formally resulting in a y-ion). In addition, a series of additional b-ions was observed. With $[M+Na]^+$ as precursor ion, the loss of 56 u (C_4H_8) and of Boc was observed, the latter formally resulting in a y-ion. With much lower abundance, a series of y-ions were observed, next to $[a–Boc]^+$- and $[b–Boc]^+$-ions [236]. Note, that the characteristic loss of *C*-terminal amino acid residues is not observed with the $[M+Na]^+$ of *C*-terminal-modified peptides [236].

7.5.6 *Oligonucleotides*

Oligonucleotides are short, single-stranded DNA or RNA molecules (typically up to 80 nucleotides). They consist of four nucleosides, i.e., the purines adenosine (A) and guanosine (G) and the pyrimidines cytosine (C), thymidine (T, in DNA), and uridine (U, in RNA), connected via a 3′-5′-phosphodiester linkage between subsequent deoxyriboses (in DNA) or riboses (in RNA). Oligonucleotides can be produced by solid-phase chemical synthesis. The role of MS in the analysis of DNA and RNA biomolecules is different from that in the analysis of proteins or oligosaccharides and glycans. For the latter, MS plays an important role in sequencing, while DNA sequencing is predominantly done by other, non-MS techniques (in DNA sequencers). As such, MS is more important for checking the identity and/or impurity profiling of synthetic oligonucleotides produced by solid-phase synthesis, eventually containing modified oligonucleotides, some of which are used in or under development for therapeutic applications, and in genotyping of short tandem repeats and single nucleotide polymorphisms in clinical applications [260].

MS analysis of oligonucleotides is predominantly based on the use of condensed-phase ionization techniques. After initial results in FDI-MS [261], PDI-MS [261] and FAB-MS [263] were applied to analyze larger oligomers (up to 13-mers). Nowadays, MALDI-MS and ESI-MS are most frequently applied in the analysis of oligonucleotides [264, 265]. In some applications, e.g., impurity profiling of synthetic oligonucleotides and the (bio) analysis of therapeutic (antisense) oligonucleotides [266], ESI-MS seems to be preferred, as it can be applied in combination with LC–MS [267].

Mononucleotides can be analyzed in either positive-ion or negative-ion using condensed-phase ionization techniques like TSI-MS, ESI-MS, and MALDI-MS. In positive-ion mode, $[M+H]^+$- and (series of) $[(M-H+Alkali)+H]^+$-ions due to $H^+/Alkali^+$-exchange are observed. In negative-ion mode, which actually is to be preferred, $[M-H]^-$- as well as (series of) $[(M-H+Alkali)-H]^-$-ions are observed. The MS analysis of nucleobases, mono-nucleosides, nucleotides, and DNA adducts has recently been reviewed [268].

The negative-ion mode is indispensable for the MS analysis of oligonucleotides [264–269]. In ESI-MS, larger oligonucleotides will show multiple-charge ions, with typically one charge per 0.7–1 kDa. In MALDI-MS, mostly single-charge ions are observed; higher charge state may only occur with very large oligonucleotides. Given the high number of acidic groups at the 3′-5′-phosphodiester linkage between the (deoxy) riboses, complicated patterns due to multiple $H^+/Alkali^+$-exchanges may be observed for oligonucleotides, unless special precautions are taken [264, 270, 271]. A variety of desalting strategies have been proposed in order to reduce the deteriorating effects of H^+/Na^+-exchange has on the spectral quality. The desalting strategies include precipitation from ethanolic ammonium acetate solution, reversed-phase or cation-exchange solid-phase extraction, microdialysis, and addition of chelating agents. In ion-pair RPLC–MS, the necessary additions of triethylamine (or another tertiary amine) as ion-pairing agent and hexafluoroisopropanol (HFIP) to the mobile phase helps in reducing the H^+/Na^+-exchange [272].

In the early days, MALDI-MS analysis of oligonucleotides seemed more difficult than ESI-MS analysis, especially due to limited mass-spectrometric resolution, problems with H^+/Na^+-exchange, and extensive fragmentation [264]. To some extent, this could be attributed to the matrices used in the early days. Oligonucleotides require different matrices than proteins or oligosaccharides [264, 273]. The most successful matrices for oligonucleotides are 3-hydroxypicolinic acid (3-HPA), THAP, 6-aza-2-thiothymine, or a mixture of 5-methoxysalicylic acid and spermine.

The fragmentation of oligonucleotides has been investigated in detail, especially for sequencing purposes [264, 274, 275]. In-source fragmentation of oligonucleotides seems to be a more prominent problem in MALDI-MS than in ESI-MS

[264]. Obviously, the selection of a characteristic precursor ion generated by ESI-MS and subsequent CID provides sequence-informative fragmentation. Several nomenclature systems for annotating MS–MS mass spectra of oligonucleotides have been proposed [264]. The system proposed by McLuckey et al. [275] seems to be adopted by most researchers nowadays [265, 274]. Oligonucleotide sequencing by MS–MS has recently been reviewed [265]. In the discussion on negative-ion fragmentation, generally little attention is paid to the fact that part of the precursor ions may actually result from H^+/Na^+-exchange processes. Perhaps, it is not relevant in negative-ion mode.

The fragmentation of oligonucleotides (2- to 6-mers) under different conditions has also been investigated using both positive-ion and negative-ion ESI-MS [276]. The data for $[M+H]^+$ as precursor ion show that the relative abundance of the base-related ions, i.e., $[BH+H]^+$, depend on the base position (3′ or 5′). This is similar to the negative-ion mode, where fragmentation of $[M–H]^-$ as precursor ion similarly yields different relative abundance of the base-related B^--ions. With $[M+Li]^+$ as precursor ion, more sequence ions were obtained [276]. Some of these findings were further investigated [276]. The fragmentation behavior of four (self-complementary) deoxydinucleotides, i.e., d(ApT), d(TpA), d(CpG), and d(GpC), as a function of the cationization agent were studied using ESI-MS on an ion-trap instrument [276]. For $[M+H]^+$, the loss of the 5′ -terminal base followed by the loss of deoxyribose was observed, i.e., the loss of A and A-deoxyribose for d(ApT) and the loss of T and T-deoxyribose for d(TpA). Similar fragmentation behavior is observed for $[M+Li]^+$, $[M+Na]^+$ and $[M+Cs]^+$ of d(ApT) and d(GpC). For d(TpA) and d(CpG), the loss of A or G is observed. This indicates that Li^+-, Na^+-, and Cs^+-ions generate H^+/Alkali$^+$-exchange products at the phosphate and show more binding interaction to pyrimidine (A and G) rather than purine bases (T and C). $[M+K]^+$ mostly shows different fragmentation behavior from the other cationized species [276].

In another MALDI-MS-related study, Stano et al. [276] demonstrated that persodiation of all phosphodiester groups in the hexameric oligonucleotides effectively suppresses all backbone cleavages. Thus, better quality spectra are obtained: simpler spectra with increased sensitivity because both fragmentation and inhomogeneities due to H^+/Na^+-exchange are no longer present. The only fragmentation still observed, is the loss of nucleobases, as this is independent of Na^+-cationization [276].

7.5.7 Synthetic Polymers

MS is also increasingly used in the characterization and analysis of synthetic polymers. Over the years, Alkali$^+$-cationization has also been frequently used to enable the MS analysis of intact synthetic oligomers and polymers, e.g., by addition of Li^+ or Na^+ to condensed-phase sample preparation to be analyzed using ionization techniques like FDI, FAB, PDI, and SIMS. More recently, Alkali$^+$-cationization is applied in the MALDI-MS and ESI-MS analysis of synthetic polymers.

Polyethylene glycols (PEG) are relatively easy to ionize using condensed-phase ionization techniques as FAB-MS, TSI-MS, ESI-MS and MALDI-MS. PEG show significant affinity to Alkali$^+$-ions. Depending on the experimental conditions and the size of the polymers, both single-charge and multiple-charge ions may be observed in ESI-MS, in most cases involving $[M+NH_4]^+$, $[M+Na]^+$ and/or $[M+K]^+$ [279, 280]. MALDI-MS, FAB-MS, and PDI-MS data for PEG-esters of (2-benzothiazolon-3-yl)acetic acid have been reported as well [281] as are data on PEGylated peptides and other (bio) molecules [282]. Similarly, $[M+NH_4]^+$- and $[M+Na]^+$-ions are frequently observed for PEG-esters of relatively nonpolar molecules, e.g., nonylphenol ethoxylates that are used as surfactants [283]. The characterization of various PEG polymers with different end groups, i.e., linear hydroxyl, amino, and/or alkyl, or cyclic crown ether end groups, using MS–MS based on FAB-MS-generated $[M+Li]^+$-ions was also reported [284]. Informative fragmentation, directed by Li^+, enables sequence determination of monomers in copolymers. For some compounds, the loss of two internal ethylene oxide units (88 Da) is observed, thus involving a Li^+-mediated rearrangement [284, 285]. Li-DHB has been applied as matrix in MALDI-MS analysis of PEGs [140].

Early accounts of the use of MALDI-MS in the analysis of synthetic polymers already indicate the potential of the addition of Alkali$^+$-salts to the sample preparations [286]. Synthetic polymers without *N*-atoms are not readily protonated. Thus, alternative cationization agents must be applied, i.e., often Alkali$^+$-ions. Chemometrics have been applied to find optimum sample preparations for MALDI-MS of synthetic polymers, by studying the effects of solvent, matrix component, and cationization agent [287]. Some recent reviews provide an excellent perspective on the importance of Alkali$^+$-cationization in MALDI-MS analysis of synthetic polymers. For instance, Rizzarelli and Carroccio [288] provide an extensive tabular overview of the MALDI matrices and cationization agents used as additive in the analysis of biodegradable synthetic polymers. Altuntaş and Schubert [289] introduce the topic of "polymeromics," comprising an integrated workflow for the characterization of synthetic polymers, involving (1) optional prefractionation of the polymer sample by LC technologies, (2) determination of the *m/z* and mass of individual components, prior to (3) MS–MS of selected polymer precursor ions using CID, and (4) the use of special data interpretation tools for faster data evaluation. Other review papers describe in-depth discussion of the fragmentation pathways involved in CID and other ion-activation methods of synthetic polymers [290, 291]. As polymer film formation can be an important issue in material sciences, the use of modern MALDI-MS imaging can be of help in that respect [292].

Similarly, Alkali$^+$-cationization has been applied in the ESI-MS analysis of synthetic polymers [288, 289]. However, ESI-MS is not as extensively used as MALDI-MS. This is mainly due to the more complex spectra obtained in ESI-MS as a result of multiple charging. Charge-state distributions may overlap chain length distributions. Although deconvolution software tools have been developed to process such complex mass spectra, generally ultra-high-resolution instruments are required to achieve useful results. A nice example of the use of Li^+-cationization in ESI-MS is the characterization of the conjugate of a poly(amidoamine) dendrimer with a PEG linear polymer. MS–MS and ion-mobility spectrometry have been used to study conformational modifications as a function of charge state (2+ to 4+) and the adducted cation (H^+ vs Li^+) [293].

Whereas MALDI-MS and ESI-MS primarily involve condensed-phase ion-attachment processes, gas-phase ion attachment may be important in APCI. A recent paper reviews the use of APCI and the closely related atmospheric-pressure photoionization (APPI) technique in the characterization of synthetic polymers [294]. Next to protonation, the formation of $[M+Na]^+$-ions is important in APCI-MS analysis of synthetic polymers.

7.6 Conclusion and Perspectives

Several condensed-phase ionization techniques have been developed over the years. The eldest of these techniques, FDI, is still used in some specific application areas, whereas techniques like PDI-MS, FAB-MS and TSI-MS are almost obsolete and have been successfully replaced by ESI-MS and MALDI-MS. These newer techniques have greatly extended the applicability range of MS and also resulted in a far broader application of MS as an analytical tool in many (new) application areas.

Whereas $[M+H]^+$- and $[M-H]^-$-ions are most generally generated using these condensed-phase ionization techniques, initial studies on the "accidentally" generated $[M+Alkali]^+$-ions have opened new ways to broader applicability of MS to compound classes not readily protonated or deprotonated. It also showed that $Alkali^+$-cationization has great potential in structure elucidation of a wide variety of compound classes. This is not only true for cationization by other $Metal^{n+}$-ions, especially Ag^+. The body of data on this is too broad to be comprehensively reviewed. By highlighting a number of exemplary applications with different and important compound classes and by providing many references to (review) papers, a broad perspective on this fascinating field has been given.

Next to facilitating the MS analysis, one of the backgrounds behind the studies on metal cationization, especially of biomolecules, is that the MS results in both liquid-phase and gas-phase interactions between metal ions and biomolecules may actually give more insight in the biochemical interaction of metal ions and biomolecules in living system. Metal ion complexes play an important role in many biological processes, for instance, in protein conformation, in protein functioning, e.g., iron in the heme group of hemoglobin or calcium in calmodulin, and in many regulatory actions at cellular level and beyond [295]. In this respect, because of its speed, sensitivity, and selectivity, it seems especially ESI-MS that is important in this area. ESI-MS enables the study of the stoichiometry of peptide-metal complexes or even mixtures of complexes. The interactions observed can be pinpointed down to the level of specific amino acids [296]. MS-based studies in this area involve the study of metal-ion interaction [296], H/D-exchange experiments [297, 298], and ion-mobility MS [299, 300]. The role MS can play in resolving protein structure has been recently reviewed [301, 302]. And of course, this type of research is not limited to proteins, but affects the conformation and function of other biomolecules as well, e.g., [303]. This dimension of the Metal$^+$-cationization studies was deliberately not highlighted, as the author of this text, being a mass spectrometrist, lacks sufficient expertise in this interesting research area.

References

1. Sleno L, Volmer DA. Ion activation methods for tandem mass spectrometry. J Mass Spectrom. 2004;39:1091–112.
2. Pittenauer E, Allmaier G. High-energy collision induced dissociation of biomolecules: MALDI-TOF/RTOF mass spectrometry in comparison to tandem sector mass spectrometry. Comb Chem High Throughput Screen. 2009;12:137–55.
3. Teesch LM, Adams J. Metal ions as special reagents in analytical mass spectrometry. Org Mass Spectrom. 1992;27:931–43.
4. Beckey HD. Field ionization mass spectrometry. Res/Dev. 1969;20(11):26.
5. Beckey HD. Principles of field ionization and field desorption mass spectrometry. Oxford: Pergamon; 1977. ISBN 0080206123.
6. Lattimer RP, Schulten H-R. Field ionization and field desorption mass spectrometry: past, present, and future. Anal Chem. 1989;61:1201A–15.
7. Linden HB. Liquid injection field desorption ionization: a new tool for soft ionization of samples including air-sensitive catalysts and non-polar hydrocarbons. Eur. J Mass Spectrom. 2004;10:459–68.
8. Smith DF, Schaub TM, Rodgers RP, Hendrickson CL, Marshall AG. Automated liquid injection field desorption/ionization for Fourier transform ion cyclotron resonance mass spectrometry. Anal Chem. 2008;80:7379–82.
9. Gross JH, Nieth N, Linden HB, Blumbach U, Richter FJ, Tauchert ME, Tompers R, Hofmann P. Liquid injection field desorption/ionization of reactive transition metal complexes. Anal Bioanal Chem. 2006;386:52–8.
10. Słomińska B, Chaładaj W, Danikiewicz W. Assessment of the various ionization methods in the analysis of metal salen complexes by mass spectrometry. J Mass Spectrom. 2014;49:392–9.
11. Gross JH. Liquid injection field desorption/ionization-mass spectrometry of ionic liquids. J Am Soc Mass Spectrom. 2007;18:2254–62.
12. Gross JH, Vékey K, Dallos A. Field desorption mass spectrometry of large multiply branched saturated hydrocarbons. J Mass Spectrom. 2001;36:522–8.
13. Qian K, Edwards KE, Siskin M, Olmstead WN, Mennito AS, Dechert GJ, Hoosain NE. Desorption and ionization of heavy petroleum molecules and measurement of molecular weight distributions. Energy Fuels. 2007;21:1042–7.
14. Schaub TM, Rodgers RP, Marshall AG, Qian K, Green LA, Olmstead WN. Speciation of aromatic compounds in petroleum refinery streams by continuous flow field desorption ionization FT-ICR mass spectrometry. Energy Fuels. 2005;19:1566–73.
15. Schulten HR, Bahr U, Monkhouse PB. Biochemical application of field desorption mass spectrometry. J Biochem Biophys Methods. 1983;8:239–69.
16. Schulten HR. Off-line combination of liquid chromatography and field desorption mass spectrometry: principles and environmental, medical and pharmaceutical applications. J Chromatogr. 1982;251:105–28.
17. Barber M, Bordoli RS, Sedgwick RD, Tyler AN, Whalley ET. Fast atom bombardment mass spectrometry of bradykinin and related oligopeptides. Biomed. Mass Spectrom. 1981;8:337–42.
18. Morris HR, Panico M, Barber M, Bordoli RS, Sedgwick RD, Tyler A. Fast atom bombardment: a new mass spectrometric method for peptide sequence analysis. Biochem Biophys Res Commun. 1981;101:623–31.

19. Bélanger J, Paré JRJ. Fast atom bombardment mass spectrometry in the pharmaceutical analysis of drugs. J Pharm Biomed Anal. 1986;4:415–41.
20. Fenselau C, Cotter RJ. Chemical aspects of fast atom bombardment. Chem Rev. 1987;87:501–12.
21. Benninghoven A, Rudenauer FG, Werner HW. Secondary ion mass spectrometry: basic concepts, instrumental aspects, applications and trends. Chichester: Wiley; 1986. ISBN 3540162631.
22. Cook KD, Todd PJ, Friar DH. Physical properties of matrices used for fast atom bombardment. Biomed Environ Mass Spectrom. 1989;18:492–7.
23. Tanaka K, Waki H, Ido Y, Akita S, Yoshida Y, Yoshida T. Protein and polymer analyses up to *m/z* 100 000 by laser ionization TOF-MS. Rapid Commun Mass Spectrom. 1988;2:151–3.
24. Karas M, Hillenkamp F. Laser desorption ionization of proteins with molecular masses exceeding 10,000 Daltons. Anal Chem. 1988;60:2299–301.
25. Karas M, Bahr U, Ingendoh A, Nordhoff E, Stahl B, Strupat K, Hillenkamp F. Principles and applications of matrix-assisted UV laser desorption ionization mass spectrometry. Anal Chim Acta. 1990;241:175–85.
26. Mann. M, Talbo G. Developments in matrix-assisted laser desorption/ionization peptide mass spectrometry. Curr Opin Biotechnol. 1996;7:11–9.
27. Karas M. Matrix-assisted laser desorption ionization mass spectrometry: a progress report. Biochem Soc Trans. 1996;24:897–900.
28. Marvin LF, Roberts MA, Fay LB. Matrix-assisted laser desorption/ionization time-of-flight mass spectrometry in clinical chemistry. Clin Chim Acta. 2003;337:11–21.
29. Knochenmuss R. Ion formation mechanisms in UV-MALDI. Analyst. 2006;131:966–86.
30. Dreisewerd K. The desorption process in MALDI. Chem Rev. 2003;103:395–426.
31. Angel PM, Caprioli RM. Matrix-assisted laser desorption ionization imaging mass spectrometry: in situ molecular mapping. Biochemistry. 2013;52:3818–28.
32. Clark AE, Kaleta EJ, Arora A, Wolk DM. Matrix-assisted laser desorption ionization-time of flight mass spectrometry: a fundamental shift in the routine practice of clinical microbiology. Clin Microbiol Rev. 2013;26:547–603.
33. Tang N, Tornatore P, Weinberger SR. Current developments in SELDI affinity technology. Mass Spectrom Rev. 2004;23:34–44.
34. Grade H, Cooks RG. Secondary ion mass spectrometry. Cationization of organic molecules with metals. J Am Chem Soc. 1978;100:5615–21.
35. MacFarlane RD, Torgerson DF. Californium-252 plasma desorption mass spectroscopy. Science. 1976;191:920–5.
36. Cotter RJ. Plasma desorption mass spectrometry: coming of age. Anal Chem. 1988;60:781A–91.
37. Sundqvist B, Macfarlane R.D. ^{252}Cf-Plasma desorption mass spectrometry. Mass Spectrom Rev. 1985;4:421–60.
38. Blakley CR, McAdams MJ, Vestal ML. Crossed-beam liquid chromatograph-mass spectrometer combination. J Chromatogr. 1978;158:261–76.
39. Blakley CR, Carmody JJ, Vestal ML. Liquid chromatograph-mass spectrometer for analysis of nonvolatile samples. Anal Chem. 1980;52:1636–41.
40. Vestal ML, Fergusson GJ. Thermospray liquid chromatograph/mass spectrometer interface with direct electrical heating of the capillary. Anal Chem. 1985;57:2373–8.
41. Arpino PJ. Combined liquid chromatography mass spectrometry. Part II. Techniques and mechanisms of thermospray. Mass Spectrom Rev. 1990;9:631–69.
42. Arpino PJ. Combined liquid chromatography mass spectrometry. Part III. Applications of thermospray. Mass Spectrom Rev. 1992;11:3–40.
43. Gáspár A, Berndt H. Thermospray flame furnace atomic absorption spectrometry (TS-FF-AAS)—a simple method for trace element determination with microsamples in the µg/l concentration range. Spectrochim Acta B. 2000;55:587–97.
44. Blakley CR, Carmody JJ, Vestal ML. A new soft ionization technique for mass spectrometry of complex molecules. J Am Chem Soc. 1980;102:5931–3.

45. Blakley CR, Vestal ML. Thermospray interface for liquid chromatography/mass spectrometry. Anal Chem. 1983;55:750–4.
46. Katta V, Rockwood AL, Vestal ML. Field limit for ion evaporation from charged thermospray droplets. Int J Mass Spectrom Ion Proc. 1991;103:129–48.
47. Dole M, Hines RL, Mack LL, Mobley RC, Ferguson LD, Alice MB. Molecular beams of macroions. J Chem Phys. 1968;49:2240–9.
48. Gieniec J, Mack LL, Nakamae K, Gupta C, Kumar V, Dole M. Electrospray mass spectroscopy of macromolecules: application of an ion-drift spectrometer. Biomed Mass Spectrom. 1984;11:259–68.
49. Yamashita M, Fenn JB. Electrospray ion source. Another variation of the free-jet theme. J Phys Chem. 1984;88:4451–9.
50. Yamashita M, Fenn JB. Negative ion production with the electrospray ion source. J Phys Chem. 1984;88:4671–5.
51. Simons DS, Colby BN, Evans CA Jr. Electrohydrodynamic ionization mass spectrometry—the ionization of liquid glycerol and non-volatile organic solutes. Int J Mass Spectrom Ion Phys. 1974;15:291–302.
52. Stimpson BP, Evans CA Jr Electrohydrodynamic ionization mass spectrometry of biochemical materials. Biomed Mass Spectrom. 1978;5:52–63.
53. Zolotai NB, Karpov GV, Tal'roze VL, Skurat VE, Ramendik GI, Basyuta YuV. Mass spectrometry of the field evaporation of ions from liquid solutions in glycerol. J Anal Chem USSR. 1980;35:937–42.
54. Zolotai NB, Karpov GV, Tal'roze VL, Skurat VE, Basyuta YuV, Ramendik GI. Mass spectrometry of the field evaporation of ions from water and aqueous solutions, aqueous sodium iodide and saccharose solutions. J Anal Chem USSR. 1980;35:1161–74.
55. Cook KD. Electrohydrodynamic mass spectrometry. Mass Spectrom Rev. 1986;5:467–519.
56. Iribarne JV, Thomson BA. On the evaporation of small ions from charged droplets. J Chem Phys. 1976;64:2287–94.
57. Thomson BA, Iribarne JV. Field-induced ion evaporation from liquid surfaces at atmospheric pressure. J Chem Phys. 1979;71:4451–63.
58. Thomson BA, Iribarne JV, Dziedzic PJ. Liquid ion evaporation/mass spectrometry/mass spectrometry for the detection of polar and labile molecules. Anal Chem. 1982;54:2219–24.
59. Iribarne JV, Dziedzic PJ, Thomson BA. Atmospheric pressure ion evaporation-mass spectrometry. Int J Mass Spectrom Ion Phys. 1983;50:331–47.
60. Fenn JB, Mann M, Meng CK, Wong SF, Whitehouse CM. Electrospray ionization—principles and practice. Mass Spectrom Rev. 1990;9:37–70.
61. Cech NB, Enke CG. Practical implications of some recent studies in electrospray ionization fundamentals. Mass Spectrom Rev. 2001;20:362–87.
62. Smith RD, Light-Wahl KJ. The observation of non-covalent interactions in solution by electrospray ionization mass spectrometry: promise, pitfalls and prognosis. Biol Mass Spectrom. 1993;22:493–501.
63. Huang MZ, Yuan CH, Cheng SC, Cho YT, Shiea J. Ambient ionization mass spectrometry. Annu Rev Anal Chem. 2010;3:43–65.
64. Mortier KA, Zhang G-F, Van Peteghem CH, Lambert WE. Adduct formation in quantitative bioanalysis: effect of ionization conditions on paclitaxel. J Am Soc Mass Spectrom. 2004;15:585–92.
65. Kelly RT, Tolmachev AV, Page JS, Tang K, Smith RD. The ion funnel: theory, implementations, and applications. Mass Spectrom Rev. 2010;29:294–312.
66. Giles K, Pringle SD, Worthington KR, Little D, Wildgoose JL, Bateman RH. Applications of a travelling wave-based radio-frequency-only stacked ring ion guide. Rapid Commun Mass Spectrom. 2004;18:2401–14.
67. Lorenzen K, van Duijn E. Native mass spectrometry as a tool in structural biology. Curr Protoc Protein Sci. 2010;62:17.12.1–17.
68. Wilm MS, Mann M. Analytical properties of the nanoelectrospray ion source. Anal Chem. 1996;68:1–8.

69. Yin H, Killeen K, Brennen R, Sobek D, Werlich M, van de Goor T. Microfluidic chip for peptide analysis with an integrated HPLC column, sample enrichment column, and nanoelectrospray tip. Anal Chem. 2005;77:527–33.
70. Bayer E, Gfrörer P, Rentel C. Coordination-ionspray-MS (CIS-MS), a universal detection and characterization method for direct coupling with separation techniques. Angew Chem Int Ed. 1997;38:992–5.
71. Carroll DI, Dzidic I, Stillwell RN, Haegele KD, Horning EC. Atmospheric pressure ionization mass spectrometry: corona discharge ion source for use in liquid chromatography-mass spectrometry-computer analytical system. Anal Chem. 1975;47:2369–73.
72. Carroll DI, Dzidic I, Horning EC, Stillwell RN. Atmospheric-pressure ionization mass spectrometry. Appl Spectrosc Re. 1981;v. 17:337–406.
73. Covey TR, Thomson BA, Schneider BB. Atmospheric pressure ion sources. Mass Spectrom Rev. 2009;28:870–97.
74. Bos SJ, van Leeuwen SM, Karst U. From fundamentals to applications: recent developments in atmospheric pressure photoionization mass spectrometry. Anal Bioanal Chem. 2006;384:85–99.
75. Robb DB, Blades MW. State-of-the-art in atmospheric pressure photoionization for LC/MS. Anal Chim Acta. 2008;627:34–49.
76. Van Berkel GJ, Pasilis SP, Ovchinnikova O. Established and emerging atmospheric pressure surface sampling/ionization techniques for mass spectrometry. J Mass Spectrom. 2008;43:1161–80.
77. Takáts Z, Wiseman JM, Cooks RG. Ambient mass spectrometry using desorption electrospray ionization (DESI): instrumentation, mechanisms and applications in forensics, chemistry. and biology. J Mass Spectrom. 2005;40:1261–75.
78. Wu C, Dill AL, Eberlin LS, Cooks RG, Ifa DR. Mass spectrometry imaging under ambient conditions. Mass Spectrom Rev. 2013;32:218–43.
79. Laiko VV, Baldwin MA, Burlingame AL. Atmospheric pressure matrix-assisted laser desorption/ionization mass spectrometry. Anal Chem. 2000;72:652–7.
80. Moyer SC, Cotter RJ. Atmospheric-pressure MALDI. Anal Chem. 2002;74:468A–476A.
81. Creaser CS, Ratcliffe L. Atmospheric pressure matrix-assisted laser desorption/ionization mass spectrometry: a review. Curr Anal Chem. 2006;2:9–15.
82. Laiko VV, Moyer SC, Cotter RJ. Atmospheric pressure MALDI/ion trap mass spectrometry. Anal Chem. 2000;72:5239–43.
83. Cody RB, Laramée JA, Durst HD. Versatile new ion source for the analysis of materials in open air under ambient conditions. Anal Chem. 2005;77:2297–302.
84. McEwen CN, McKay RG, Larsen BS. Analysis of solids, liquids, and biological tissues using solids probe introduction at atmospheric pressure on commerical LC/MS instruments. Anal Chem. 2005;77:7826–31.
85. Hirabayashi A, Sakairi M, Koizumi H. Sonic spray ionization method for atmospheric pressure ionization mass spectrometry. Anal Chem. 1994;66:4557–9.
86. Hirabayashi A, Sakairi M, Koizumi H. Sonic spray mass spectrometry. Anal Chem. 1995;67:2878–82.
87. Haddad R, Sparrapan R, Eberlin MN. Desorption sonic spray ionization for (high) voltage-free ambient mass spectrometry. Rapid Commun Mass Spectrom. 2006;20:2901–5.
88. Arpino PJ, Guiochon G. Optimization of the instrumental parameters of a combined LC-MS, coupled by an interface for DLI. III. Why the solvent should not be removed in LC-MS interfacing methods. J Chromatogr. 1982;251:153–64.
89. Vestal ML. Ionization techniques for nonvolatile molecules. Mass Spectrom Rev. 1983;2:447–80.
90. Busch KL. Desorption ionization mass spectrometry. J Mass Spectrom. 1995;30:233–40.
91. Amad MH, Cech NB, Jackson GS, Enke CG. Importance of gas-phase proton affinities in determining the electrospray ionization response for analytes and solvents. J Mass Spectrom. 2000;35:784–9.
92. Bursey MM. Comment to readers: style and the lack of it. Mass Spectrom Rev. 1991;19:1–2.

93. Jemal M, Almond RB, Teitz DS. Quantitative bioanalysis utilizing high-performance liquid chromatography/electrospray mass spectrometry via selected-ion monitoring of the sodium ion adduct [M+Na]+. Rapid Commun Mass Spectrom. 1997;11:1083–8.
94. Suzuki H, Kameyama A, Tachibana K, Narimatsu H, Fukui K. Computationally and experimentally derived general ru1es for fragmentation of various glycosyl bonds in sodium adduct oligosaccharides. Anal Chem. 2009;81:1108–20.
95. Rodriquez CF, Fournier R, Chu IK, Hopkinson AC, Siu KWM. A possible origin of $[M-nH+mX]^{(m-n)+}$ ions (X=alkali metal ion) in electrospray ionization mass spectrometry of peptides. Int J Mass Spectrom. 1999;192:303–17.
96. Newton KA, McLuckey SA. Gas-phase peptide/protein cationizing agent switching via ion/ion reactions. J Am Chem Soc. 2003;125:12404–5.
97. Newton KA, McLuckey SA. Generation and manipulation of sodium cationized peptides in the gas phase. J Am Soc Mass Spectrom. 2004;15:607–15.
98. van Kampen JJ, Burgers PC, de Groot R, Gruters RA, Luider TM. Biomedical application of MALDI mass spectrometry for small-molecule analysis. Mass Spectrom Rev. 2011;30:101–20.
99. Mohr MD, Bomsen KO, Widmer HM. Matrix-assisted laser desorption/ionization mass spectrometry: improved matrix for oligosaccharides. Rapid Commun Mass Spectrom. 1995;9:809–14.
100. Kamel AM, Brown PR, Munson B. Effects of mobile-phase additives, solution pH, ionization constant, and analyte concentration on the sensitivities and electrospray ionization mass spectra of nucleoside antiviral agents. Anal Chem. 1999;71:5481–92.
101. Stefansson M, Sjoberg PJR, Markides KE. Regulation of multimer formation in electrospray mass spectrometry. Anal Chem. 1996;68:1792–7.
102. Mann M, Meng CK, Fenn JB. Interpreting mass spectra of multiply charged ions. Anal Chem. 1989;61:1702–8.
103. Covey TR, Bonner RF, Shushan BI, Henion JD. The determination of protein, oligonucleotide and peptide molecular weights by ion-spray mass spectrometry. Rapid Commun Mass Spectrom. 1988;2:249–56.
104. Ferrige AG, Seddon MJ, Jarvis S. Maximum entropy deconvolution in electrospray mass spectrometry. Rapid Commun Mass Spectrom. 1991;5:374–7.
105. Reinhold BB, Reinhold VN. Electrospray ionization mass spectrometry: deconvolution by an entropy-based algorithm. J Am Soc Mass Spectrom. 1992;3:207–15.
106. Ferrige AG, Seddon MJ, Green BN, Jarvis SA, Skilling J. Disentangling electrospray spectra with maximum entropy. Rapid Commun Mass Spectrom. 1992;6:707–11.
107. Kelly MA, Vestling MM, Fenselau CC, Smith PB. Electrospray analysis of proteins: A comparison of positive-ion and negative-ion mass spectra at high and low pH. Org Mass Spectrom. 1992;27:1143–7.
108. Loo JA, Loo RR, Light KJ, Edmonds CG, Smith RD. Multiply charged negative ions by electrospray ionization of polypeptides and proteins. Anal Chem. 1992;64:81–8.
109. Potier N, Van Dorsselaer A, Cordier Y, Roch O, Bischoff R. Negative electrospray ionization mass spectrometry of synthetic and chemically modified oligonucleotides. Nucleic Acids Res. 1994;22:3895–903.
110. Lin ZJ, Li W, Dai G. Application of LC-MS for quantitative analysis and metabolite identification of therapeutic oligonucleotides. J Pharm Biomed Anal. 2007;44:330–41.
111. Zaia J. Mass spectrometry of oligosaccharides. Mass Spectrom Rev. 2004;23:161–227.
112. Röllgen FW, Borchers F, Giessmann U, Levsen K. Collisional activation of ions formed by $[Li]^{+}$ion attachment. Org Mass Spectrom. 1977;12:541–3.
113. Wu Y, Zhao J, Henion JD, Korfmacher WA, Lpaiguera AP, Lin C-C. Microsample determination of lovastatin and its hydroxy acid metabolite in mouse and rat plasma by liquid chromatography-ionspray tandem mass spectrometry. J Mass Spectrom. 1997;32:379–87.
114. Zhao JJ, Xie IH, Yang AY, Roadcap BA, Rogers JD. Quantitation of simvastatin and its beta-hydroxy acid in human plasma by liquid-liquid cartridge extraction and liquid chromatography-tandem mass spectrometry. J Mass Spectrom. 2000;35:1133–43.

115. Zhao JJ, Yang AY, Rogers JD. Effects of liquid chromatography mobile phase buffer contents on the ionization and fragmentation of analytes in liquid chromatography-ionspray tandem mass spectrometric determination. J Mass Spectrom. 2002;37:421–33.
116. Nozaki K, Tarui A, Osaka I, Kawasaki H, Arakawa R. Elimination technique for alkali metal ion adducts from an electrospray ionization process using an on-line ion suppressor. Anal Sci. 2010;26:715–8.
117. Bruggink C, Maurer R, Herrmann H, Cavalli S, Hoefler F. Analysis of carbohydrates by anion exchange chromatography and mass spectrometry. J Chromatogr A. 2005;1085:104–9.
118. Bruggink C, Wuhrer M, Koeleman CA, Barreto V, Liu Y, Pohl C, Ingendoh A, Hokke CH, Deelder AM. Oligosaccharide analysis by capillary-scale high-pH anion-exchange chromatography with on-line ion-trap mass spectrometry. J Chromatogr B. 2005;829:136–43.
119. Li XF, Ma M, Scherban K, Tam YK. Liquid chromatography-electrospray mass spectrometric studies of ginkgolides and bilobalide using simultaneous monitoring of proton, ammonium and sodium adducts. Analyst. 2002;127:641–6.
120. Hua W, Ierardi T, Lesslie M, Hoffman BT, Mulvana D. Development and validation of a HILIC-MS/MS method for quantification of decitabine in human plasma by using lithium adduct detection. J Chromatogr B. 2014;969:117–22.
121. Eichhorn P, Knepper TP. Electrospray ionization mass spectrometric studies on the amphoteric surfactant cocamidopropylbetaine. J Mass Spectrom. 2001;26:677–84.
122. Kamel AM, Brown PR, Munson B. Electrospray ionization mass spectrometry of tetracycline, oxytetracycline, chlorotetracycline, minocycline, and methacycline. Anal Chem. 1999;71:968–77.
123. Cerny RL, MacMillan DK, Gross ML, Mallams AK, Pramanik BN. Fast-atom bombardment and tandem mass spectrometry of macrolide antibiotics. J Am Soc Mass Spectrom. 1994;5:151–8.
124. Chang TT, Lay JO. Direct analysis of thin-layer chromatography spots by fast atom bombardment mass spectrometry. Anal Chem. 1984;56:109–11.
125. Siegel MM, McGahren WJ, Tomer KB, Chang TT. Applications of fast atom bombardment mass spectrometry and fast atom bombardment mass spectrometry-mass spectrometry to the maduramicins and other polyether antibiotics. Biomed Environ Mass Spectrom. 1987;14:29–38.
126. Volmer DA, Lock CM. Electrospray ionization and collision-induced dissociation of antibiotic polyether ionophores. Rapid Commun Mass Spectrom. 1998;12:157–64.
127. Wang J, Sporns P. MALDI-TOF MS quantification of coccidiostats in poultry feeds. J Agric Food Chem. 2000;48:2807–11.
128. Grimalt S, Pozo OJ, Marín JM, Sancho JV, Hernández F. Evaluation of different quantitative approaches for the determination of noneasily ionizable molecules by different atmospheric pressure interfaces used in liquid chromatography tandem mass spectrometry: abamectin as case of study. J Am Soc Mass Spectrom. 2005;16:1619–30.
129. Kamel A, Munson B. Collision induced dissociation studies of alkali metal adducts of tetracyclines and antiviral agents by electrospray ionization, hydrogen/deuterium exchange and multiple stage mass spectrometry. Eur J Mass Spectrom. 2008;14:281–97.
130. Fredenhagen A, Derrien C, Gassmann E. An MS/MS library on an ion-trap instrument for efficient dereplication of natural products, different fragmentation patterns for $[M + H]^+$ and $[M + Na]^+$ ions. J Nat Prod. 2005;68:385–91.
131. Rivera SM, Christou P, Canela-Garayoa R. Identification of carotenoids using mass spectrometry. Mass Spectrom Rev. 2014;33:353–72.
132. van Breemen RB, Dong L, Pajkovic ND. Atmospheric pressure chemical ionization tandem mass spectrometry of carotenoids. Int J Mass Spectrom. 2012;312:163–72.
133. Bijttebier SK, D'Hondt E, Hermans N, Apers S, Voorspoels S. Unravelling ionization and fragmentation pathways of carotenoids using orbitrap technology: a first step towards identification of unknowns. J Mass Spectrom. 2013;48:740–54.
134. van Breemen RB. Innovations in carotenoid analysis using LC-MS. Anal Chem. 1996;68:299A–304A.

135. Weesepoel Y, Vincken J-P, Pop RM, Liu K, Gruppen H. Sodiation as a tool for enhancing the diagnostic value of MALDI-TOF/TOF-MS spectra of complex astaxanthin ester mixtures from *Haematococcus pluvialis*. J Mass Spectrom. 2013;48:862–74.
136. Rentel C, Strohschein S, Albert K, Bayer E. Silver-plated vitamins: a method of detecting tocopherols and carotenoids in LC/ESI-MS coupling. Anal Chem. 1998;70:4394–400.
137. Ma Y-C, Kim H-Y. Determination of steroids by liquid chromatography-mass spectrometry. J Am Soc Mass Spectrom. 1997;8:1010–20.
138. Pozo OJ, Van Eenoo P, Deventer K, Delbeke FT. Ionization of anabolic steroids by adduct formation in liquid chromatography electrospray mass spectrometry. J Mass Spectrom. 2007;42:497–516.
139. Kim SH, Cha EJ, Lee KM, Kim HJ, Kwon OS, Lee J. Simultaneous ionization and analysis of 84 anabolic androgenic steroids in human urine using liquid chromatography-silver ion coordination ionspray/triple-quadrupole mass spectrometry. Drug Test Anal. 2014;6:1174–85.
140. Cvacka J. Svatos A. matrix-assisted laser desorption/ionization analysis of lipids and high molecular weight hydrocarbons with lithium 2, 5-dihydroxybenzoate matrix. Rapid Commun Mass Spectrom. 2003;17:2203–7.
141. Horká P, Vrkoslav V, Hanus R, Pecková K, Cvačka J. New MALDI matrices based on lithium salts for the analysis of hydrocarbons and wax esters. J Mass Spectrom. 2014;49:628–38.
142. Roussis SG, Proulx R. Probing the molecular weight distributions of non-boiling petroleum fractions by Ag^+ electrospray ionization mass spectrometry. Rapid Commun Mass Spectrom. 2004;18:1761–75.
143. Grewal RN, Rodriquez CF, Shoeib T, Chu IK, Tu Y-P, Hopkinson AC, Siu KWM. Elimination of AgR (R=H, CH_3, C_6H_5) from collisionally-activated argentinated amines. Eur J Mass Spectrom. 2000;6:187–92.
144. Shi T, Zhao J, Shoeib T, Siu KWM, Hopkinson AC. Fragmentation of singly charged silver/α,ω-diaminoalkane complexes: competition between the loss of H_2 and AgH molecules. Eur J Mass Spectrom. 2004;10:931–40.
145. Schäfer M, Dreiocker F, Budzikiewicz H. Collision-induced loss of AgH from Ag^+ adducts of alkylamines, aminocarboxylic acids and alkyl benzyl ethers leads exclusively to thermodynamically favored product ions. J Mass Spectrom. 2009;44:278–84.
146. Martha CT, van Zeist W-J, Bickelhaupt FM, Irth H, Niessen WMA. Tandem mass spectrometry of silver-adducted ferrocenyl catalyst complexes in continuous-flow reaction detection systems. J Mass Spectrom. 2010;45:1332–43.
147. Dreifuss PA, Wood GE, Roach JA, Brumley WC, Andrzejewski D, Sphon JA. Field desorption mass spectrometry of cyanogenic glycosides. Biomed Mass Spectrom. 1980;7:201–4.
148. Schulten H-R, Games DE. High resolution field desorption mass spectrometry. II—glycosides. Biomed Mass Spectrom. 1974;1:120–3.
149. Stobiecki M. Application of mass spectrometry for identification and structural studies of flavonoid glycosides. Phytochem. 2000;54:237–56.
150. Cuyckens F, Claeys M. Mass spectrometry in the structural analysis of flavonoids. J Mass Spectrom. 2004;39:1–15.
151. de Rijke E, Out P, Niessen WMA, Ariese F, Gooijer C, Brinkman UATh. Analytical separation and detection methods for flavonoids. J Chromatogr A. 2006;1112:31–63.
152. March R, Brodbelt J. Analysis of flavonoids: tandem mass spectrometry, computational methods, and NMR. J Mass Spectrom. 2008;43:1581–617.
153. Vukics V, Guttman A. Structural characterization of flavonoid glycosides by multi-stage mass spectrometry. Mass Spectrom Rev. 2010;29:1–16.
154. Ma YL, Vedernikova I, Van den Heuvel H, Claeys M. Internal glucose residue loss in protonated O-diglycosyl flavonoids upon low-energy collision-induced dissociation. J Am Soc Mass Spectrom. 2000;11:136–44.
155. Brüll LP, Kovácik V, Thomas-Oates JE, Heerma W, Haverkamp J. Sodium-cationized oligosaccharides do not appear to undergo 'internal residue loss' rearrangement processes on tandem mass spectrometry. Rapid Commun Mass Spectrom. 1998;12:1520–32.

156. Harvey DJ, Mattu TS, Wormald MR, Royle L, Dwek RA, Rudd PM. "Internal residue loss": rearrangements occurring during the fragmentation of carbohydrates derivatized at the reducing terminus. Anal Chem. 2002;74:734–40.
157. Kite GC, Veitch NC. Identification of common glycosyl groups of flavonoid O-glycosides by serial mass spectrometry of sodiated species. Rapid Commun Mass Spectrom. 2011;25:2579–90.
158. Hofmeister GE, Zhou Z, Leary JA. Linkage position determination in lithium-cationized disaccharides: tandem mass spectrometry and semiempirical calculations. J Am Chem Soc. 1991;113:5964–70.
159. Lemoine J, Strecker G, Leroy Y, Fournet B, Ricart G. Collisional-activation tandem mass spectrometry of sodium adduct ions of methylated oligosaccharides: sequence analysis and discrimination between alpha-NeuAc-(2-3) and alpha-NeuAc-(2-6) linkages. Carbohydr Res. 1991;221:209–17.
160. Asam MR, Glish GL. Tandem mass spectrometry of alkali cationized polysaccharides in a quadrupole ion trap. J Am Soc Mass Spectrom. 1997;8:987–95.
161. Song F, Cui M, Liu Z, Yu B, Liu S. Multiple-stage tandem mass spectrometry for differentiation of isomeric saponins. Rapid Commun Mass Spectrom. 2004;18:2241–8.
162. Madhusudanan KP, Mathad VT, Raj SK, Bhaduri AP. Characterization of iridoids by fast atom bombardment mass spectrometry followed by collision-induced dissociation of $[M + Li]^+$ ions. J Mass Spectrom. 2000;35:321–9.
163. Es-Safi NE, Kerhoas L, Ducrot PH. Fragmentation study of iridoid glucosides through positive and negative electrospray ionization, collision-induced dissociation and tandem mass spectrometry. Rapid Commun Mass Spectrom. 2007;21:1165–275.
164. Ricci A, Fiorentino A, Piccolella S, Golino A, Pepi F, D'Abrosca B, Letizia M, Monaco P. Furofuranic glycosylated lignans: a gas-phase ion chemistry investigation by tandem mass spectrometry. Rapid Commun Mass Spectrom. 2008;22:3382–92.
165. Ricci A, Fiorentino A, Piccolella S, D'Abrosca B, Pacifico S, Monaco P. Structural discrimination of isomeric tetrahydrofuran lignan glucosides by tandem mass spectrometry. Rapid Commun Mass Spectrom. 2010;24:979–85.
166. Satterfield M, Brodbelt JS. Enhanced detection of flavonoids by metal complexation and electrospray ionization mass spectrometry. Anal Chem. 2000;72:5898–906.
167. Pikulski M, Brodbelt JS. Differentiation of flavonoid glycoside isomers by using metal complexation and electrospray ionization mass spectrometry. J Am Soc Mass Spectrom. 2003;14:1437–53.
168. Satterfield M, Brodbelt JS. Structural characterization of flavonoid glycosides by collisionally activated dissociation of metal complexes. J Am Soc Mass Spectrom. 2001;12:537–49.
169. Zhang J, Wang J, Brodbelt JS. Characterization of flavonoids by aluminum complexation and collisionally activated dissociation. J Mass Spectrom. 2005;40:350–63.
170. Zhang J, Brodbelt JS. Silver complexation and tandem mass spectrometry for differentiation of isomeric flavonoid diglycosides. Anal Chem. 2005;77:1761–70.
171. Davis BD, Brodbelt JS. LC-MS^n methods for saccharide characterization of monoglycosyl flavonoids using postcolumn manganese complexation. Anal Chem. 2005;77:1883–90.
172. Pikulski M, Aguilar A, Brodbelt JS. Tunable transition metal-ligand complexation for enhanced elucidation of flavonoid diglycosides by electrospray ionization mass spectrometry. J Am Soc Mass Spectrom. 2007;18:422–31.
173. Dell A, Carman HH, Tiller PR, Thomas-Oates JE: Fast atom bombardment mass spectrometric strategies for characterizing carbohydrate-containing biopolymers. Biomed Environ Mass Spectrom. 1988;16:19–24.
174. Fukuda M, Dell A, Fukuda MN. Structure of fetal lactosaminoglycan. The carbohydrate moiety of Band 3 isolated from human umbilical cord erythrocytes. J Biol Chem. 1984;259:4782–91.
175. Aduru S, Chait BT. Californium-252 plasma desorption mass spectrometry of oligosaccharides and glycoconjugates: control of ionization and fragmentation. Anal Chem. 1991;63:1621–5.

176. Harvey DJ. Matrix-assisted laser desorption/ionisation mass spectrometry of oligosaccharides and glycoconjugates. J Chromatogr A. 1996;720:429–46.
177. Harvey DJ. Matrix-assisted laser desorption/ionization mass spectrometry of carbohydrates. Mass Spectrom Rev. 1999;18:349–450.
178. Harvey DJ. Analysis of carbohydrates and glycoconjugates by matrix-assisted laser desorption/ionization mass spectrometry: an update covering the period 1999–2000. Mass Spectrom Rev. 2006;25:595–662.
179. Harvey DJ. Analysis of carbohydrates and glycoconjugates by matrix-assisted laser desorption/ionization mass spectrometry: an update covering the period 2001–2002. Mass Spectrom Rev. 2008;27:125–201.
180. Harvey DJ. Analysis of carbohydrates and glycoconjugates by matrix-assisted laser desorption/ionization mass spectrometry: An update for 2003–2004. Mass Spectrom Rev. 2009;28:273–361.
181. Harvey DJ. Analysis of carbohydrates and glycoconjugates by matrix-assisted laser desorption/ionization mass spectrometry: an update for the period 2005–2006. Mass Spectrom Rev. 2011;30:1–100.
182. Harvey DJ. Analysis of carbohydrates and glycoconjugates by matrix-assisted laser desorption/ionization mass spectrometry: an update for 2007–2008. Mass Spectrom Rev. 2012;31:183–311.
183. Harvey DJ. Analysis of carbohydrates and glycoconjugates by matrix-assisted laser desorption/ionization mass spectrometry: an update for 2009–2010. Mass Spectrom Rev. 34, 2015 (in press). doi:10.1002/mas.21411.
184. Wuhrer M, de Boer AR, Deelder AM. Structural glycomics using hydrophilic interaction chromatography (HILIC) with mass spectrometry. Mass Spectrom Rev. 2009;28:192–206.
185. Ongay S, Boichenko A, Govorukhina N, Bischoff R. Glycopeptide enrichment and separation for protein glycosylation analysis. J Sep Sci. 2012;35:2341–72.
186. Harvey DJ. Electrospray mass spectrometry and fragmentation of N-linked carbohydrates derivatized at the reducing terminus. J Am Soc Mass Spectrom. 2000;11:900–15.
187. Domon B, Costello CE. A systematic nomenclature for carbohydrate fragmentations in FAB-MS/MS spectra of glycoconjugates. Glycoconjugate J. 1988;5:397–409.
188. Orlando R, Bush CA, Fenselau C. Structural-analysis of oligosaccharides by tandem mass-spectrometry — Collisional activation of sodium adduct ions. Biomed Environ Mass Spectrom. 1990;19:747–54.
189. Fahy E, Subramaniam S, Brown HA, Glass CK, Merrill AH Jr, Murphy RC, Raetz CR, Russell DW, Seyama Y, Shaw W, Shimizu T, Spener F, van Meer G, VanNieuwenhze MS, White SH, Witztum JL, Dennis EA. A comprehensive classification system for lipids. J Lipid Res. 2005;46:839–61.
190. Griffiths WJ. Tandem mass spectrometry in the study of fatty acids, bile acids, and steroids. Mass Spectrom Rev. 2003;22:81–152.
191. Murphy RC, Axelsen PH. Mass spectrometric analysis of long-chain lipids. Mass Spectrom Rev. 2011;30:579–99.
192. Cajka T, Fiehn O. Comprehensive analysis of lipids in biological systems by liquid chromatography-mass spectrometry. Trends Anal Chem. 2014;61:192–206.
193. Blanksby SJ, Mitchell TW. Advances in mass spectrometry for lipidomics. Annu Rev Anal Chem. 2010;3:433–65.
194. Li M, Yang L, Bai Y, Liu H. Analytical methods in lipidomics and their applications. Anal Chem. 2014;86:161–75.
195. Murray KE, Schulten H-R. Field desorption mass spectrometry of lipids. I. The application of field desorption mass spectrometry to the investigation of natural waxes. Chem Phys Lipids. 1981;29:11–21.
196. Lehmann WD, Kessler M. Characterization and quantification of human plasma lipids from crude lipid extracts by field desorption mass spectrometry. Biol Mass Spectrom. 1983;10:220–6.

197. Puzo G, Tissie G, Lacave C, Aurelle H, Prome JC. Structural determination of 'cord factor' from a *Corynebacterium diphtheriae* strain by a combination of mass spectral ionization methods: field desorption cesium cationization and electron impact mass spectrometry studies. Biomed Mass Spectrom. 1978;5:699–703.
198. Matsubara T, Hayashi A. FAB/mass spectrometry of lipids. Prog Lipid Res. 1991;30:301–22.
199. Fredrickson HL, De Leeuw JW, Tas AC, Van der Greef J, La Vos GF, Boon JJ. Fast atom bombardment (tandem) mass spectrometric analysis of intact polar ether lipids extractable from the extremely halophilic archaebacterium *Halobacterium cutivubrum*. Biomed Environ Mass Spectrom. 1989;18:96–105.
200. Adams J, Gross ML. Energy requirement for remote charge site ion decompositions and structural information from collisional activation of alkali metal cationized fatty alcohols. J Am Chem Soc. 1986;108:6915–21.
201. Contado MJ, Adams J. Collision-induced dissociations and B/E linked scans for structural determination of modified fatty acid esters. Anal Chim Acta. 1991;246:187–97.
202. Crockett, J, S,; Gross ML, Christie WW, Holman RT. Collisional activation of a series of homoconjugated octadecadienoic acids with fast atom bombardment and tandem mass spectrometry. J Am Soc Mass Spectrom. 1990;1:183–91.
203. Gross ML. Charge-remote fragmentation: an account of research on mechanisms and applications. Int J Mass Spectrom. 2000;200:611–24.
204. Zirrolli JA, Davoli E, Bettazzoli L, Gross ML, Murphy RC. Fast atom bombardment and collision-induced dissociation of prostaglandins and thromboxanes: Some examples of charge remote fragmentation. J Am Soc Mass Spectrom. 1990;1:325–35.
205. Ann Q, Adams J. Structure determination of ceramides and neutral glycosphingolipids by collisional activation of $[M+Li]^+$ ions. J Am Soc Mass Spectrom. 1992;3:260–3.
206. Hsu F-F, Turk J, Stewart ME, Downing DT. Structural studies on ceramides as lithiated adducts by low energy collisional-activated dissociation tandem mass spectrometry with electrospray ionization. J Am Soc Mass Spectrom. 2002;13:680–95.
207. Levery SB, Toledo MS, Doong RL, Straus AH, Takahashi HK. Comparative analysis of ceramide structural modification found in fungal cerebrosides by electrospray tandem mass spectrometry with low energy collision-induced dissociation of Li^+adduct ions. Rapid Commun Mass Spectrom. 2000;14:551–63.
208. Hsu F-F, Turk J. Distinction among isomeric unsaturated fatty acids as lithium adducts by ESI-MS using low energy CID on a triple stage quadrupole instrument. J Am Soc Mass Spectrom. 1999;10:600–12.
209. Byrdwell WC. Atmospheric-pressure chemical ionization mass spectrometry for analysis of lipids. Lipids. 2001;36:327–46.
210. Hsu F-F, Turk J. Structural characterization of triacylglycerols as lithiated adducts by electrospray ionization mass spectrometry using low-energy collisionally activated dissociation on a triple stage quadrupole instrument. J Am Soc Mass Spectrom. 1999;10:587–99.
211. Domingues P, Domingues MR, Amado FM, Ferrer-Correia AJ. Characterization of sodiated glycerol phosphatidylcholine phospholipids by mass spectrometry. Rapid Commun Mass Spectrom. 2001;15:799–804.
212. Hsu FF, Turk J. Electrospray ionization with low-energy collisionally activated dissociation tandem mass spectrometry of glycerophospholipids: mechanisms of fragmentation and structural characterization. J Chromatogr B. 2009;877:2673–95.
213. Kushi Y, Handa S. Application of field desorption mass spectrometry for the analysis of sphingoglycolipids. J Biochem. 1982;91:923–31.
214. Kushi Y, Handa S, Kambara H, Shizukuishi K. Comparative study of acidic glycosphingolipids by field desorption and secondary ion mass spectrometry. J Biochem. 1983;94:1841–1150.
215. Haynes CA, Allegood JC, Park H, Sullards MC. Sphingolipidomics: methods for the comprehensive analysis of sphingolipids. J Chromatogr B. 2009;877:2696–708.
216. Ann Q, Adams J. Structure determination of sphingolipids by mass spectrometry. Mass Spectrom Rev. 1993;12:51–85.

217. Ann Q, Adams J. Structure-specific collision-induced fragmentations of ceramides cationized with alkali-metal ions. Anal Chem. 1993;65:7–13.
218. Park T, Park YS, Rho JR, Kim YH. Structural determination of cerebrosides isolated from *Asterias amurensis* starfish eggs using high-energy collision-induced dissociation of sodium-adducted molecules. Rapid Commun Mass Spectrom. 2011;25:572–8.
219. Fuchs B. Analysis of phospolipids and glycolipids by thin-layer chromatography-matrix-assisted laser desorption and ionization mass spectrometry. J Chromatogr A. 2012;1259:62–73.
220. Fuchs B, Schiller J. Application of MALDI-TOF mass spectrometry in lipidomics. Eur J Lipid Sci Technol. 2009;111:83–98.
221. Gode D, Volmer DA. Lipid imaging by mass spectrometry—a review. Analyst. 2013;138:1289–315.
222. Rujoi M, Estrada R, Yappert MC. In situ MALDI-TOF MS regional analysis of neutral phospholipids in lens tissue. Anal Chem. 2004;76:1657–63.
223. Ho YP, Huang PC, Deng KH. Metal ion complexes in the structural analysis of phospholipids by electrospray ionization tandem mass spectrometry. Rapid Commun Mass Spectrom. 2003;17:114–21.
224. Christie WW. Separation of molecular species of triacylglycerols by HPLC with a silver ion column. J Chromatogr A. 1988;454:273–84.
225. Sandra P, Medvedovici A, Zhao Y, David F. Characterization of triglycerides in vegetable oils by silver-ion packed-column supercritical fluid chromatography coupled to mass spectroscopy with atmospheric pressure chemical ionization and coordination ion spray. J Chromatogr A. 2002;974:231–41.
226. Lísa M, Velínská H, Holčapek M. Regioisomeric characterization of triacylglycerols using silver-ion HPLC/MS and randomization synthesis of standards. Anal Chem. 2009;81:3903–30.
227. Havrilla CM, Hachey DL, Porter NA. Coordination (Ag^+) ion spray—mass spectrometry of peroxidation products of cholesterol linoleate and cholesterol arachidonate: high-performance liquid chromatography—mass spectrometry analysis of peroxide products from polyunsaturated lipid autoxidation. J Am Chem Soc. 2000;122:8042–55.
228. Seal JR, Porter NA. Liquid chromatography coordination ion-spray mass spectrometry (LC-CIS-MS) of docosahexaenoate ester hydroperoxides. Anal Bioanal Chem. 2004;378:1007–13.
229. Yin H, Brooks JD, Gao L, Porter NA, Morrow JD. Identification of novel autoxidation products of the omega-3 fatty acid eicosapentaenoic acid in vitro and in vivo. J Biol Chem. 2007;282:29890–1.
230. Roepstorff P, Nielsen PF, Klarskov K, Højrup P. Applications of plasma desorption mass spectrometry in peptide and protein chemistry. Biomed Environ Mass Spectrom. 1988;16:9–18.
231. Biemann K. Contributions of mass spectrometry to peptide and protein structure. Biomed Environ Mass Spectrom. 1988;16:99–111.
232. Desiderio DM, Sabbatini JZ. Field desorption collision activation linked scanning mass spectrometry of underivatized oligopeptides. Biol Mass Spectrom. 2005;8:565.
233. Smith RD, Loo JA, Edmonds CG, Barinaga CJ, Udseth HR. New developments in biochemical mass spectrometry: electrospray ionization. Anal Chem. 1990;62:882–99.
234. Smith RD, Loo JA, Ogorzalek Loo RR, Busman M, Udseth HR. Principles and practice of electrospray ionization-mass spectrometry for large polypeptides and proteins. Mass Spectrom Rev. 1991;10:359–452.
235. Grese RP, Cerny RL, Gross ML. Metal ion-peptide interactions in the gas phase: A tandem mass spectrometry study of alkali metal cationized peptides. J Am Chem Soc. 1989;111:2835–42.
236. Sabareesh V, Balaram P. Tandem electrospray mass spectrometric studies of proton and sodium ion adducts of neutral peptides with modified N- and C-termini: synthetic model peptides and microheterogeneous peptaibol antibiotics. Rapid Commun Mass Spectrom. 2006;20:618–28.

237. Deutsch J, Gilon C, Chorev M. Field desorption mass spectrometry. II. Potassium cationization field desorption mass spectrometry of some penta- and hexapeptides derived from substance P. Int J Pept Protein Res. 1981;18:203–7.
238. Draper WM, Xu D, Perera SK. Electrolyte-induced ionization suppression and microcystin toxins: ammonium formate suppresses sodium replacement ions and enhances protiated and ammoniated ions for improved specificity in quantitative LC-MS-MS. Anal Chem. 2009;81:4153–60.
239. Paizs B, Suhai S. Fragmentation pathways of protonated peptides. Mass Spectrom Rev. 2005;24:508–48.
240. Mouls L, Aubagnac J-L, Martinez J, Enjalbal C. Low energy peptide fragmentations in an ESI-Q-TOF type mass spectrometer. J Proteome Res. 2007;6:1378–91.
241. Roepstorff P, Fohlmann J. Proposal for a common nomenclature for sequence ions in mass spectra of peptides. Biomed Mass Spectrom. 1984;11:601.
242. Hunt DF, Yates JR III, Shabanowitz J, Winston S, Hauer CR. Protein sequencing by tandem mass spectrometry. Proc Natl Acad Sci U S A. 1986;83:6233–7.
243. Papayannopoulos IA. The interpretation of collision-induced dissociation tandem mass spectra of peptides. Mass Spectrom Rev. 1995;14:49–73.
244. Jensen ON, Podtelejnikov AV, Mann M. Identification of the components of simple protein mixtures by high-accuracy peptide mass mapping and database searching. Anal Chem. 1997;69:4741–50.
245. Yates JR III, Eng JK, McCormack AL, Schieltz D. Methods to correlate tandem mass spectra of modified peptides to amino acid sequences in the protein database. Anal Chem. 1995;67:1426–38.
246. Liska AJ, Shevchenko A. Combining mass spectrometry with database interrogation strategies in proteomics. Trends Anal Chem. 2003;22:291–8.
247. Yates JR III. Mass spectrometry and the age of proteome. J Mass Spectrom. 1998;33:1–19.
248. Kinter M, Sherman NE. Protein sequencing and identification using tandem mass spectrometry. New York: Wiley Interscience; 2000. ISBN 978-0-47132-249–8.
249. Zhang Y, Fonslow BR, Shan B, Baek MC; Yates JR 3rd. Protein analysis by shotgun/bottom-up proteomics. Chem Rev. 2013;113:2343–94.
250. Liebler DC. Introduction to proteomics: tools for the new biology. Totowa: Humana Press Inc; 2002. ISBN 978-0-89603-991–9.
251. Shevchenko A, Jensen ON, Podtelejnikov AV, Sagliocco F, Wilm M, Vorm O, Mortesen P, Shevchenk A, Boucherie H, Mann M. Linking genome and proteome by mass spectrometry: large-scale identification of yeast proteins from two-dimensional gels. Proc Natl Acad Sci U S A. 1996;93:14440–5.
252. Cramer R, Gobom J, Nordhoff E. High-throughput proteomics using matrix-assisted laser desorption/ionization mass spectrometry. Expert Rev Proteomics. 2005;2:407–20.
253. Hardouin J. Protein sequence information by matrix-assisted laser desorption/ionization in-source decay mass spectrometry. Mass Spectrom Rev. 2007;26:672–82.
254. Rodrigo MA, Zitka O, Krizkova S, Moulick A, Adam V, Kizek R. MALDI-TOF MS as evolving cancer diagnostic tool: a review. J Pharm Biomed Anal. 2014;95:245–55.
255. Zhurov KO, Fornelli L, Wopdrich MD, Laskay ÜA, Tsybin YO. Principles of electron capture and transfer dissociation mass spectrometry applied to peptide and protein structure analysis. Chem Soc Rev. 2013;42:5014–30.
256. Kulik W, Heerma W, Terlouw JK. A novel fragmentation process in the fast-atom bombardment/tandem mass spectra of peptides cationized with Na^+, determining the identity of the C-terminal amino acid. Rapid Commun Mass Spectrom. 1989;3:276–9.
257. Lin T, Glish GL. C-terminal peptide sequencing via multistage mass spectrometry. Anal Chem. 1998;70:5162–5.
258. Feng WY, Gronert S, Fletcher KA, Warres A, Lebrilla CB. The mechanism of C-terminal fragments in alkali metal ion complexes of peptides. Int J Mass Spectrom. 2003;222:117–34.

259. Anbalagan V, Silva ATM, Rajagopalachary S, Bulleigh K, Talaty ER, Van Stipdonk MJ. Influence of "Alternative" C-terminal amino acids on the formation of $[b_3+17+Cat]^+$ products from metal cationized synthetic tetrapeptides. J Mass Spectrom. 2004;39:495–504.
260. Tost J, Gut IG. Genotyping single nucleotide polymorphisms by MALDI mass spectrometry in clinical applications. Clin Biochem. 2005;38:335–50.
261. Schulten HR, Beckey HD. High resolution field desorption mass spectrometry-I: nucleosides and nucleotides. Org Mass Spectrom. 1973;7:861–7.
262. McNeil CJ, Macfarlaine RD. Observation of a fully protected oligonucleotide dimer at *m/z* 12637 by californium-252 plasma desorption mass spectrometry. J Am Chem Soc. 1981;103:1609–10.
263. Grotjahn L, Taylor LCE. The use of signal averaging techniques for the quantitation and mass measurement of high molecular weight compounds using fast atom bombardment mass spectrometry. Org Mass Spectrom. 1985;20:146–52.
264. Nordhoff E, Kirpekar F, Roepstorff P. Mass spectrometry of nucleic acids. Mass Spectrom Rev. 1996;15:67–138.
265. Schürch S. Characterization of nucleic acids by tandem mass spectrometry—The second decade (2004-2013): from DNA to RNA and modified sequences. Mass Spectrom Rev. 2015 (in press). doi:10.1002/mas.21442.
266. van Dongen WD, Niessen WMA. Bioanalytical LC-MS of therapeutic oligonucleotides. Bioanalysis. 2011;3:541–64.
267. Huber CG, Oberacher H. Analysis of nucleic acids by on-line liquid chromatography-mass spectrometry. Mass Spectrom Rev. 2001;20:310–43.
268. Dudley E, Bond L. Mass spectrometry analysis of nucleosides and nucleotides. Mass Spectrom Rev. 2014;33:302–31.
269. Bleicher K, Bayer E. Various factors influencing the signal intensity of oligonucleotides in electrospray mass spectrometry. Biol Mass Spectrom. 1994;23:320–2.
270. Castleberry CM, Rodicio LP, Limbach PA. Electrospray ionization mass spectrometry of oligonucleotides. Curr Protoc Nucleic Acid Chem. 2008. doi:10.1002/0471142700.nc1002s35.
271. Castleberry CM, Chou CW, Limbach PA. Matrix-assisted laser desorption/ionization time-of-flight mass spectrometry of oligonucleotides. Curr Protoc Nucleic Acid Chem. 2008. doi:10.1002/0471142700.nc1001s33.1.
272. Apffel A, Chakel JA, Fischer S, Lichtenwalter K, Hancock WS. Analysis of oligonucleotides by HPLC-electrospray ionization mass spectrometry. Anal Chem. 1997;69:1320–5.
273. Sauer S. The essence of DNA sample preparation for MALDI mass spectrometry. J Biochem Biophys Methods. 2007;70:311–8.
274. McLuckey SA, Van Berkel GJ, Glish GL. Tandem mass spectrometry of small, multiply charged oligonucleotides. J Am Soc Mass Spectrom. 1992;3:60–70.
275. Murray KK. DNA sequencing by mass spectrometry. J Mass Spectrom. 1996;31:1203–15.
276. Xiang Y, Abliz Z, Takayama M. Cleavage reactions of the complex ions derived from self-complementary deoxydinucleotides and alkali-metal ions using positive ion electrospray ionization with tandem mass spectrometry. J Am Soc Mass Spectrom. 2004;15:689–96.
277. Boschenok J, Sheil MM. Electrospray tandem mass spectrometry of nucleotides. Rapid Commun Mass Spectrom. 1996;10:144–9.
278. Stano M, Flosadottir HD, Ingolfsson O. Effective quenching of fragment formation in negative ion oligonucleotide matrix-assisted laser desorption/ionization mass spectrometry through sodium adduct formation. Rapid Commun Mass Spectrom. 2006;20:3498–502.
279. Wong SF, Meng CK, Fenn JB. Multiple charging in electrospray ionization of poly(ethylene glycols). J Phys Chem. 1988;92:546–50.
280. Varray S, Aubagnac J-L, Lamaty F, Lazaro R, Martinez J, Enjalbal C. Poly(ethyleneglycol) in electrospray ionization (ESI) mass spectrometry. Analusis. 2000;28:263–8.
281. Mincheva Z, Hadjieva P, Kalcheva V, Seraglia R, Traldi P, Przybylski M. Matrix-assisted laser desorption/ionization, fast atom bombardment and plasma desorption mass spectrometry of polyethylene glycol esters of (2-benzothiazolon-3-yl)acetic acid. J Mass Spectrom. 2001;26:626–32.

282. González-Valdez J, Rito-Palomares M, Benavides J. Advances and trends in the design, analysis, and characterization of polymer-protein conjugates for "PEGylaided" bioprocesses. Anal Bioanal Chem. 2012;403:2225–35.
283. Ayorinde FO, Eribo BE, Johnson JH Jr, Elhilo E. Molecular distribution of some commercial nonylphenol ethoxylates using matrix-assisted laser desorption/ionization time-of-flight mass spectrometry. Rapid Commun Mass Spectrom. 1999;13:1124–8.
284. Lattimer RP. Tandem mass spectrometry of lithium-attachment ions from polyglycols. J Am Soc Mass Spectrom. 1992;3:225–34.
285. Lattimer RP. Tandem mass spectrometry of poly(ethylene glycol) lithium-attachment ions. J Am Soc Mass Spectrom. 1994;5:1072–80.
286. Bahr U, Deppe. A, Karas M, Hillenkamp F, Giessmann U. Mass spectrometry of synthetic polymers by UV-matrix-assisted laser desorption/ionization. Anal Chem. 1992;64:2866–9.
287. Brandt H, Ehmann T, Otto M. Toward prediction: using chemometrics for the optimization of sample preparation in MALDI-TOF MS of synthetic polymers. Anal Chem. 2010;82:8169–75.
288. Rizzarelli P, Carroccio S. Modern mass spectrometry in the characterization and degradation of biodegradable polymers. Anal Chim Acta. 2014;808:18–43.
289. Altuntaş E, Schubert US. "Polymeromics": mass spectrometry based strategies in polymer science toward complete sequencing approaches: a review. Anal Chim Acta. 2014;808:56–69.
290. Wesdemiotis C, Solak N, Polce MJ, Dabney DE, Chaicharoen K, Katzenmeyer BC. Fragmentation pathways of polymer ions. Mass Spectrom Rev. 2011;30:523–59.
291. Crecelius AC, Baumgaertel A, Schubert US. Tandem mass spectrometry of synthetic polymers. J Mass Spectrom. 2009;44:1277–86.
292. Crecelius AC, Vitz J, Schubert US. Mass spectrometric imaging of synthetic polymers. Anal Chim Acta. 2014;808:10–7.
293. Tintaru A, Chendo C, Wang Q, Viel S, Quéléver G, Peng L, Posocco P, Pricl S, Charles L. Conformational sensitivity of conjugated poly(ethylene oxide)-poly(amidoamine) molecules to cations adducted upon electrospray ionization—a mass spectrometry, ion mobility and molecular modeling study. Anal Chim Acta. 2014;808:163–74.
294. Terrier P, Desmazières B, Tortajada J, Buchmann W. APCI/APPI for synthetic polymer analysis. Mass Spectrom Rev. 2011;30:854–74.
295. Flick TG, Merenbloom SI, Williams ER. Effects of metal ion adduction on the gas-phase conformations of protein ions. J Am Soc Mass Spectrom. 2013;24:1654–62.
296. Carlton DD Jr, Schug KA. A review on the interrogation of peptide-metal interactions using electrospray ionization-mass spectrometry. Anal Chim Acta. 2011;686:19–39.
297. Jaswal SS. Biological insights from hydrogen exchange mass spectrometry. Biochim Biophys Acta. 1834;2013:1188–201.
298. Balasubramaniam D, Komives EA. Hydrogen-exchange mass spectrometry for the study of intrinsic disorder in proteins. Biochim Biophys Acta. 1834;2013:1202–9.
299. Uetrecht C, Rose RJ, van Duijn E, Lorenzen K, Heck AJ. Ion mobility mass spectrometry of proteins and protein assemblies. Chem Soc Rev. 2010;39:1633–55.
300. Lanucara F, Holman SW, Gray CJ, Eyers CE. The power of ion mobility-mass spectrometry for structural characterization and the study of conformational dynamics. Nat Chem. 2014;6:281–94.
301. Vandermarliere E, Stes E, Gevaert K, Martens L. Resolution of protein structure by mass spectrometry. Mass Spectrom Rev. 33, 2014, doi:10.1002/mas.21450.
302. Konermann L, Vahidi S, Sowole MA. Mass spectrometry methods for studying structure and dynamics of biological macromolecules. Anal Chem. 2014;86:213–32.
303. Seo Y, Schenauer MR, Leary JA. Biologically relevant metal-cation binding induces conformational changes in heparin oligosaccharides as measured by ion mobility mass spectrometry. Int J Mass Spectrom. 2011;303:191–8.Author Query

Chapter 8
Direct Analysis Mass Spectrometry

Murray J McEwan

8.1 Introduction

Ever since mass spectrometers have been used to analyze trace amounts of chemicals in the environment, there has been a demand for simpler and faster methods of analysis. This is true for applications ranging from analysis of trace chemicals in the environment to bioanalytical analysis of proteins. Many of the early methods required significant effort and time to be expended in sample preparation, often with a potential loss of accuracy in identifying the analytes of interest. Traditionally, gas chromatography mass spectrometry (GC-MS) methods (introduced originally in the 1950s and 1960s) were chosen as the gold standard of analytical analysis, and many of the methods required by regulatory bodies for analysis of air samples have been written in terms of GC-MS methodologies [1–3]. Ideally the mass spectrometric analytical method applied should be fast, selective, and have high sensitivity. In recent times, there have been several newer technologies that have helped change the traditional landscape of mass spectrometric analysis using what might be called more direct methods of analysis. Additional emphasis has been given in this chapter on the technologies that monitor volatile organic compounds (VOCs).

In the chapter although we will include some of the more established methods for direct analysis, we will also be examining some of the newer methods that have been introduced for analysis of samples, and that have simplified the analysis process either by reducing or removing the need for sample preparation and/or providing results in real time. The essential feature of these newer techniques is that the analysis of trace gases either in the gas phase or released from surfaces can be obtained directly avoiding sample collection or modification. These technologies not only reduce the time for analysis but in many cases are easier to use. Clearly, in situations such as environmental and breath analysis of trace volatiles in air, a rapid or real-time response offers a considerable advantage. Here we will focus on those

M. J. McEwan (✉)
Department of Chemistry, University of Canterbury, PB 4800, Christchurch 8140, New Zealand
e-mail: murray.mcewan@canterbury.ac.nz

T. Fujii (ed.), *Ion/Molecule Attachment Reactions: Mass Spectrometry*,
DOI 10.1007/978-1-4899-7588-1_8

technology advances that have been successful in quantitative analysis of volatiles. Among these methods is atmospheric pressure chemical ionization (APCI), ion–molecule reaction mass spectrometry (IMR-MS), proton transfer mass spectrometry (PTR-MS), selected ion flow tube mass spectrometry (SIFT-MS), and the ionization techniques of direct analysis in real time (DART) and desorption electrospray ionization (DESI) and related methods. The main emphasis in this chapter is to focus on those techniques that have been successful in providing direct analyses of VOCs.

To aid in a comparison of the ways the different techniques have been applied to monitor volatiles, a short commentary is presented at the end of the discussion of each method summarizing their applications mainly in three different areas: environment, food science, and medicine/health. The importance of VOCs to each of these areas is briefly summarized next.

VOCs have a very important role in the atmosphere. They are emitted by natural and anthropogenic processes and their presence and distribution can have a marked influence on the environment. In the troposphere, VOCs from anthropogenic sources play a major role in the formation of photochemical smog. However, biogenic VOCs (BVOCs) from the world's forests and oceans dominate over VOCs from anthropogenic sources except in the densely populated areas. The role of VOCs in the workplace environment is another growing area of interest. In recent times, there has been a concerted effort in the application of analytical techniques that can directly monitor VOCs in the environment, that are easy to apply and that are both fast and accurate.

VOCs also have an influential role in food science and technology. They are present in foods and drinks and their presence is intimately involved in aroma and flavor. Both of these factors influence the choice and preference of consumers. Further, VOCs emitted by foods can provide evidence of the shelf life of a product, of the ripening stage, and whether product adulteration has occurred. The human olfactory threshold can be as low as the pptv range for key aroma compounds, so sensitive analytical techniques are required for interrogation.

VOCs also play a very important role in medicine. They are produced by metabolic processes in the body and they are present on breath. The distribution of these VOCs can be indicative of disease and even the onset of organ failure. There has been a growing emphasis on searching for VOC markers of disease using breath, but so far without the return researchers had hoped for although inroads are now being made with some promising discoveries. VOCs are also present in the headspace above body fluids such as blood and urine; these continue to be objects for interrogation by direct methods of mass spectrometry. Bacterial cultures also provide a source of VOCs and these too are subjects of investigation to see whether the traditional methods of identification can be shortened.

8.2 Atmospheric Pressure Chemical Ionization (APCI)

APCI is an ionization technique that relies on gas phase ion–molecule reactions that take place at or near atmospheric pressure. It was originally developed in the mid-1970s by Horning and coworkers [4, 5] and has used a variety of different ion generation techniques including ^{63}Ni, corona discharge, photoionization, and glow discharge that all operate at atmospheric pressure. Ions generated in an APCI source are therefore introduced to the mass spectrometer at higher pressures than from the traditional electron impact source with a corresponding increase in sensitivity. The most widely used source for ionization is a direct current corona discharge needle as it is simple to install, robust, and operates with high efficiency. At the time of its introduction APCI was primarily used for air analysis as it provided high sensitivity for polar volatiles and semivolatile species [6, 7].

Generally in an APCI source, multiple ion species are generated from air (e.g., N_2^+, O_2^+) and these ions then undergo a series of ion–molecule reactions ultimately reacting with water vapor that is also present in the sample. At atmospheric pressure, the mean free path between collisions is sufficiently short for a large number of gas-phase reactions to occur. The end result is that the most stable (or least reactive) ions in air predominate and it is these ions that serve as the reagent ions for the APCI source. In the ensuing ion–molecule chemistry in moist air, water cluster ions are generated and these may react with the analyte. Many of these same processes occur in the Earth's upper atmosphere as a result of ionizing radiation from the sun impacting the Earth's outer atmosphere.

8.2.1 *Typical Reaction Sequence for An APCI Source (Somewhat Simplified) Is*

$$N_2 + e^- \rightarrow N_2^+ + 2e^- \tag{8.1}$$

$$N_2^+ + 2N_2 \rightarrow N_4^+ + N_2 \tag{8.2}$$

$$O_2 + e^- \rightarrow O_2^+ + 2e^- \tag{8.3}$$

$$N^{4+} + H_2O \rightarrow H_2O^+ + 2N_2 \tag{8.4}$$

$$N_2^+ + H_2O \rightarrow H_2O^+ + N_2 \tag{8.5}$$

$$O_2^{+} * + H_2O \rightarrow H_2O^+ + O_2 \tag{8.6}$$

$$H_2O^+ + H_2O \rightarrow H_3O^+ + OH \tag{8.7}$$

$$H_3O^+ + H_2O + M \rightarrow H_3O^+ \cdot (H_2O) + M \tag{8.8}$$

$$H_3O^+ \cdot (H_2O) + H_2O + M \rightarrow H_3O^+ \cdot (H_2O)_2 + M \tag{8.9}$$

In this scheme, M is any molecule that can remove energy from the ion-neutral collision. The water cluster ions are not strongly bound and can dissociate in collisions and do not necessarily react with all analytes that may be present in a sample. Many of the designs of APCI sources are directed at reducing the number of cluster ions in the source. Several different designs have been followed [6]. One design consists of a small heated closed volume source (~ 1 cm^3) that is connected to the vacuum through a small 25 μm orifice. It is flushed with a pure carrier gas and is then coupled to a gas or liquid chromatograph. This configuration copes best with a single analyte at a time. A second design was developed at the University of Toronto and was later manufactured by Sciex [6]. In this design, the connection of the ion source to the vacuum is via a larger 100 μm orifice and is separated by a stream of high-purity nitrogen from where the ion–molecule reactions with the analyte takes place. This arrangement enables collisional dissociation to reduce the number of cluster ions.

Early APCI sources used an axial configuration with the ions produced on the axis of the orifice. Now orthogonal configurations are used in many ion sources particularly in liquid chromatography mass spectrometry (LC-MS) as in this configuration only the ions and not the neutral gas are directed towards the inlet of the mass spectrometer avoiding some of the contamination. Declustering (or desolvation) of the ions generated in an APCI source can also be improved by accelerating the ions in the interface region between the ion source and the mass spectrometer at a pressure of around 10^{-3} Torr so that collisions with the residual gas molecules break up some of the weakly bound cluster ions.

The sensitivity of analyte detection by a chemical ionization (CI) source is directly proportional to the concentration of analyte in favorable cases [8, 9]. Sensitivities have been reported of up to 5000 counts per second (cps) per ppbv [7]. Sensitivities can also be enhanced by operating the APCI at elevated temperatures (up to 400 °C) and an increase in sensitivity for $(CH_3)_2S$ of 2×10^6 has been reported by the temperature increase [10].

8.2.2 *Current Trends for APCI*

Although APCI began as a tool for analysis of VOCs in air, it is now more commonly used as the ionization source in many mass spectrometry (MS) techniques. This is because the APCI source can be simply adapted for direct coupling of separation techniques to the mass spectrometer. Although the emphasis in this chapter is on air samples, there has been widespread integration of APCI into LC-MS where a heated nebulizer vaporizes the LC solvent and creates the reagent ions from the

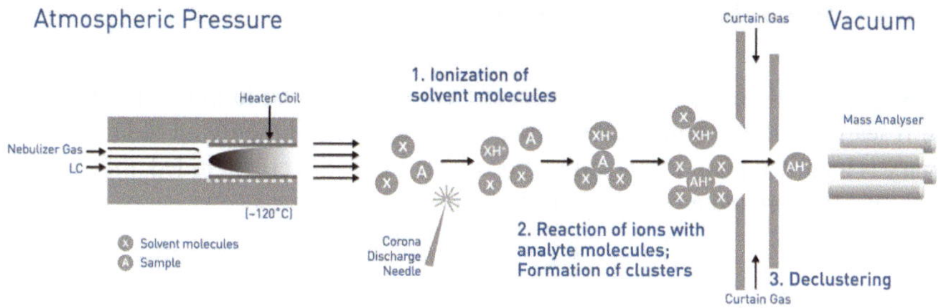

Fig. 8.1 Schematic diagram of an APCI source in use in LC-MS

resultant vapor with a high-voltage corona discharge [11]. LC-MS is now the largest user of APCI. The APCI source has also been adapted for examining solid surfaces by thermal evaporation, laser ablation, and desorption [12] (Fig. 8.1).

Generally APCI is applied to both polar and nonpolar analytes with a molecular weight up to around 1500 Da, and it invariably gives singly charged ions. In LC-MS, the analyte is a solution and the eluate from the liquid chromatograph is introduced to a nebulizer where it is converted into small droplets by a nitrogen nebulizer gas. The droplets are transmitted through a heated quartz tube which vaporizes the solvent *X* and sample analyte *A*. After the desolvation step they are carried along past the corona discharge where ionization of the analyte takes place at atmospheric pressure. Two modes of operation are possible. In the positive ion mode, proton transfer and association of the reagent ions produce diagnostic ions from the analyte. In the negative ion mode, negative ions of the analyte are generally produced by either proton abstraction or adduct formation of negative ions from the discharge. The primary positive ions are generally N_2^+ and O_2^+, which then react with the solvent molecules to form secondary reagent ions. As soon as a molecule more basic than water such as ammonia or acetonitrile is introduced, the solvated H_3O^+ ions form $NH_4^+(H_2O)_n$ ions or $CH_3CNH^+.(H_2O)_m$ and these ions then react with the analyte. The primary negative ion is mainly O_2^- which rapidly hydrates to $O_2^-.(H_2O)_n$ and these ions then react with the solvent molecules to form secondary reagent ions. The solvated ions react with the analyte A:

$$O_2^- \cdot (H_2O)_n + A \rightarrow (A - H)^- + HO_2 + nH_2O \tag{8.10}$$

$$\rightarrow AO_2^- + nH_2O \tag{8.11}$$

$$\rightarrow A^- + O_2 + nH_2O \tag{8.12}$$

The reactant ions spectrum generated from an APCI source is therefore dominated by solvent-derived cluster ions [13].

8.2.3 Some Applications of APCI

Gas-phase analysis using ACPI sources have been applied to many different areas of research with the emphasis on direct VOC analysis and only an indicative summary is included here. If the applications from LC-MS were included, their applications would demand a full chapter in themselves.

8.2.3.1 Environmental

APCI combined with mass spectrometric detection has been shown to be a valuable and useful technique for monitoring atmospheric VOCs [14]. An instrument that provides instantaneous, on-site environmental analysis with the acronym of trace atmospheric gas analyzer (TAGA). The TAGA instrument also utilized an APCI source was developed and installed in vehicles to provide real-time tracking of industrial emissions [15]. The instrument was successfully used to provide a fingerprint of industrial emissions including ammonia, dimethylamine, aniline, benzothiazole, diphenylamine, acetone, and methyl isobutyl ketone [15]. TAGA instruments were used to respond to the World Trade Center disaster, to monitor anthrax found in the 2001 incident in the Hart Senate office building in the USA and also to monitor chlorine dioxide (ClO_2) emissions in ambient air near residences [16].

Direct APCI in combination with tandem mass spectrometry was applied to the real-time monitoring of benzene, toluene, ethylbenzene, and xylenes in ambient air. The method was linear over four orders of magnitude with a sensitivity of 0.5 ppbv (2 μg m^{-3}) [17]. A similar study using a hand-held mass spectrometer and APCI source achieved limits of detection for benzene, toluene, and ethylbenzene of around 0.5 ppbv [18]. Another study utilizing an APCI source coupled to a triple quadrupole mass spectrometer and linked to a smog chamber examined the products of hydrocarbon oxidation in the environment [19]. A similar study examined the products from α-pinene ozonolysis using an APCI source coupled to an ion trap mass spectrometer [20]. In another investigation, an APCI source linked with a tandem mass spectrometer and utilizing a Cl^- reagent ion was used to monitor fluorinated phenols $C_6H_{(5-x)}F_xOH$, where $x = 0-5$ in nitrogen. The product ions were MCl^- through association reactions or $(M\text{-}H)^-$ through proton abstraction [21].

8.2.3.2 Food and Food Technology

Most of the applications of APCI to the area of food science are in conjunction with LC-MS and high-performance liquid chromatography-mass spectrometry (HPLC-MS) and its variations. A few samples of investigations using these techniques are also included to get the "flavor" of what has been achieved.

A useful review of APCI-MS applied to volatile flavor release examines the successes and limitations [22], and an even more comprehensive review of the whole area of food and flavor is in the book edited by Taylor and Linforth [23]. The profile of Stilton cheese was examined by APCI-MS which provided rapid discrimination of the different cheese profiles although some interference from ammonia was found [24]. APCI sources were developed for real-time volatile release from different foods looking at the VOCs ethyl butyrate, 2-methylbutanal, 2,5-dimethylpyrazine, and 1-hexanol [25]. The feasibility of using APCI-MS for food classification was successfully demonstrated by monitoring 15 VOCs from a number of monovarietal apple juices in conjunction with chemometrics [26]. The motivation for the study was to provide a faster and simpler alternative than the more time-consuming methods of conventional techniques.

In the wider area of LC-APCI-MS, a large number of investigations have focused on identifying particular contaminants [27–29] or quantification of ingredients [30, 31].

8.2.3.3 Medical

APCI-MS has not been applied to problems in the medical area for analysis of VOCs in gas mixtures as frequently as some of the other direct analysis techniques that are discussed in this chapter. One early study utilizing an APCI source plus mass spectrometer system examined individual breath exhalations of human subjects using both positive and negative ion modes [32]. As with the classification of food technology, the majority of work utilizing APCI in the medical area is from LC-MS studies and electrospray ESI-MS. Although APCI does not have the versatility of some of the other direct techniques for interrogation and quantification of VOCs in the medical area, it still has an important role when integrated with established analytical techniques such as LC-MS and electrospray. Two illustrative examples follow: APCI was found to be the most effective analytical technique when coupled with LC-MS in the detection and quantification of polydimethylsiloxanes when used in cosmetic surgery under uncontrolled conditions [33]. In another investigation, it was effective in demonstrating medical misadventure in an overdose of a drug used in cancer chemotherapy [34].

8.3 Ion–Molecule Reaction Mass Spectrometry (IMR-MS)

Another variant of a technique that utilizes specific ion–molecule reactions to determine analyte concentrations is ion–molecule reaction mass spectrometry (IMR-MS). IMR-MS instruments utilize designated reagent ions to monitor selected analytes. A substantial effort of the early flowing afterglow studies initiated by Ferguson et al. [35] focused on understanding the processes occurring in the Earth's upper atmosphere using both positive and negative reagent ions. In several cases, dedicated laboratory experiments have been established to monitor molecules of interest at trace levels. Arnold and coworkers have continued this early work using flowing afterglows to examine reactions of ions that could be used for specific atmospheric trace species [36, 37]. These flow tube instruments were in the main large laboratory-based instruments and although the chemistry studied using them can, in principle, be classed as IMR-MS they will not be discussed further in this chapter.

At least two commercial IMR-MS instruments have been produced [38, 39]. A commercial IMR-MS instrument was manufactured by V&F Analyse that began as a spin-off company out of the University of Innsbruck in 1985 by Villinger and Federer [38]. The instrument has been marketed under a number of different guises that are dependent on the particular application of the instrument. Some of these instruments use conventional electron impact ion sources (EI Sense) and others use a combination of electron impact and ion–molecule reactions. The "Air Sense" instrument was designed to measure trace gas components in industrial applications such as in fuel cells or reformer gas manufacture and in monitoring of the work place environment. It is the "Air Sense" instrument that we discuss further here (Fig. 8.2).

In this instrument, ions are generated via electron impact in a low pressure source using three reagent gases/vapors: krypton, xenon, and mercury. The atomic reagent ions formed Kr^+, Xe^+, and Hg^+; all have different ionization energies (IE) (Kr = 13.997 eV, Xe = 12.13 eV, Hg = 10.437 eV); and these differences can be utilized

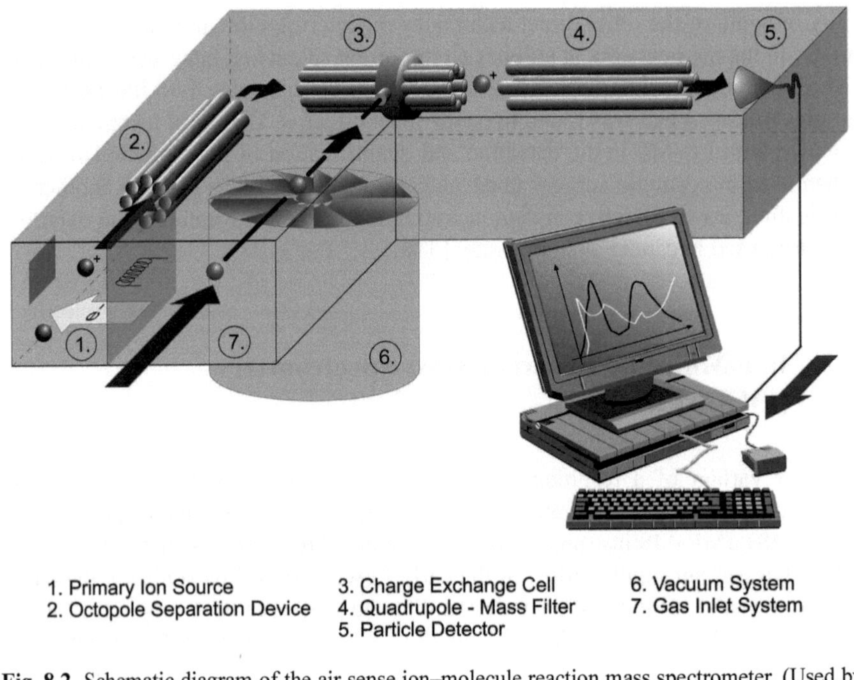

Fig. 8.2 Schematic diagram of the air sense ion–molecule reaction mass spectrometer. (Used by permission from V&F Analyse)

in some situations where isobaric ions are produced. The two octopole assemblies are used to store reagent and product ions in a confined volume against their coulombic repulsion before releasing the ions to the quadrupole mass spectrometer for mass analysis. The first octopole (2) acts as an ion guide for the primary ion beam and also as a high-pass mass filter. The second octopole (3) acts as a charge exchange cell. The pressure in the reaction region is low—typically 8×10^{-5} Torr—and those pressures remove the possibility of association reactions occurring. Switching time between the three reagent ions is about 400 ms. The response time of the instrument to changes in concentration is reported to be around 50 ms [40].

8.3.1 Reaction Chemistry in IMR-MS

The three atomic reagent ions in IMR-MS (V&F variety) react with an analyte by electron transfer and dissociative electron transfer.

$$Xe^{+} + C_2H_6 \rightarrow C_2H_6{}^{+} + Xe(0.44) \quad (8.13)$$

$$\rightarrow C_2H_5{}^{+} + Xe + H(0.03) \quad (8.14)$$

$$\rightarrow C_2H_4{}^{+} + Xe + H_2(0.53) \quad (8.15)$$

Reactions (8.13–8.15) have a combined rate coefficient of $k=8.5 \times 10^{-10}$ cm^3 s^{-1} and the reaction branching ratios are given in parenthesis. The advantage of the mercury ion is that it does not react with the ambient air matrix for samples in air.

$$Hg^{+} + C_6H_6 \rightarrow C_6H_6{}^{+} + Hg \quad (8.16)$$

Here a fast electron transfer reaction occurs between Hg^{+} and benzene.

8.3.2 Some Applications of IMR-MS

IMR-MS has been used in quite a wide range of different applications where the lack of sample preparation required by the technique was found to be a distinct advantage.

8.3.2.1 Environmental

The environmental area has been one of the major areas for application of IMR-MS. Dearth used it to monitor ten hydrocarbon species and H_2S in the exhaust gas of motor vehicles in real time. The results compared favorably with offline gas chromatography flame ionization detection (GC-FID) methods [41]. The technology has also been utilized to monitor in real time, a number of inorganic compounds including O_2, H_2O, CO_2, HCl, SO_2, and NO in the exhaust gas of a municipal solid waste pilot plant. In that investigation, CF_3I^+ was also included with Xe^+ and Kr^+ as a source of reagent ions [42]. The applicability of the technology was further demonstrated when the IMR-MS technique was applied to the detection of volatiles present in cigarette smoke in which 20 VOCs were identified [43].

8.3.2.2 Medical

Recently there has been a growing interest in the applications of IMR-MS in the medical area. Two studies using IMR-MS have examined monitoring the anesthetic propofol (2,6-diisopropylphenol) using the Hg^+ ion as the ionization source. In the first study, comparisons between blood propofol, and propofol in the exhaled breath of patients undergoing intravenous anesthesia were monitored and the conclusion was reached that IMR-MS may allow continuous noninvasive monitoring of propofol levels in patients undergoing general anesthesia [40]. A second study also on propofol found that IMR-MS can detect propofol on human breath within seconds after a bolus injection [44]. Another investigation using IMR-MS examined breath gas from patients who were mechanically ventilated and compared the results from IMR-MS with a conventional electron impact mass spectrometer for monitoring CO_2. The study reported measurements of acetaldehyde, acetone, ethanol, and isoprene in alveolar breath [45]. IMR-MS was also investigated for identifying breath gas marker candidates in liver disease [46] and in developing a more rapid test for distinguishing between gram-positive bacteria by monitoring the VOCs produced from cultures [47].

8.4 Proton Transfer Mass Spectrometry (PTR-MS)

Proton transfer mass spectrometry (PTR-MS) was developed in 1995 at the University of Innsbruck by Werner Lindinger and his collaborators [48]. PTR-MS was able to utilize the ability of the H_3O^+ ion to transfer its proton ($PA(H_2O) = 691\ kJ\ mol^{-1}$) to any analyte in air that has a higher proton affinity. Proton transfer reactions are invariably fast when exoergic and occur at the collision rate [49]. H_3O^+ is readily formed from a discharge in moist air; it does not react with the bulk constituents of air and is therefore a good indicator of the presence of any volatile species that may be present in a sample. Many of the organic molecules that are important analytes in medicine, foods, and the environment have proton affinities (PAs) higher than that of H_2O. Notable exceptions are the small alkanes, ethane, and acetylene.

8.4.1 PTR-MS Operation

A schematic diagram of a PTR-MS instrument is shown in Fig. 8.3.

A very good summary of PTR-MS has been presented in a recent book by Ellis and Mayhew 2014 [50]. The PTR-MS instrument consists of the hollow cathode (HC) ion source, a drift tube, a mass spectrometer, and an ion detector. There are several instrument manufacturers producing instruments that can be broadly classed as PTR-MS units. Those discussed here have generally been manufactured by Ionicon in Austria.

8.4.2 PTR-MS Ion Source

The ion source in most PTR-MS instruments is a hollow cathode that operates off a DC voltage, and generates H_3O^+ from a discharge of water vapor at a typical pressure of ~0.75 Torr. In this process the ion fragments arising from the dissociation of H_2O in the discharge and other ions generated from air molecules in the discharge undergo a series of ion–molecule reactions such that the fragment ions OH^+, N_2^+, O^+, H_2^+, and H_2O^+ are converted into H_3O^+. The units for k, the rate coefficient are 10^{-9} cm^3 s^{-1} [51]

$$H_2O^+ + H_2O \rightarrow H_3O^+ + OH \quad k = 1.85 \times 10^{-9} \tag{8.17}$$

$$OH^+ + H_2O \rightarrow H_3O^+ + O \quad k = 1.3 \times 10^{-9} \tag{8.18}$$

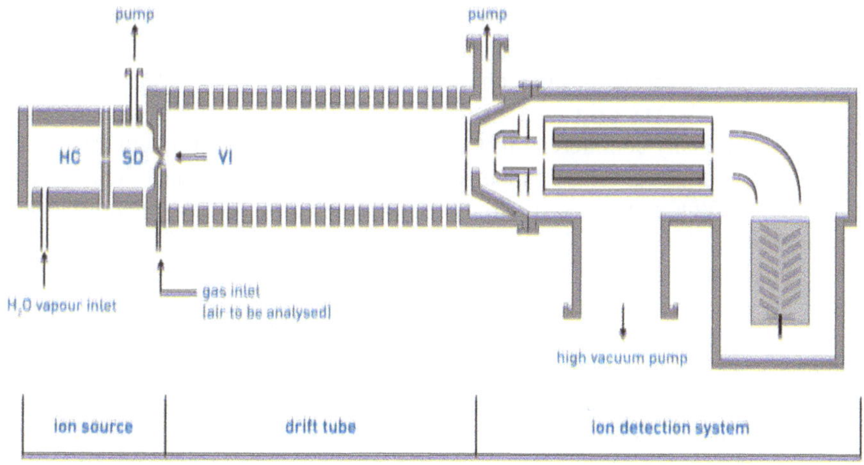

Fig. 8.3 A schematic diagram of a PTR-MS instrument showing the hollow cathode (HC) ion source, the source drift region (SD), the Venturi inlet (VI) and the drift tube

$$\rightarrow H_2O^+ + OH \quad k = 1.6 \times 10^{-9} \tag{8.19}$$

$$O^+ + H_2O \rightarrow H_2O^+ + O \quad k = 2.6 \times 10^{-9} \tag{8.20}$$

$$H_2{}^+ + H_2O \rightarrow H_3O^+ + H \quad k = 3.4 \times 10^{-9} \tag{8.21}$$

$$\rightarrow H_2O^+ + H_2 \quad k = 3.9 \times 10^{-9} \tag{8.22}$$

$$H^+ + H_2O \rightarrow H_2O^+ + H \quad k = 6.9 \times 10^{-9} \tag{8.23}$$

$$N_2{}^+ + H_2O \rightarrow H_2O^+ + N_2 \quad k = 1.9 \times 10^{-9} \tag{8.24}$$

The absence of any mass spectrometric filtering of the ions is not a significant problem from the HC discharge as most of the ion–molecule reactions in the HC region terminate in H_3O^+. The principle contaminant ions generated in the discharge are NO^+ and $O_2{}^+$ and although they are formed, they do not have a readily available pathway to H_3O^+. By careful monitoring of the discharge conditions the hollow cathode can deliver a signal of H_3O^+ that approaches 99 % purity.

8.4.3 *PTR-MS Drift Tube*

The drift tube consists of a series of equally spaced rings separated by insulators to establish a voltage on each of the rings. A voltage gradient is then established in the drift tube that accelerates the ions in the direction of the flow. An electric field strength in a PTR-MS drift tube of around 50 V cm^{-1} is typical of the operating conditions for many commercial instruments. For a drift tube of length 10 cm, the voltage gradient along the drift tube is 500 V. In addition, to reduce the degree of adsorption of VOCs on the walls from a sample, the drift tube is generally heated between 40 and 100°C.

An important parameter in PTR-MS is the energy provided by collisions between the ions and the neutral gas in the drift tube. This energy depends on the applied electric field, *E*, and on the neutral gas density, *N* (number of molecules per cubic centimeter), in the drift tube. The ratio *E/N* is called the reduced electric field. For a typical drift tube at a pressure of 1.5 Torr, the gas density is 4.4×10^{16} mol/cm^3. The air flow into the instrument (that may also contain the sample in some cases), at a pressure of about 1.5 Torr is around 20 cm^3 min^{-1} at standard temperature and pressure (STP; 273 K, 1 atmosphere = 760 Torr). The electric field strength is 500 V in 10 cm.

So,

$$E = 500/10 = 50\,\mathrm{Vcm}^{-1} \tag{8.25}$$

From which the reduced electric field may be calculated

$$E/N = 50/(4.4\times10^{16})\,\mathrm{Vcm}^2 = 1.14\times10^{-15}\,\mathrm{Vcm}^2 \tag{8.26}$$

To reduce the difficulty of working with such small numbers, another unit has been defined called the Townsend (Td), where

$$1Td = 10^{-17}\,\mathrm{Vcm}^2 \tag{8.27}$$

In the PTR-MS drift tube described here, the reduced electric field strength

$$E/N = 1.14\times10^{-15}/1\times10^{-17} = 114Td \tag{8.28}$$

8.4.4 PTR-MS Neutral Gas Residence Time

The mean residence time of an ion in a drift tube is quite different from the residence time for a neutral gas molecule. For a drift tube with the dimensions considered here, viz., 10 cm long, diameter of 4 cm, and an STP flow rate of 20 $cm^3\,min^{-1}$ the average residence time of the neutral gas can be calculated from the volumetric flow rate F $cm^3\,s^{-1}$.

$$F = F_{STP}\times(P_{std}/P_{drift})\times(T_{drift}/T_{std}) \tag{8.29}$$

$$= (20/60)\times(760/1.5)\times(373/273)\,\mathrm{cm}^3\mathrm{s}^{-1} = 200\,\mathrm{cm}^3\mathrm{s}^{-1} \tag{8.30}$$

For an internal drift tube volume of 126 cm^3, the gas should therefore have a residence time in the flow tube of

$$126/200\mathrm{s} = 0.63\mathrm{s} \tag{8.31}$$

8.4.5 Ion Residence Times

The presence of the voltage gradient in the drift tube accelerates the ions until a constant drift velocity is established. The constant drift velocity is the result of multiple collisions between the ions and molecules generating an opposing force that results in a steady drift velocity v_{drift}. The ion drift velocity is linearly dependent on the electric field (E) at low electric fields and can be written as [52]

$$v_{drift} = KE, \tag{8.32}$$

where the coefficient K is called the ion mobility and is defined here as the proportionality factor of an ion's drift velocity v_{drift} in a gas and an electric field of strength E. K is a function of temperature and pressure and also depends on the cross section for ion-neutral collisions. A more useful ion mobility for comparison that does not depend on pressure and temperature is the reduced mobility K_0 is defined

$$K = (760/P)(T/273)K_0 \tag{8.33}$$

Here, the pressure P is expressed in Torr and the temperature is in Kelvin. Once K_0 is known for a particular combination of ions and neutral molecules, the ion mobility may be found at any other temperature and pressure. Substituting K_0 for K in Eq. (8.32) gives

$$v_{drift} = (760/P)(T/273)K_0E \tag{8.34}$$

If the number density N_0 is defined as the gas number density under STP conditions (N_0=2.687 × 10^{19} cm^{-3}), then

$$v_{drift} = N_0K_0(E/N). \tag{8.35}$$

The velocity of the ion swarm can be measured by gating the ions and then measuring their arrival times at the end of the drift tube. Tables of reduced mobilities have been published for H_3O^+ ions among others in nitrogen [53, 54]. Once knowledge of the mobility has been found, the transit time through the drift tube can be estimated. For a drift tube that is 10 cm long and has an *E/N* value of 120 Td, where the H_3O^+ ions exhibit minimal clustering with H_2O remaining largely as individual H_3O^+ ions, the reduced mobility of H_3O^+ from [7] is ~2.81 cm^2 V^{-1}. From Eq. (8.35), the average drift velocity is

$$v_{drift} = 2.69\times10^{19}\times2.81\times1.14\times10^{-15}\ \text{cm s}^{-1} = 862\,\text{ms}^{-1}. \tag{8.36}$$

For a drift tube of length L = 10 cm, the average drift time can be estimated as

$$t_{drift} = (\text{L}/v_{drift}) = 116\,\mu\text{s}.$$

The calculation shown here is only approximate because of the assumptions made. However, it is clear that the drift time for ions is more than two orders of magnitude shorter than the residence time taken for the neutral gas in the flow tube. A compromise is reached between the electric field applied and the drift tube sensitivity desired. Higher drift fields lead to shorter residence times but also provide less opportunity for the reaction and therefore lower sensitivity. The drift field on the other hand helps to remove cluster ions formed by association of H_3O^+ with water in the sample which can also react with an analyte.

$$H_3O^+ + H_2O + M \rightarrow H_3O^+ \cdot H_2O + M \tag{8.8}$$

In Eq. (8.8), M represents any gaseous species that can remove energy from an ion-neutral collision. Another complicating factor is that the reagent H_3O^+ ions may have different mobilities in the drift tube than the product protonated ions of an analyte. Larger ions will travel at a slower velocity through the drift field.

8.4.6 Mass Spectrometers

In the first PTR-MS instruments, quadrupole mass spectrometers were used. On the plus side they are relatively low in cost but they have limited mass range and unit mass resolution. Although other mass analyzers have been used including quadrupole ion traps [55, 56], the major alternative in commercial instruments is the time-of-flight PTR-TOF-MS. This instrument has good mass resolution without the mass discrimination problems exhibited by quadrupole instruments [57, 58]. A schematic diagram of a commercial PTR-TOF-MS is shown in Fig. 8.4.

The substitution of a TOF MS for the quadrupole MS provides increased mass resolution but it also takes up more space in the instrument and has larger data han-

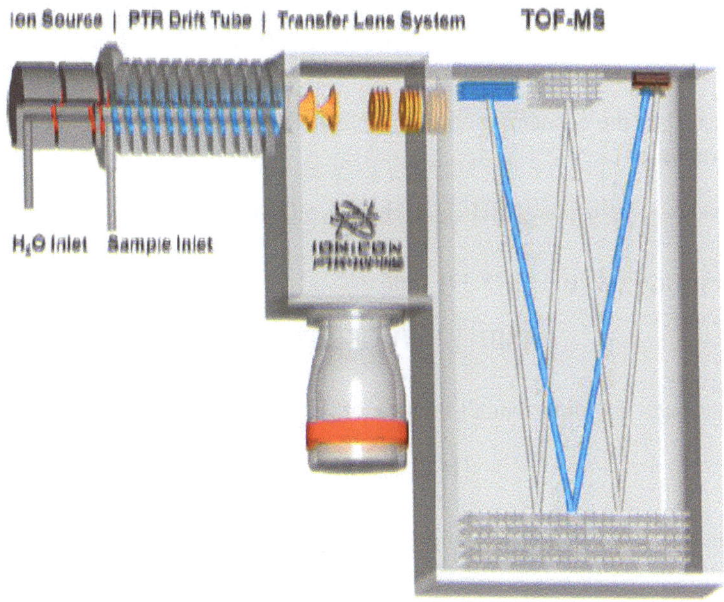

Fig. 8.4 Schematic diagram of a commercial PTR-TOF-MS instrument incorporating a reflectron time-of-flight mass spectrometer. (Reprinted from Int J Mass Spectrom. Authors Jordan A et al. 2009, 286:122 with permission from Elsevier)

dling requirements. Also as the flight time of the ions is measured, it is necessary to gate the ions into the flight tube to scan their arrival times and this process imposes a duty cycle that can limit detection sensitivity. However, some impressive claims have been made of a mass resolution up to 6000 m/Δm and sub ppbv detection within a second [56, 57]. (A later version is commented on in Sect. 8.8.4)

8.4.7 *Quantitative Analysis*

An advantage that PTR-MS and the other techniques discussed in this chapter have over conventional analytical methods is that they are real-time techniques and can provide an immediate analysis of a sample. In PTR-MS for a reaction between H_3O^+ and an analyte M in an air sample, we have

$$H_3O^+ + M \rightarrow MH^+ + H_2O \tag{8.37}$$

The reaction is a pseudo first-order reaction as [M] >> $[H_3O^+]$ even when [M] is at trace levels in a sample. The rate equation may be expressed as

$$-d[H_3O^+]/dt = k[H_3O^+][M] \tag{8.38}$$

,that when integrated becomes

$$[H_3O^+]_t = [H_3O^+]_0 e^{-k[M]t} \tag{8.39}$$

Here $[H_3O^+]_t$ is the concentration of H_3O^+ ions at the end of the drift tube. As the concentration of

$$[MH^+]_t = [H_3O^+]_0 - [H_3O^+]_t \tag{8.40}$$

then substituting (8.40) into (8.39) yields

$$[MH^+]_t = [H_3O^+]_0 (1 - e^{-k[M]t}) \tag{8.41}$$

If it is assumed that only a small fraction of $[H_3O^+]_0$ is removed, then

$$[H_3O^+]_0 \sim [H_3O^+]_t \tag{8.42}$$

Substituting (8.42) into (8.41) and taking the first term of the Taylor expansion gives

$$[MH^+]_t/[H_3O^+] = I_{MH^+}/I_{H3O^+} = k[M]t \tag{8.43}$$

where I_{MH+} and $I_{H_3O}^{+}$ are the respective ion signals of the protonated analyte M and the reagent ion H_3O^+. In principle, Eq. (8.43) defines an absolute method independent of calibration (providing the rate coefficient k and reaction time t are known), simply by recording the ratio of the MH^+ ion counts to those of the H_3O^+ reagent ion.

However, although PTR-MS is loosely based on established measurements of proton transfer rate coefficients these cannot be applied directly as an absolute method for quantification of an analyte because of the presence of the electric field in the drift tube. For exothermic proton transfer, a reaction rate coefficient of 2×10^{-9} cm^3 s^{-1} is often used in PTR-MS to estimate trace analyte concentrations. However, the rate coefficients for proton transfer have been measured under thermal conditions (e.g., SIFT) and vary between 1×10^{-9} and 8×10^{-9} cm^3 s^{-1}. It is therefore necessary to resort to calibration procedures using permeation tubes or calibrated mixtures of analytes at the operational field strength of the instrument to determine the absolute analyte concentration in a sample.

One other influence of the drift field is the effect that it has on association reactions. These are used commonly to help distinguish between isobaric compounds in SIFT-MS instruments. But in PTR-MS, the influence of the electric field can disrupt the more weakly bound cluster ions and this is very evident in the reagent ion mass spectrum generated from H_3O^+.

$$H_3O^+ + H_2O + M \rightarrow H_3O^+ \cdot H_2O + M \tag{8.8}$$

m/z19 m/z37

M is any neutral molecule that can participate in the reaction to remove energy during an ion–molecule collision. In a drift tube, the formation of cluster ions is suppressed as weakly bound cluster ions break up in the more energetic collisions with the neutral gas. There are advantages in this is as it makes the kinetics simpler but there are also disadvantages as it reduces the diagnostic possibilities of quadrupole-based PTR-MS instruments when reactions producing product ions formed during association reactions are used to differentiate isobaric products.

8.4.8 Switchable Reagent Ions

Although PTR-MS has been widely used in many applications, it is limited to those analytes in a sample that can be identified by whether proton transfer from H_3O^+ is exoergic under drift tube conditions. In order to expand the range of analytes available for investigation, additional reagent ions other than H_3O^+ have been used. Clean sources of the reagent ions NH_4^+, NO^+, and O_2^+ were generated without mass selection in a PTR-MS instrument and their reactions were examined [59, 60]. Later, a further modification with online switching between the reagent ions H_3O^+,

NO^+, and O_2^+ was developed with sensitivities of up to 800 cps/ppbv for aromatics, aldehydes, and ketones and this instrument has been denoted as PTR-SRI-MS [61]. The H_3O^+ source in this instrument was again water vapor, the source gas for NO^+ was ambient air that is passed through a charcoal filter and the source gas for O_2^+ was pure O_2. The switching time in the PTR-SRI-MS instrument between reagent ions is around 10 s which when combined with the drift field does not allow the same separation of isobaric ions that are found using SIFT-MS instruments.

8.4.9 Some Applications of PTR-MS

Since its introduction in 1995, a large number of applications using PTR-MS instruments have appeared. A conference devoted to applications of the technique was held in 2003 and it has now become a regular biennial event. A summary of these conferences can be found online from the conference website [62].

Only a very brief summary of some of the interesting applications or reviews of PTR-MS are presented here with several of the more recent investigations.

8.4.9.1 Environmental

Quite a number of the applications of PTR-MS fall in the general environmental area. Useful reviews on applications of PTR-MS to monitoring VOCs in the atmosphere have been presented [63–66]. The technique is now an established monitoring method of analysis for real time appraisal of atmospheric VOCs and numerous studies have been completed. The applications presented using PTR-MS cover a wide range of volatiles arising from widely different sources and processes. These include monitoring normal atmospheric VOC variations, anthropogenic VOCs, biogenic VOCs, and pollution monitoring. Different platforms have been utilized including field site measurements [67, 68], vehicles [69], ships [70], and aircraft [71–73]. Some of these variations in the applications of PTR-MS are demonstrated in the snap shot of publications that follow: (i) measurements of VOCs of anthropogenic origin above a large city [69]; (ii) analysis of syngas in an industrial Fischer Tropsch process [74]; (iii) a comparison of PTR-MS with GC-MS of environmental analytes [75]; (iv) monitoring the atmosphere for pollutants near intensive farm animal production [76].

Biological processes in both the marine and terrestrial environments emit a substantial fraction of their assimilated carbon into the atmosphere as biological volatile organic compounds (BVOCs) [77]. Quite a number of BVOCs have been monitored in different studies using PTR-MS. One study used BVOCs to estimate the OH radical concentration [78]; PTR-MS was also used to measure the profile of several BVOCs from 35 m to 10 km and their diurnal variations in order to estimate emission fluxes from the local forest canopy [79].

8.4.9.2 Food and Food Technology

The versatility and ease of applicability of PTR-MS to the monitoring of VOCs in food related areas is seen in the wide range of studies undertaken in food research [80]. Most of these studies have utilized the appearance and concentration levels of VOCs in foods and processed food products and linked them to quality, stability with age and degradation, process control, and classification. A good summary of PTR-MS applications to the food sciences is included in the book by Ellis and Mayhew, who list entries for more than 130 investigations of food and related issues up until 2012 [50]. A review of PTR-MS applications initiated in Italy was also presented by Cappellin et al. [66]. After 2012, many more PTR-MS investigations in food science utilized higher resolution PTR-TOF-MS instruments and several of these investigations are included here. Zardin et al. [81] used the higher resolving power of the TOF based instrument to distinguish between four pairs of isobaric compounds including cis-3-hexanol and 2,3-pentanedione, and benzaldehyde and m-xylene. The volatiles in coffee aroma as perceived by the nose were examined using PTR-MS and compared with sensory panel analysis [82]. Another study on coffee beans from different origins examined the effect of roasting on the VOCs emitted during the roasting process using a PTR-TOF-MS [83]. Another study examined the production of ethylene in apples with a view to extending the postharvest storage [84]. A second study on apples examined the process of apple ripening and VOC release during postharvest storage [85].

PTR-TOF-MS was also used to probe the effect of heat and high hydrostatic pressure treatment of milk base in the presence or absence of a transglutaminase protein and how it influences the flavor development of yoghurt [86]. The influence of pig diet on the VOC profile of dry-cured Iberian ham was examined with differences found in aldehydes, ketones, and sulfur-containing compounds [87].

8.4.9.3 Medical

One of the primary objectives of instruments like PTR-MS is to use its ease of VOC monitoring to provide options for diagnosis that avoid invasive medical procedures. The number of studies published using the technique in the medical area demonstrate its widespread versatility. Quite a number of these until 2012 have been included in the review by Ellis and Mayhew [50]. As with the previous application of food and food technology (8.4.9.2), many of the more recent applications have employed the higher resolution PTR-TOF-MS instrument.

As an aid for diagnosis, a PTR-TOF-MS was used in the investigation of liver cirrhosis and the severity of the disease was assessed by direct analysis of exhaled breath [88]. Kohl et al. examined the breath of 96 patients shortly before and after kidney transplantation using PTR-MS. They found creatine in urine and breath showed good correlation [89]. Aprea et al. used a PTR-TOF-MS to investigate metabolites in the exhaled breath of patients affected by coeliac disease who were on a gluten free diet [90]. A PTR-TOF-MS was also used to calibrate an acetone sensor applied to monitor breath acetone levels [91]. Another study also using a PTR-TOF-MS examined VOCs in the breath of medical staff and mechanically ventilated patients monitoring isoprene and the anesthetic sevofluorane [92]. One quite different application of PTR-TOF-MS was to monitor anthropogenic VOCs in a football stadium to assess human emissions [93].

8.5 Selected Ion Flow Tube-Mass Spectrometry (SIFT-MS)

SIFT-MS has its origins in the pioneering work of Spanel and Smith in 1996 [94]. The advantage the technology had is that it offered a direct route for analyzing samples for their VOC content in real time without the need for chromatographic separation and extensive calibration procedures. SIFT-MS technology was a follow on from selected ion flow tube (SIFT) technology introduced about 20 years earlier by Adams and Smith [95]. SIFT in turn was an extension of the early work of adapting flow tubes to investigate ion-molecule reactions begun in the late 1960s by Ferguson and collaborators [96]. SIFT and SIFT-MS can be thought of as reversible applications of the same core methodology. In SIFT the kinetic parameters of an ion-molecule reaction are derived from known reactant concentrations whereas in SIFT-MS, the known kinetic parameters are used to derive the analyte concentrations.

Many reviews of SIFT-MS methodology have been written with an emphasis on applications in medical areas [97–100]. What places SIFT-MS technology apart from all the other methods discussed in this chapter on direct analysis mass spectrometric methods is that the reagent (or precursor) ions are well characterized, are mass-selected, and undergo known ion-molecule reactions with the analytes. In principle, providing the reagent ion chemistry with the analyte is known, all that is necessary to determine an analyte's concentration is already known without the need for analyte calibration. Interfacing the technology to different applications is straightforward and the capability of switching between reagent ions in < 10 ms offers a substantial benefit in increased selectivity for analysis in mixtures containing multiple analytes.

8.5.1 SIFT-MS Instrumentation

A schematic outline showing the principle of operation of a SIFT-MS instrument is shown in Fig. 8.5 and Fig 8.6.

The SIFT-MS instrument shown here may be viewed as consisting of four distinct regions. The ion source region is a microwave discharge of moist air. As noted in Sect. 8.1.2, the dominant terminal ions from a discharge in air are H_3O^+, NO^+, and O_2^+. This mixture of ions is transmitted to the lower pressure upstream quadrupole

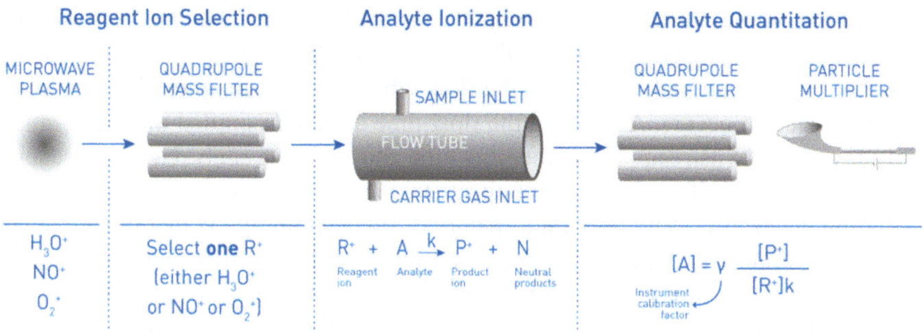

Fig. 8.5 A schematic representation of the essential regions in a SIFT-MS instrument. (Reprinted by permission from Syft Technologies Ltd [101])

mass filter where mass selection takes place at a typical pressure in the 10^{-4} Torr range. The selected reagent ion (one of H_3O^+, NO^+, or O_2^+) is then transmitted into the flow tube where the reagent ion-analyte reaction occurs under controlled conditions. The flow tube pressure is typically 0.6 Torr so the ions from the upstream quadrupole enter the flow tube against a pressure gradient. The entry is assisted by means of a Venturi orifice [97, 102] which facilitates the transmission of ions

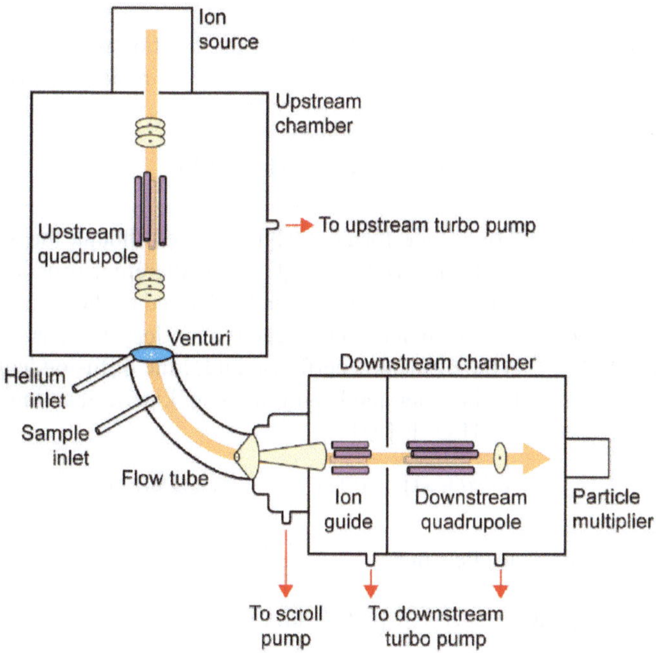

Fig. 8.6 Schematic of a commercial SIFT-MS instrument from Syft Technologies Ltd showing the arrangement of the various components for their Voice200 unit. (Reprinted by permission from Syft Technologies Ltd [101])

against the pressure gradient. The reagent ions are then carried along the flow tube by the carrier gas (usually helium although nitrogen is being used in an increasing number of applications), and it is in the flow tube where the diagnostic reagent ion-analyte reaction occurs. All ions in the flow tube are then sampled through a small orifice at the downstream end of the flow tube and are mass analyzed by the second quadrupole mass filter. The ion number densities are then counted by the pulse counting electronics. The important feature of this technology is that three reagent ions are available and can be interchanged within a few milliseconds to obtain a complete analysis in real time of analytes from all three reagent ions. Most of the commercial units made by Syft Technologies Ltd. (Voice200 and Voice200*Ultra* SIFT-MS instruments) operate at a flow tube temperature of 110 °C and at a carrier gas pressure of 0.6 Torr. In some reactions where association reactions compete with electron transfer, the ratio of the product ion peaks are influenced by the conditions of temperature and pressure selected.

8.5.2 Data Analysis

For the chemical reaction

$$H_3O^+ + A \rightarrow AH^+ + H_2O \tag{8.44}$$

providing the PA (A) > PA (H_2O = 691 kJ mol^{-1}), the reaction will occur at the collision rate [103]. As discussed in Sect. 8.3.7, the rate of removal of H_3O^+ can be expressed

$$-d[H_3O^+]/dt = k[H_3O^+][A] \tag{8.45}$$

In a SIFT or SIFT-MS flow tube for an analyte A, even though the analyte A is at trace levels, its concentration [A] >> [H_3O^+] and therefore pseudo first order kinetics can be applied. The reaction time t can be defined as the reaction length l, divided by the ion velocity v. In addition to reactive ion loss, there is loss of H_3O^+ by diffusion. If D_{H3O+} is the diffusion coefficient and Λ the diffusion length (which depends on the flow tube diameter and length) then the kinetic equation expressing the loss of the reagent ion H_3O^+ is [97]

$$-d[H_3O^+]/dt = [H_3O^+]D_{H3O+}/\Lambda^2 + k[H_3O^+][A] \tag{8.46}$$

On integration, Eq. (8.46) becomes

$$[H_3O^+]_t = [H_3O^+]_0 \exp(-k[A]t - (D_{H3O+}/\Lambda^2)t) \tag{8.47}$$

It can also be shown that [97]

$$\left[AH^+\right]_t = \left[H_3O^+\right]_t k[A]t \frac{\exp\left(k[A]t + \frac{(D_{H_3O^+} - D_{AH^+})}{\Lambda^2} \cdot t\right) - 1}{\left(k[A]t + \frac{(D_{H3O^+} - D_{AH^+})}{\Lambda^2} \cdot t\right)} \tag{8.48}$$

In the SIFT mode of operation, Eq. (8.47) indicates that the semilogarithmic decay of $[H_3O^+]$ (which is proportional to the ion count of H_3O^+, I_{H3O^+}) against the reaction concentration [A] is linear, with the slope defining the reaction rate constant k.

In the SIFT-MS mode of operation, as long as the rate coefficient for a particular reagent ion reaction with an analyte is known, Eq. (8.48) suggests in principle that the ratio of reagent ion count to product ion count can be used to determine the concentration of an analyte [A]. However a distinguishing feature of SIFT-MS is that when a mixture of analytes in a sample is added to the flow tube, the reduction in the reagent ion represents the sum of all reactions of that reagent ion with all analytes. The important factor in this situation is that different product ions are formed from each analyte and it is the identity and amplitude of these product ion signals that enable the analyte to be quantified. It is therefore important in SIFT-MS that accurate measurements are made of both reagent ion count and all product ion counts for each analyte.

Provided both the flow rate of analyte gas and concentration of analyte in the gas sample is small, the reduction in the reagent count is also small (a fractional reduction of $\leq 10\%$ is required for good quantitation) then the ratio of product ion counts to reagent ion counts gives the analyte concentration in the mixture. Smith and Spanel have shown the analytical solution to the set of differential equations for $[AH^+]$ in the limit as $[A] \approx 0$, to be [98]

$$[AH^+]_t = [H_3O^+]k[A]tDe \tag{8.49}$$

In Eq. (8.49), k is the rate coefficient for reaction, the quantities in square brackets are the number densities of the ions and analyte A, t is the reaction time and De is a differential diffusion enhancement coefficient that accommodates the difference in diffusion rate to the walls between reagent ions and product ions. Ions with larger *m/z* generally have slower rates of diffusion than smaller ions. Another important factor when using quadrupole mass filters is to account for the reduced transmission of heavier ions. The two opposing effects of diffusion and transmission oppose each other. A proper analysis will account for both effects.

When multiple product ion channels from an analyte are present in a gas mixture with the product ions represented by p_1, p_2, etc., and analyte reactions occur with both the primary H_3O^+ reagent ion and its water clusters $H_3O^+.H_2O$, then [A] is given by [97]

$$[A] = \frac{1}{t} \frac{(I_{p1} + I_{p2} + \ldots)}{\left(I_{r1}k_1 + I_{r2}\left(\frac{k_1 + k_2}{2}\right) + \ldots\right)} \tag{8.50}$$

In Eq. (8.50) I_{p1}, I_{r1}, are the signal amplitudes of product ions and reagent ions (H_3O^+, $H_3O^+.H_2O$, etc.) corrected for differential diffusion in the flow tube and mass discrimination in the downstream quadrupole mass filter. Additional parameters f_{r1}, f_{r2}, may also be included in the reactions of H_3O^+ reagent ions to account for analytes where not all water cluster ion reactions with the analyte are reactive [98].

8.5.3 *Analyte Concentration*

Equation (8.50) provides a concentration of an analyte A in units of molecules cm^{-3} inside the flow tube. The conversion to the more useful parameters of parts per billion by volume, ppbv, is readily made. If ϕ_{car} and ϕ_{sample} are the gas flows of the carrier gas and sample gas through the capillary, P_{FT} is the pressure in the flow tube, then:

$$\phi_{anal} = (\phi_{car} + \phi_{sample})\cdot(P_{analyte}/P_{FT}) \tag{8.51}$$

where ϕ_{anal} is the gas flow of the analyte in the same units as those of the carrier and sample gas. In a commercial Voice200 instrument these are typically $\phi_{car}=3.5$ Torr L s^{-1} and $\phi_{sample}=0.25$ Torr L s^{-1}. Converting [A] from molecules cm^{-3} to a partial pressure in Torr, we have (R is the gas constant, T is the temperature in Kelvin, and N_A is Avogadro's number):

$$P_{analyte} = (10^6[A]RT)/N_A = 1.035\times10^{-19}[A]T \tag{8.52}$$

From Eq. (8.51), the conversion from molecules cm^{-3} to ppbv is then simply

$$[Analyte]_{ppbv} = (\phi_{anal}/\phi_{sample})\times10^9 \tag{8.53}$$

In summary, the SIFT-MS instrument provides a value for [A] in molecules cm^{-3} on the basis of the ratio of product ion counts to reagent ion counts with appropriate corrections for mass discrimination and differential diffusion and the known reagent-analyte ion-molecule reaction kinetics (8.50). This value is converted to the analyte concentration in ppbv, $[Analyte]_{ppbv}$, by combining Eqs. (8.51), (8.52), and (8.53).

$$[Analyte]_{ppbv} = \frac{(1.035\times10^{-10}[A]T)}{(P_{FT})}\left(1+\frac{(\phi_{car})}{(\phi_{sample})}\right). \tag{8.54}$$

8.5.4 *Data Acquisition*

The SIFT-MS instrument has two modes of operation: (i) a mass scan mode (MS) in which a mass scan is made of all reagent ions and product ions over a selected mass-to-charge ratio, m/z. In the mass scan mode of operation, ion counts are recorded for typically 50–100 ms at each m/z value. The mass spectra are interpreted by relating the product ion peaks to the trace gases that are present in the sample. This mode of operation is useful for determining what trace analytes might be present in a sample. The concentrations of analytes recorded using this mode of operation are less precise than that in the second mode of operation because a relatively short time is spent depleting product ions at each mass. (ii) A selected ion mode (SIM) scan where only the reagent ion and selected product ions are monitored. This mode of operation is used when known analytes are monitored and therefore the identity of the product ions is known. It provides a more precise measure of analyte concentrations as the residence time for acquisition for the product ion counts may be 100 ms or more averaged over multiple cycles of counting and allows for measurement of temporal variations and the ability to evaluate trends.

8.5.5 *Ion Chemistry in SIFT-MS*

This has been well summarized in several articles [97, 104]. As was mentioned earlier in this chapter, the terminal ions arising from a discharge in moist air are H_3O^+, NO^+, and O_2^+. These ions have been utilized in APCI. The advantage that SIFT-MS has is that the reagent ions of H_3O^+, NO^+, and O_2^+ are mass selected, so only the product ions of an analyte from a specific reagent ion are present during analysis together with unreacted reagent ions. It is also important that the reagent ions do not react with the major components of air. Almost all analyses undertaken using SIFT-MS have been restricted to the three reagent ions H_3O^+, NO^+, and O_2^+. However other ions including negative ions may be used for monitoring specific analytes. One example of an alternative positive ion is $CH_3OCH_2^+$ which was found to be unreactive with air and small saturated hydrocarbons but reactive with molecules containing sulfur [105]. The use of $CH_3OCH_2^+$ in this situation opens the way for monitoring odorant compounds containing sulfur that have been added to propane and butane mixtures for safety reasons.

8.5.5.1 H_3O^+ Reactions

In SIFT-MS when the H_3O^+ ion is injected into the flow tube, clusters of water ions are also formed. These often react in the same way as the H_3O^+, ion does with the VOC providing an exoergic pathway for proton transfer exists.

$$H_3O^+ + H_2O + He \rightarrow H_3O^+ \cdot H_2O + He \tag{8.55}$$

$$H_3O^+ \cdot H_2O + H_2O + He \rightarrow H_3O^+ \cdot (H_2O)_2 + He \tag{8.56}$$

The predominant reaction of H_3O^+ reagent ions with VOCs is exothermic proton transfer

$$H_3O^+ + A \rightarrow AH^+ + H_2O \quad (8.44)$$

When proton transfer is quite exothermic (more than about 1 eV), then dissociative proton transfer may occur from $[AH^+]^*$ as in some reactions with alcohols, aldehydes, and carboxylic acids leading to H_2O elimination [106].

$$H_3O^+ + C_5H_{11}CHO \rightarrow C_6H_{13}O^+ + H_2O \quad 0.50 \quad (8.57)$$

$$\rightarrow C_6H_{11}^+ + 2H_2O \quad 0.50 \quad (8.58)$$

Another reaction of H_3O^+, that does occur is association when H_3O^+ simply adds to the analyte molecule either as a single product or in conjunction with other product channels

$$H_3O^+ + A + He \rightarrow H_3O^+ \cdot A + He \quad (8.59)$$

In some cases the weakly bound association complexes, $H_3O^+.A$, may switch out the H_3O^+, to water and if this happens the use of H_3O^+ as a diagnostic reagent ion is lost.

$$H_3O^+ \cdot A + H_2O \rightarrow H_3O^+ \cdot H_2O + A \quad (8.60)$$

More detailed discussion of H_3O^+ chemistry in SIFT-MS is given in the reviews by Smith and Spanel [97, 104].

8.5.5.2 NO⁺ Chemistry

There is more variety in the reactions of NO^+ compared to H_3O^+ and there are usually only one or two product ions in its reactions with an analyte. This makes NO^+ a very useful and important reagent ion in SIFT-MS analysis.

Electron transfer can occur when the IE of the analyte A is less than IE(NO = 9.26 eV). Quite a number of VOCs come into this category of lower IEs than NO with electron transfer a common reaction pathway.

$$NO^+ + C_6H_5OCH_3 \rightarrow C_7H_8O^+ + NO \quad (8.61)$$

H^- transfer is also a common pathway of reactions of NO^+ with aldehydes, ethers, and alcohols with the exception of tertiary alcohols when hydroxide ion transfer occurs [104].

$$NO^+ + CH_3CHO \rightarrow CH_3CO^+ + HNO \quad (8.62)$$

$$NO^+ + (CH_3)_3COH \rightarrow (CH_3)_3C^+ + HNO_2 \quad (8.63)$$

Association reactions of NO^+ also provide a very useful diagnostic reaction for analytes, especially ketones. A good example of the usefulness of NO^+ as a diagnostic reagent ion is the analysis of a mixture containing the three isobaric compounds 3-methylbutanal, 2-pentanone, and diacetyl (2,3-butanedione). The reagent ion H_3O^+ gives mainly proton transfer at m/z 87 for all three analytes but NO^+ gives complete separation.

$$\underset{\text{3-methylbutanal}}{NO^+ + C_5H_{10}O} \rightarrow \underset{m/z\ 85}{C_5H_9O^+ + HNO} \quad H^-\text{transfer} \quad (8.64)$$

$$\underset{\text{2-pentanone}}{NO^+ + (CH_3)_3COH} \rightarrow \underset{m/z116}{NO^+.(CH_3)_3COH} \quad \text{association} \quad (8.65)$$

$$\underset{\text{2,3-butanedione}}{NO^+ + (CH_3CO)_2} \rightarrow \underset{m/z\ 86}{C_4H_6O_2{}^+ + NO} \quad \text{electron transfer} \quad (8.66)$$

$$\rightarrow \underset{m/z\ 43}{C_2H_3O^+} + C_2H_3O + NO \quad \text{dissociative electron transfer} \quad (8.67)$$

The usefulness of NO^+ as a diagnostic reagent ion in SIFT-MS is shown later in this chapter in Sect. 8.8.4 in the comparison of PTR-MS and SIFT-MS.

8.5.5.3 O_2^+ Chemistry

O_2^+ is the most energetic of the usual SIFT-MS reagent ions having an IE of 12.06 eV. This is larger than the IEs of most VOCs with the result that O_2^+ generally reacts rapidly either by electron transfer (ET) or dissociative electron transfer (DET) and produces multiple product fragment ions in some cases. O_2^+ is particularly useful in monitoring analytes that are unreactive with H_3O^+ and NO^+ reagent ions such as the small hydrocarbons [106]. The O_2^+ reaction with isoprene is a typical example in that there are three product ion channels [107].

$$O_2^+ + CH_2 = C(CH_3)CH = CH2 \rightarrow C_5H_8^+ + O_2\ 0.45\ \text{ET} \quad (8.68)$$

$$\rightarrow C_5H_7^+ + (H + O_2)\ 0.45 \quad \text{DET} \quad (8.69)$$

$$\rightarrow C_4H_5^+ + (CH_3 + O_2)\ 0.10 \quad \text{DET} \quad (8.70)$$

Another useful reaction of O_2^+ occurs with methane which is unreactive with NO^+ and H_3O^+ reagent ions [106]. The rate coefficient for this reaction is small

($k = 5 \times 10^{-12}$ cm^3 s^{-1}) meaning that the detection limit for methane is larger than for most other analytes where the rate coefficients are typically 3×10^{-9} cm^3 s^{-1}. The small rate coefficient for methane means that the limits of detection for methane are in the mid to high ppbv range.

$$O_2^+ + CH_4 \rightarrow CH_3O_2^+ + H \tag{8.71}$$

Possibly the greatest value of O_2^+ as a reagent ion in SIFT-MS is in monitoring small molecules and those analytes where the number of product ions from their reaction of O_2^+ is limited to 1 or 2 as in the aromatic hydrocarbons and a number of heterocyclic molecules.

As noted earlier in Sect. 8.5.1, SIFT-MS instruments can operate at different conditions of temperature and pressure. For example, most of the Syft Technologies Ltd Voice200 SIFT-MS instruments operate at a flow tube temperature of 110 °C and a carrier gas pressure of 0.6 Torr. In some reactions where association reactions compete with electron transfer, the product ion peak ratios of association/electron transfer are sensitive to temperature and pressure. For example, the reaction of the terpenoid carvone, $C_{10}H_{14}O$, with NO^+ has product ions of reaction $C_{10}H_{14}O^+$ and $C_{10}H_{14}O.NO^+$. The ratio of ($C_{10}H_{14}O.NO^+/C_{10}H_{14}O^+$) is somewhat sensitive to temperature and pressure.

The advantage of having clearly defined reagent ions at close to thermal energies reacting via identified reactions is seen clearly in Fig. 8.7 by monitoring the VOC carvone ($C_{10}H_{14}O^+$). When carvone is ionized using conventional GC-MS technology with 70 eV electrons, a large number of fragment peaks are observed requiring chromatographic separation of the sample containing carvone before identification can occur. With SIFT-MS a much simpler product ion spectrum is found allowing identification in real time without chromatographic separation using both H_3O^+ and NO^+ reagent ions.

8.5.6 *Some Applications of SIFT-MS*

Due to the ease of application of SIFT-MS to research in the medical area and particularly for those studies that require direct breath analysis in real time there has been a major focus on medical applications. Quite a number of reviews of the medical area have been undertaken and some of these will be referenced in the appropriate section. For example, slightly more than two thirds of all SIFT-MS investigations published in 2013 had a medical focus. This does not mean that there are fewer opportunities for SIFT-MS in say, the environment and food technology areas but the preponderance of medical studies is merely a legacy of the early research focus of the technology. Most commercial SIFT-MS instruments are in fact operating in the environmental and food areas doing routine analyses.

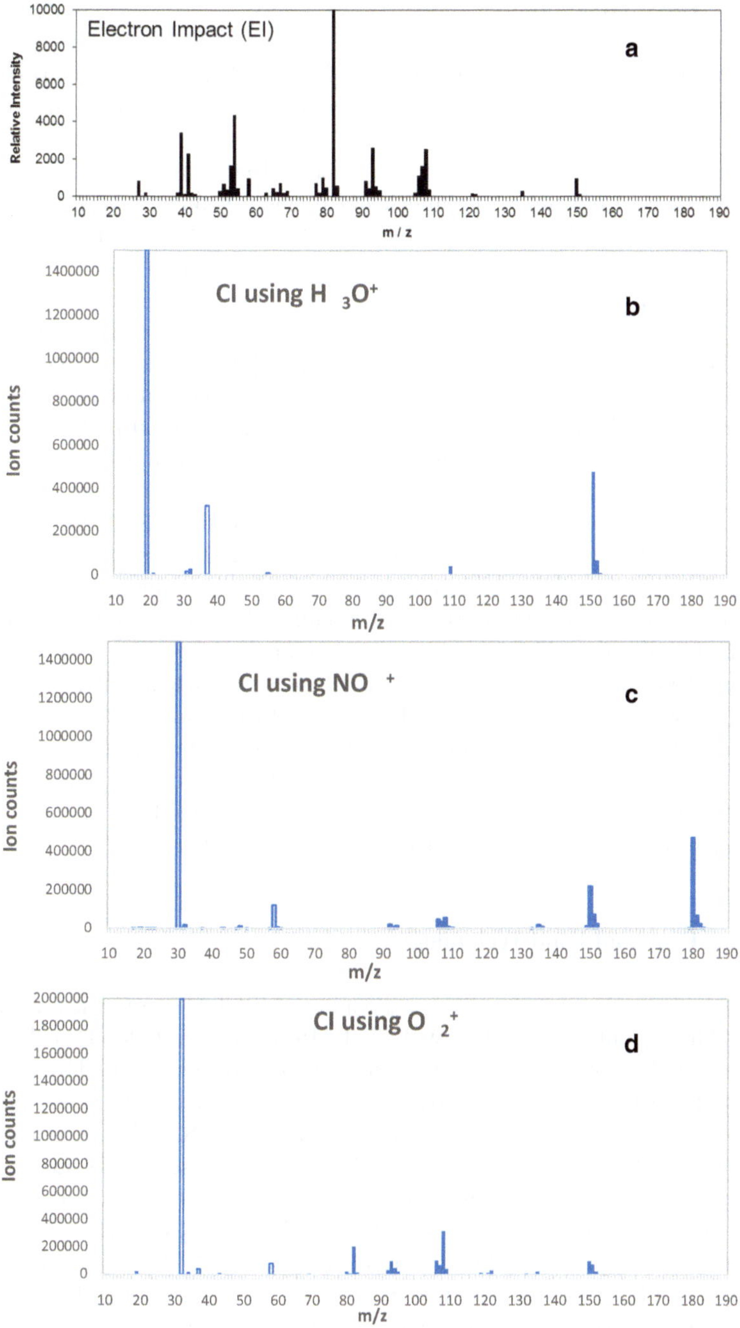

Fig. 8.7 A comparison of the mass spectrum of carvone ($C_{10}H_{14}O$) $M_r = 150$ generated by **a** 70 eV electrons with the three reagent ions in SIFT-MS as found in a commercial Voice200 Syft instrument operating with a nitrogen carrier gas. **b** Chemical ionization (CI) by H_3O^+ reagent ion. **c** CI by NO^+ reagent ion. **d** CI by O_2^+ reagent ion. In **b**, **c**, and **d** the reagent ion signals for H_3O^+ at m/z 19, NO^+ at m/z 30 and O_2^+ at m/z 32 are represented by hollow bars and the product ions by filled bars

8.5.6.1 Environmental

The real-time monitoring of atmospheric VOCs has been a major focus of direct analysis mass spectrometric techniques without quite the same emphasis for SIFT-MS until around 2010. Diesel engine exhaust gases were analyzed using SIFT-MS to monitor NO, NO_2, HNO_2, aldehydes, and ketones [108]. Peroxyacetyl nitrate (PAN) is known to be a precursor to photochemical smog. A method was developed to monitor PAN using SIFT-MS with a limit of detection of 20 pptv in 10 s [109]. The ability of SIFT-MS methodology for real-time monitoring of atmospheric VOCs was demonstrated when a number of VOCs including ethene, ethanol, 1,3-butadiene, benzene, and toluene were monitored over a 4-day period. The concentration of 1,3-butadiene was measured down to 9 pptv [110]. Volatile selenium compounds such as H_2Se, CH_3Se, $(CH_3)_2Se$, and $(CH_3)_2Se_2$ have been monitored in real time using SIFT-MS [111]. The kinetic parameters for 17 hazardous air pollutants (HAPs) were measured and limits of quantitation were found for them using SIFT-MS that were well below the limits required by legislating bodies for monitoring the work place environment [112]. Methods were also developed for SIFT-MS to monitor the presence of 15 chemical warfare agent precursors. Thirteen of the fifteen compounds were found to have atmospheric detection limits in the pptv concentration range for real-time measurements [113]. Landfills have been and are a common feature of civilization. The increasing presence of siloxanes from a variety of personal products dumped in landfills in modern times has led to a growing production of anthropogenic siloxanes from landfill sites that have posed problems when gas emissions from landfills are combusted. First methods were developed using SIFT-MS to monitor siloxanes and these were then applied to monitor concentrations of four siloxanes at three landfill sites [114, 115]. VOCs arising from intensive animal farming have also been examined using SIFT-MS [116, 117].

8.5.6.2 Food and Food Technology

Although numerous reviews of the application of SIFT-MS to medical areas have been published, only one has dealt with SIFT-MS applications to food [118]. The first SIFT-MS study applied to the real-time monitoring of flavors released by food products was published in 1999 [119], but since that time, more than 45 investigations of food and food products have been published using SIFT-MS, with most related to the ease with which the technology monitors VOC release. A series of studies on olive oil demonstrated that the dominant volatiles were methanol and ethanol, with the VOCs that had been the object of numerous studies in the past such as E-2-hexenal being present at significantly lower concentrations [120]. A total oxyradical scavenging capacity (TOSC) assay based on SIFT-MS was also developed for monitoring antioxidants in olive oil [121]. Barringer and coworkers have focused on key volatiles generated by enzymes, especially lipoxygenase (LOX) which catalyses the oxidation of polyunsaturated fatty acids [118]. Linoleic acid and linolenic acids are the major precursors of many volatile acids, alcohols, aldehydes and esters. Detection of these

VOCs is monitored in real-time SIFT-MS measurements and it provides input for a number of qualities, including flavor, ripening stage, food quality and freshness. The effect of chewing and concomitant release of VOCs has also been examined. The application to flavors includes tomatoes [122–124], tomatilloes [125], carrots [126], pumpkin seeds [127], jalapenos [128], bell peppers [129, 130], strawberries [131], garlic [132, 133], cocoa [134, 135], almonds [136] and cashews [137].

Several studies of different unifloral (monofloral) honeys have been undertaken to investigate VOC profiles of honey and the effect of location and botanical origin [138–140]. Several studies of meat and meat products have also been done, where SIFT-MS has been used to quantify aroma compounds in comparison to conventional analytical techniques [141–143]. Methods have also been developed using SIFT-MS to examine the spoilage and storage effects on fish and fish products, including cod [144], grey shrimp [145, 146], packaged fillets [147] and brown shrimp [148]. In a very different study, the VOCs associated with macaroni were quantified to assess any differences in flavor associated with the addition of a sodium-reduced cheese sauce [149]. Several investigations have also been used to assess various qualities of cheese. The VOC profile was used to distinguish between Italian parmesan and New Zealand parmesan [150] and, in conjunction with odor activity values, to differentiate Swiss cheese production from five different factories [151]. In another investigation the VOC profile examined using SIFT-MS aided the evaluation of parameters responsible for split defects in Swiss cheese [152].

8.5.6.3 Medical

There have been a number of reviews of applications of SIFT-MS to the medical area by Smith and Spanel; they give a good summary of the advances that have been made [104, 153, 155]. A special issue of the *Journal of Breath Research* (2014, 8(3)) has highlighted some of the contributions of David Smith and other users of SIFT-MS to breath research. SIFT-MS has a considerable advantage over many techniques in that individual breaths can be examined because of the rapid response time of the instrument. Four major objectives can be stated of what direct breath analysis might achieve [156]: (a) identify and quantify VOCs in exhaled breath; (b) identify abnormal concentrations of common metabolites in exhaled breath; (c) identify biomarker compounds related to disease conditions; (d) identify and track the prophylactic effect of medication that might be administered from monitoring biomarkers. An important factor that has been identified in monitoring exhaled breath is that there is a difference between breath exhaled through the mouth and breath exhaled through the nose as the oral cavity can be a major source of some trace gases [153, 157]. It is important therefore to distinguish between trace VOCs generated systemically from those generated in the oral cavity due to bacterial and enzymatic activity. Reference ranges for a number of metabolites in mouth-exhaled breath in healthy populations have been established, and it has been shown for example, that mouth-exhaled ammonia clearly increases with age [158]. Because so many of the medical aspects of SIFT-MS have been reviewed well, we will only briefly mention here some of the more recent activities in (b) and (c) of the objectives for breath analysis. Hydrogen cyanide has been identified in the head space of plate cultures of the bacterium *Pseudomonas aeruginosa* derived from sputum and cough swab samples from cystic fibrosis patients [104, 159]. Differences in 7 VOCs were found in the headspace of vapor from urine samples in patients with gastro-esophageal cancer [160, 161]. Pentane has been examined as a potential biomarker of bowel disease [162]. A careful analysis of changes in breath metabolites associated with advanced chronic kidney disease (CKD) has been undertaken and outlined in a review paper [163]. Attention has been given to changes in ammonia concentration, including other biomarkers such as isoprene, ethane, and pentane. The potential of using acetone to monitor starvation and metabolic stress in intensive care units in hospital has been examined [164].

Dweik and coworkers have also made a number of interesting discoveries monitoring breath volatiles. They report a study in which they monitor six VOCs, including trimethylamine, acetone, and pentane whose levels increased in patients with liver disease and found they could predict diagnosis and severity of the disease in a sample of 80 patients who developed alcoholic hepatitis [165]. In another study of 21 patients with juvenile idiopathic arthritis (JIA) and 55 healthy controls in which they monitored 13 VOCs using SIFT-MS, and concluded that the exhaled breath prints allowed them to distinguish those with JIA [166]. The team also compared changes in the VOC levels in exhaled breath in 60 obese children and 55 lean controls. They concluded that obese children have a unique pattern of exhaled VOCs that may provide some insight into the pathophysiological processes and pathways leading to childhood obesity [167]. A similar methodology was used to distinguish patients with pulmonary arterial hypertension (PAH). In a study of 31 patients with PAH and 34 healthy controls, a comparison of the breath prints of the two groups found statistically significant differences in eight volatile compounds. Exhaled ammonia levels correlated with the severity of the disease [168]. In another investigation of 24 patients with pancreatic cancer and 72 patients with benign biliary conditions, the team examined the headspace from 96 bile samples. They found that the concentrations of six VOCs had increased in those patients with pancreatic cancer and concluded that the simple test of monitoring VOCs in the biliary fluid headspace may help in distinguishing malignant from benign biliary strictures [169].

8.6 DART and DESI

Two further techniques that can be included under the title of direct analysis are DART and DESI. The focus of these two techniques is slightly different from those already presented in this chapter. What distinguishes DART and DESI is that they do not focus on gas phase VOCs but instead extract or “wipe off” analyte molecules from samples by exposing the surface to the ionizing gas or aerosol. The ionization still occurs at atmospheric pressure. This enables the direct examination of completely unprocessed samples or even entire objects. Both DESI and DART sources are available commercially. The item for analysis is simply positioned at the entrance of the mass spectrometer and analyte molecules released from the sample after ionization are transferred to the mass spectrometer for analysis. There are now many variants of these techniques that directly remove analyte molecules from surfaces but DART and DESI are the most established of them. Due to the different nature of DART and DESI, the applications of these techniques will not be classified under the same headings as the methods already discussed in this chapter.

8.6.1 DART

DART had its origin in the work of Cody and Laramee in 2005 [170]. A general representation of the DART source is shown in Fig. 8.8. The ionization source relies on Penning ionization from helium 2^3 S metastable atoms produced in the corona-to-glow discharge and the ensuing gas phase reactions that occur are very similar to those in the APCI sources. Molecular nitrogen is the dominant molecule in air

$$He^* + N_2 \rightarrow He + N_2^+ + e^- \tag{8.72}$$

The sequences of reactions that can follow have already been discussed in reactions (8.1) through (8.9) for APCI sources. The DART ion source may also be operated in the negative ion mode when the electrons generated in reaction (8.72) attach to molecular oxygen creating negative ions

$$O_2 + e^- \rightarrow O_2^- \tag{8.73}$$

The O_2^- ions can associate with the analyte molecule A to form adducts (reaction (8.74)), undergo direct electron transfer (8.75), or undergo the solvation reactions with H_2O shown in reactions 8.10 through 8.12.

$$O_2^- + A \rightarrow AO_2^- \tag{8.74}$$

$$\rightarrow A^- + O_2 \tag{8.75}$$

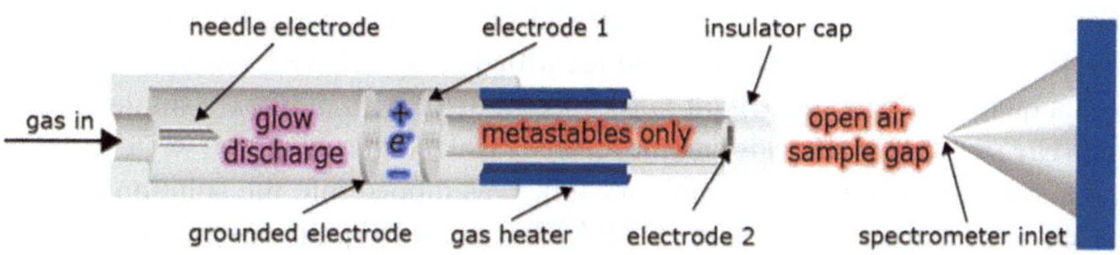

Fig. 8.8 A schematic diagram of a DART ion source. (Used by permission from Jeol USA, Inc.)

The DART ion source has many possible different configurations that depend on the sample being analyzed, and these are discussed in several reviews [171–173]. A flow of helium gas is normally used with flows through the ion source at around 1–4 L min^{-1}. A potential of 1–5 kV on the needle electrode relative to a grounded perforated disk electrode causes a corona-to-glow discharge [173] that produces the metastable helium atoms, ions, and electrons. Ionization of the analyte occurs when the DART gas plasma makes contact with the sample in the open-air gap between the DART source outlet and the mass spectrometer sampling orifice. The role of cluster ions and helium metastables in APCI sources and DART sources has been the subject of a recent study that seeks to examine further the role of cluster ions in the generation of analyte ions [174].

8.6.2 Applications of DART-MS

An indication of the acceptance of the DART ion source is seen in the growth in research papers that have been published since its introduction in 2005. More than 530 articles and conference papers have been published using the technique until mid-2014. The application of DART ion sources has become an established technique for rapid analysis of many widely differing sample types, particularly for analytes deposited or adsorbed on surfaces. In the area of analysis of foodstuffs there have been a number of widely variant studies. A DART-time-of-flight system was used to examine and classify olive oil triacyl glycerols and olive oil authenticity [175]. The chemical changes in vegetable oils were also monitored during thermally induced oxidation [176]. An analysis of several flavors and fragrances in complex mixtures was performed using DART which was applied to examine 12 chewing gum samples for the presence of cooling agents merely by placing the gum stick or tablet in the ion beam. The release of fragrance from fabric and hair was also examined [177]. A method using DART was developed to monitor 32 mycotoxins in beer with some sample preparation by adding a beer sample to acetonitrile [178]. In another study, investigations were made of organosulfur species in the aroma of crushed garlic, onion, and other alliums by homogenizing the garlic bulbs in a food blender followed by extraction in ether, filtration, and evaporation [179]. The DART-MS has also been used to authenticate animal fats in lard and beef tallow and also minced meat prepared from pork and beef [180]. Milk and milk-based foods were examined using DART-MS and high-resolution mass spectrometry, allowing discrimination using statistical analysis between milk from cows, sheep, and goats and also for cow's milk from conventional farming compared to organic farming [181]. A number of saccharides were identified using a DART ion source coupled with a quadrupole time-of-flight mass spectrometer with minimal sample preparation. The standard saccharide samples were dissolved in a methanol–water mixture, and a standard melting point capillary tube was dipped into the solution and then placed in the helium stream of the DART ion source [182]. Six phytohormones were analyzed simultaneously using a DART ion source and single drop liquid–liquid microextraction in three different kinds of freshly prepared fruit juice in less than 30 min [183]. Another study on fruit juice using a DART ion source also demonstrated high-throughput screening and detection of polysorbate 20/80 micelles in apple juice [184]. A DART ionization source coupled with a high-resolution mass spectrometer was used for screening aflatoxins from different surfaces [185]. A review of DART applications in food quality and safety analysis covers qualitative confirmation of chemical identity, metabolomics fingerprinting/profiling, quantification of low molecular weight food components, including trace organic contaminants [186]. A summary of many of the applications of DART to food and food-related adulterants and contaminants is available from a commercial supplier [187].

In the medical area, one study using DART has been applied to human breath. Diagnosis through online breath analysis was found to be difficult due to the dispersion of the breath samples in open air. A confined DART ion source (cDART) was developed and this was tested on four standard compounds—ethanol, acetone, 2-hexanone, and limonene—with detection limits approaching 1 ppbv after calibration [188].

Other applications of DART include forensics where differences in writing inks were identified [189]; screening for trace explosives [190]; pharmaceuticals [191]; and chemical analysis [192–194] to name just a few of the many investigations.

8.7 DESI

Like DART, DESI has received widespread acceptance as evidenced by more than 750 papers and conference presentations till mid-2014, referring to the technique since its introduction in 2004 by Cooks and coworkers [195, 196]. The technique makes use of electrospray ionization (ESI) that is widely used in the mass spectrometry of larger molecules in which a solution is nebulized to create a fine spray of very small droplets. In DESI, a standard electrospray of charged droplets hits the surface where the molecules of interest are present or adsorbed (including larger biomolecules), detaches them from the surface, and delivers them as desolvated ions in the mass spectrometer. DESI is thus similar to DART (Sect. 8.4) where the gaseous plasma of ions from the ion source desorbs molecules from a surface. A schematic diagram of the main aspects of a DESI ion source is shown in Fig. 8.9.

Charged microdroplets with a velocity of around 120 ms^{-1} and average diameter ~5 μm impact on the surface and release dozens of microdroplets of size ~0.8–3.3 μm [197]. The secondary droplet velocities depend on the size of the primary droplet, the take-off angle, and the distance from the point of impact. The performance of DESI is also influenced by the choice of solvent. As an example, a solution of dimethylformamide with water or methanol enhances the signal of low molecular weight compounds [198]. Nonaqueous solvents are better at extracting lipids from tissue sections without causing major changes in morphology on a microscopic scale. Further details on the set-up and solvents are contained in a review of the technique [198].

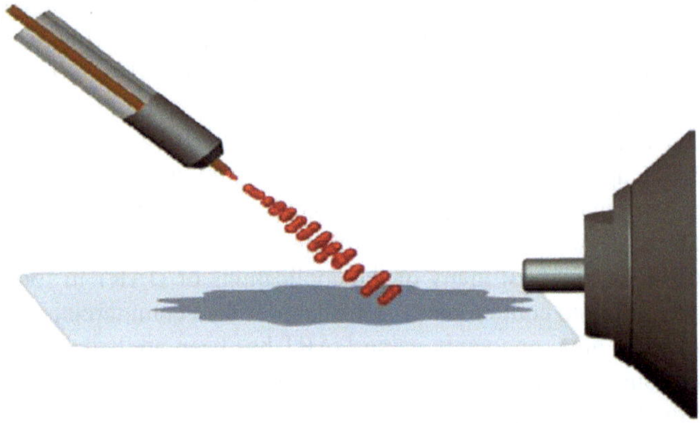

Fig. 8.9 DESI ion source showing an electrospray of charged droplets impacting the surface where the molecules of interest are present. Once detached, they enter the ion source shown of the mass spectrometer. (Used with permission of Prosolia, Inc., Indianapolis, IN, USA)

8.7.1 Applications of DESI-MS

As already noted, both DESI and DART techniques can readily extract molecules from surfaces. A number of the applications for this type of analysis are accessible to both techniques providing "direct analysis with little sample preparation." DESI can however be applied to larger molecules giving it great potential in bioanalysis. Surface imaging is a major area of application of DESI. The use of differential mobility was applied to image biological tissue in sea algae and rat brain tissue [199]; flat thin layer chromatographic imprints of leaves and petals were imaged using DESI to provide knowledge of varietal differences, toxic metabolite production, and pest/pathogen attack [200]; mouse uterine sections of implantation sites on day 6 of pregnancy were analyzed by imaging without sample pretreatment [201] and nicotine has been imaged in rat brain tissue following in vivo drug administration [202].

DESI as with DART has found a number of applications in forensics. The lack of sample preparation and ease of surface examination makes it suited to examination for screening for explosives [203–205], forensic applications such as document ink analysis [206], latent fingerprints [207] and counterfeit pharmaceuticals [208, 209].

In the area of food and food contaminants, DESI has been applied to the rapid analysis of pesticides on fruits and vegetables and to the adulteration of food [210]. In medicine, there have been quite a number of different applications. A statistical analysis of DESI-MS imaging data could distinguish between normal cells and tumor cells in canine and human bladder cancer [211], and it has been used to assess surgical resection margins of gastric cancer [212]. The penetration of drugs into the skin has been examined using DESI [213]; it has also been used to discriminate between individual oocytes and blastocysts to aid the understanding of events that can trigger obesity and type 2 diabetes later in life [214].

8.8 Comparison of Direct Ionization Methods

Each of the direct analysis methods discussed in this chapter has its advantages in particular applications. The first four techniques, viz. APCI, IMR-MS, PTR-MS, and SIFT-MS have direct applications for monitoring VOCs and BVOCs in gas samples and the emphasis in this section is on them. The strengths of DART and DESI and related techniques are particularly aimed at direct ways of extracting samples from surfaces and surface imaging, and they will be excluded from this comparison.

8.8.1 MS-e-nose

One other technique that may also be included in gas analysis is the MS-e-nose. An electronic nose or (e-Nose) is a detector that uses an array of chemical sensors which does not generally identify specific VOCs but has in addition an appropriate pattern recognition system that is capable of recognizing simple or complex odors [215, 216]. With the MS-e-nose, the mass sensor provides a fingerprint of the sample being analyzed. The signal then requires chemometric processing before sample identification can occur. The problem with direct identification without statistical processing is that the degree of fragmentation from electron impact of the sample makes direct identification impossible. Some attempts have been made to improve the performance using other ionization methods, and although the MS-e-nose has been used successfully in monitoring some aroma properties [217], it is not as versatile or quantitative as the techniques to be considered next.

8.8.2 APCI

APCI is a versatile and effective means of ionization but it is not an inherently quantitative technique simply because it has multiple simultaneous reactions from all the ions generated in the ion source. Calibration reactions are required for quantitation. Because large ion signals can be generated at atmospheric pressure, the APCI source provides good sensitivity, but ionization does not necessarily occur with all analytes. In the ion generation scheme presented for APCI in reactions (8.1) through (8.9), a large amount of the ionization will occur through water cluster ions $H_3O^+.(H_2O)_n$, although some ionization in the positive ion mode may also occur from O_2^+ and in the negative ion mode from O_2^-. For positive ion formation, an analyte needs to have a proton affinity greater than that of H_2O (691 kJ mol^{-1}) for proton transfer H_3O^+ to occur. APCI sources for analysis work best in conjunction with separation devices such as GC/MS and LC/MS. Even then it is sometimes necessary for sample preparation. For example, van Leeuwen et al. [218] monitored aldehydes and ketones in car exhausts and cigarette smoke with both APCI and atmospheric

Table 8.1 List of carbonyl groups detected with signal-to-noise≥3 in automobile exhausts using an APCI source from ref [218]

Fuel	A	B	C	D	E
Reg Fuel 1	C1–C6	C1	C3–C7	C7–C8	C8
Reg Fuel 2	C1–C6	C1–C2	C3–C7	C7–C8	C8

A saturated, *B* saturated hydroxycarbonyls or carboxylic acids, *C* one double bond or saturated ring, *D* aromatic, *E* phenolic

pressure photoionization (APPI) sources and an LC/MS by derivatization of the samples with 2,4-dinitrophenylhydrazine (DNPH). The first DNPH derivative standards of several aldehydes and ketones in acetonitrile and water solvents were prepared. They then sampled the exhaust gases from the car exhaust pipe and cigarette smoke with DNPH-coated silica gel cartridges and were able to detect, using the APCI source, the diphenylhydrazones of formaldehyde, acetaldehyde, acetone, propanal, and benzaldehyde at limits of quantitation around 2 ppb and elution times out to 16 min. A list of carbonyls detected after derivatization in the exhaust gases of vehicles running on different fuels using an APCI source is shown in Table 8.1 from reference [218].

The APPI source in this application was found to give better sensitivity than the APCI source. An interesting comparison can be made here with the techniques of PTR-MS and SIFT-MS. With these techniques, and particularly with SIFT-MS, the exhaust gases of motor vehicles can be sampled directly in real time without any sample preparation as shown in Fig. 8.10a and b on a commercial instrument (Syft Technologies Ltd).

8.8.3 IMR-MS

The strength of the IMR-MS system is that it uses reagent ions that have a large range of energies that enable them to undergo reactions with most analytes. (The ion energies are Kr 13.997, Xe 12.13, and Hg 10.437 eV.) However, for analysis of samples in air, only Hg^+ can be used for monitoring trace analytes as Kr^+ and Xe^+ react with O_2. On the other hand, the advantage that Kr and Xe ions bring is reactivity with analytes such as CO, SO_2, N_2O, SF_6, HF, and CF_4 that instruments using reagent ions generated from terminal processes in air (H_3O^+, NO^+, and O_2^+) cannot measure. Wang et al. [219], using IMR-MS, applied the methodology to monitor volatiles in cigarette smoke but in order to do this it was necessary for them to establish the ion products of the reactions of the various reagent ions with analytes that might be present in cigarette smoke. This was achieved by making mixtures with known concentrations at the ppm level in nitrogen of individual hydrocarbons, aldehydes, nitriles, HCN, H_2S, and chloromethane. The products of each individual analyte/nitrogen mixture were examined with each of the reagent ions. For most analytes, multiple product ions were found. Equipped with this knowledge, they examined the filtered gas from burning cigarettes and were able to identify the

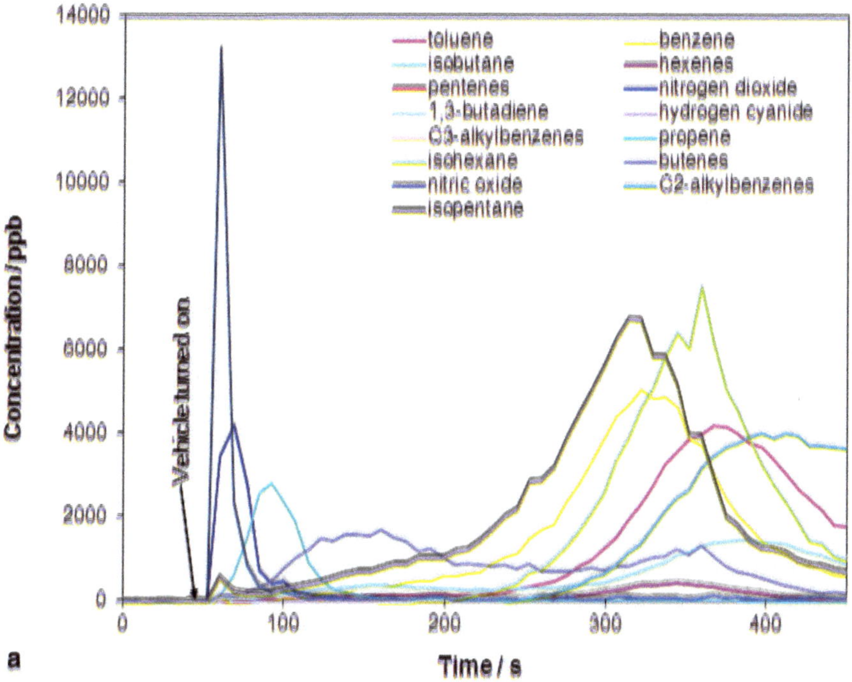

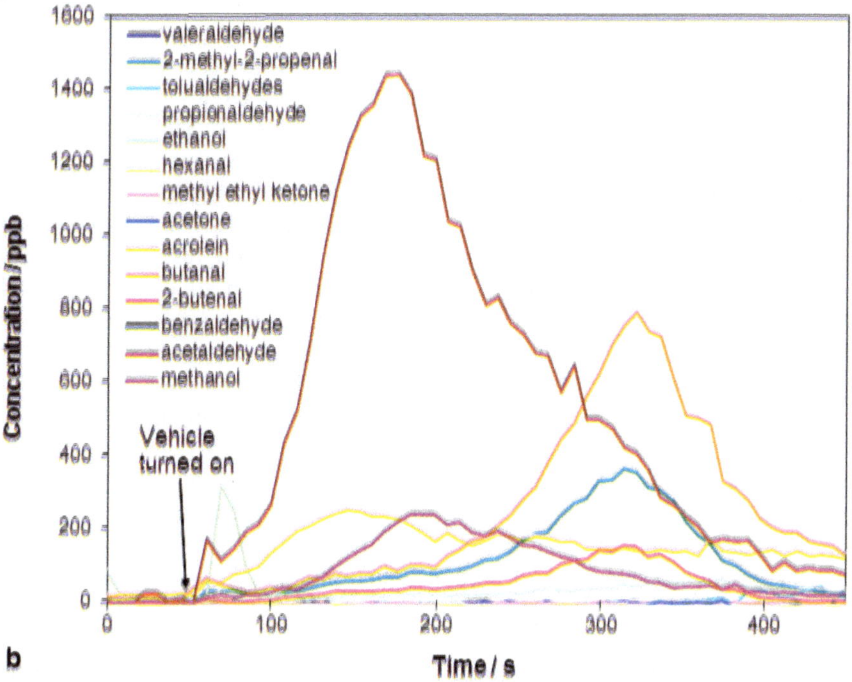

Fig. 8.10 **a** Aldehydes and **b** hydrocarbons monitored (using a Syft Technologies Voice200 SIFT-MS instrument) in a car exhaust that had been fitted with a catalytic converter. The concentrations shown in the figure were based on the known reagent ion kinetics with the analytes listed

most abundant VOCs in cigarette smoke. Although the methodology allowed direct examination of cigarette smoke in real time without any sample preparation, they could not identify a number of analytes with the quadrupole mass spectrometer because of isobaric ion overlap.

8.8.4 Comparison of PTR-MS and SIFT-MS

PTR-MS and SIFT-MS have a clear advantage over most other techniques for direct analysis of gas mixtures as no derivatization, adsorption onto traps followed by desorption steps, or other sample pretreatment steps are required before the sample is admitted to the analytical instrument. A comparison between the two techniques relevant to breath-sampling has been presented recently [220]. Both technologies allow immediate quantification of an analyte in a gas mixture in real time whether it be a breath sample or an environmental sample. As discussed in Sect. 8.4 (PTR-MS) and 8.5 (SIFT-MS), there are three main differences between these two techniques: (i) PTR-MS operates with a single H_3O^+ reagent ion generated by a hollow cathode ion source that is not mass selected (although in some instruments, a valve allows mechanical switching to gases that produce NO^+ or O_2^+ ions in addition to H_3O^+, that are also generated by the hollow cathode ion source, with a switching delay of 10 s or so); SIFT-MS operates with a microwave discharge and mass-selected reagent ions (usually H_3O^+, NO^+, and O_2^+) that can be switched within a few ms; (ii) In PTR-MS, the reagent ion-analyte reactions take place in a drift tube that noticeably modifies the thermal reagent ion–analyte chemistry; in SIFT-MS, the reagent ion–analyte reactions take place at close to thermal energies; (iii) the PTR-MS generally uses air as the carrier gas whereas helium is used in SIFT-MS (although recently some commercial SIFT-MS instruments have successfully operated using nitrogen as the carrier gas [221]).

An advantage of the PTR-MS instrument is that high ion densities can be transmitted from the ion source to the flow tube without the more restricting influence of mass selection by a quadrupole mass filter as occurs with SIFT-MS. But the advantage of the increase in reagent ion density in a PTR-MS instrument is mitigated somewhat by several other factors. In PTR-MS, the axial electric field produces an ion drift time of typically 116 μs at 120 Td (Sect. 8.4.5) as opposed to a SIFT-MS ion residence time of several ms. Although the shorter drift time in PTR-MS lowers diffusive ion loss, it also lowers the number of reagent ions that are converted to product ions. Perhaps one of the most telling factors of difference between the two techniques is the influence that the drift fields in PTR-MS have on the reagent ion–analyte chemistry. As mentioned earlier in Sect. 8.4.7, a common database of reagent ion–molecule reactions is used by both techniques. This database is derived from thermal energy reagent ions with known rate coefficients for reaction with analytes, and these values are not necessarily valid for the conditions of a PTR-MS drift tube. In PTR-MS, a generic value of 2×10^{-9} cm^3 s^{-1} is often selected as many of the reactions have not been investigated at the particular E/N conditions used. What is also significant is that the ion-neutral collision energies imposed by the

Table 8.2 Product masses and branching ratios of ions formed from several compounds in the same gas stream using H_3O^+ in SIFT-MS and PTR-MS instruments (at 165 Td)

Compound (molar mass; $gmol^{-1}$)	SIFT-MS product masses, m/z (branching ratio in %)	PTR-MS product masses, m/z (branching ratio in %)[a]
Acetaldehyde [44]	45(100 %)[b]	61(26 %), 45(47 %), 43(20 %), 41(6 %), 39(1 %)
Ethyl acetate [88]	89(100 %)[c]	89(5 %), 61(74 %), 43(21 %)
Hexanal [100]	101(50 %), 83(50 %)[b]	83(57 %), 55(43 %)

[a] Ref [222]
[b] Ref [223]
[c] Ref [224]

drift field in PTR-MS cause much more product ion fragmentation from collision-induced dissociation and many of the more weakly bound cluster ions dissociate. These differences between the two techniques can be clearly seen in the comparison of the product ions generated in the three common analytes: acetaldehyde, ethylacetate, and hexanal, and these are presented in Table 8.2.

Another interesting comparison of relative ion energies between SIFT-MS and PTR-MS is the measurement of chloroform, $CHCl_3$. This comparison is shown in Fig. 8.11 (a, b and c). In SIFT-MS, only a very slow reaction occurs between the reagent ion H_3O^+ and $CHCl_3$ ($k = 5 \times 10^{-11}$ cm^3 s^{-1}) with the products of association of H_3O^+, $H_3O^+.CHCl_3$ occurring at m/z 137, 139, and 141. The mass spectrum observed on the Voice200 SIFT-MS is shown in Fig. 8.11a up to m/z 125. A rapid reaction of the reagent ion O_2^+ with $CHCl_3$ in SIFT-MS occurs ($k = 1.8 \times 10^{-9}$ cm^3 s^{-1}) producing the product ion $CHCl_2^+$ at m/z 83, 85, and 87 as shown in Fig. 8.11b. But when the same sample of $CHCl_3$ is introduced to a PTR-MS instrument operating at 127 Td drift field, the spectrum from H_3O^+ shown in Fig. 8.11c is found which is essentially identical to the SIFT-MS spectrum from O_2^+. This emphasizes the fact that the presence of the drift field in PTR-MS does cause additional excitation in reactions of H_3O^+ producing product ions that are not present in libraries of reagent ions under thermal conditions.

SIFT-MS, with its capability for rapid switching between reagent ions, has another advantage in that the diagnostic capability of three reagent ions enables some isobaric compounds to be monitored in real time. For example, the VOCs acetone (CH_3COCH_3) and propionaldehyde (CH_3CH_2CHO) both have a molar mass of 58.08 g mol^{-1} but they can readily be monitored in the same sample by utilizing the different reagent ion diagnostics of H_3O^+ and NO^+. Similarly, 3-methyl butanal ($M_r = 86.13$), 2-pentanone ($M_r = 86.13$), and diacetyl (2,3-butanedione) ($M_r = 86.09$) are almost identical in mass and require a high-resolution mass spectrometer such as a TOF to resolve the VOCs if present together in a sample. Even then it would be difficult. With a single H_3O^+ reagent ion alone, it would not be possible to distinguish between them. However, the three reagent ions of the SIFT-MS instrument can readily monitor each analyte because of the different reagent ion chemistries. This same separation in real time cannot be achieved using a conventional PTR-MS instrument with switchable reagent ions because of the time needed for switching

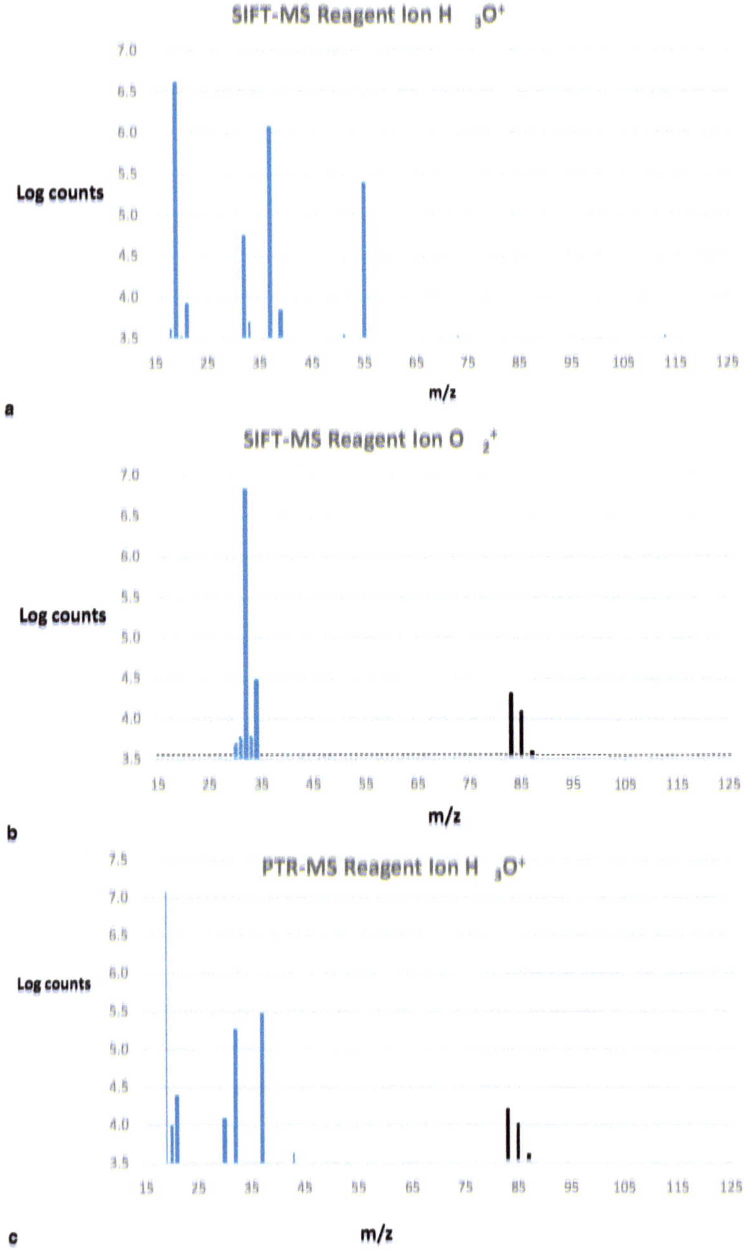

Fig. 8.11 SIFT-MS and PTR-MS analysis comparison of chloroform, $CHCl_3$, mixture. **a** SIFT-MS reagent ion H_3O^+. **b** SIFT-MS reagent ion O_2^+. **c** PTR-MS mass spectrum at 127 Td

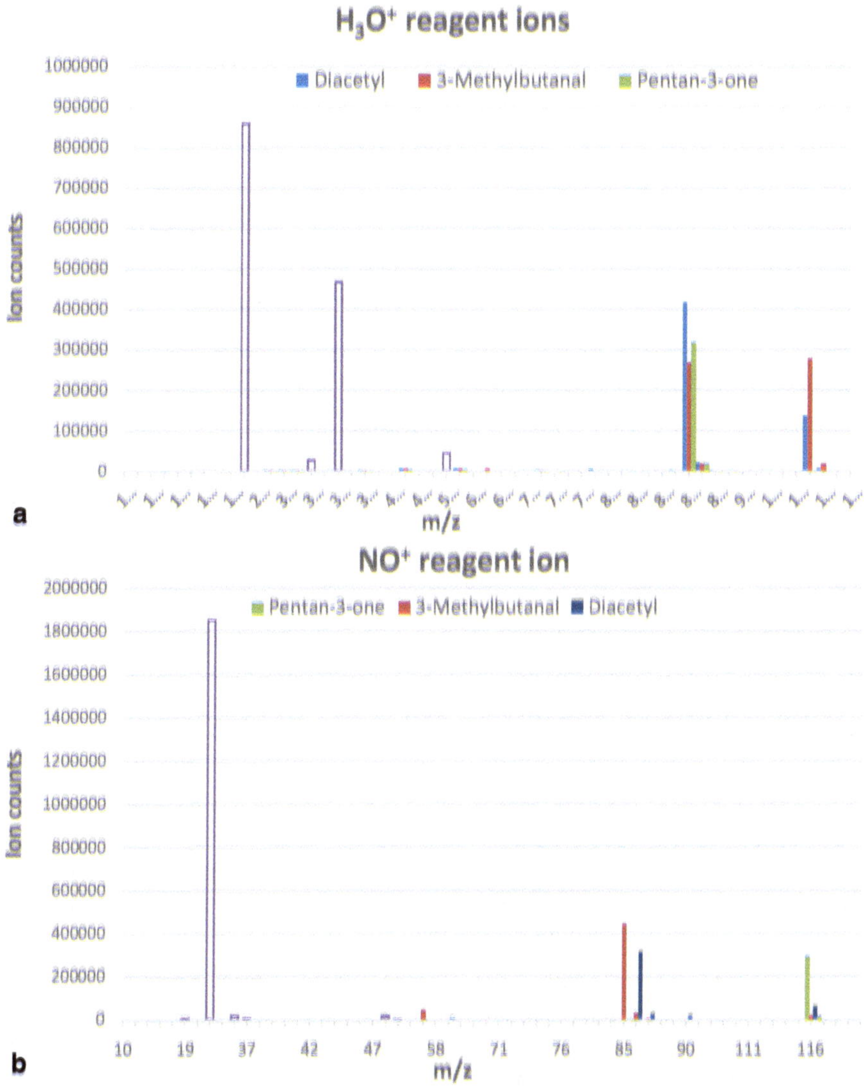

Fig. 8.12 SIFT-MS mass spectra on a Syft Technologies Ltd Voice200 instrument with a helium carrier gas. The unfilled bars represent reagent ions and the filled bars are product ions. Figure 8.12a is the mass spectra generated by H_3O^+ ions in which the product ion peaks are superimposed. Figure 8.12b is the mass spectra generated by NO^+ ions which show clear separation of the isobaric ions

between reagent ions and the product ion distribution incurred by the drift field. In SIFT-MS, the isobaric separation is clearly seen in Fig. 8.12 which shows the reagent ion spectra for H_3O^+ (Fig. 8.12a) and NO^+ (Fig. 8.12b) of these three VOCs with the separation and identification that can be achieved of these isobaric compounds using SIFT-MS.

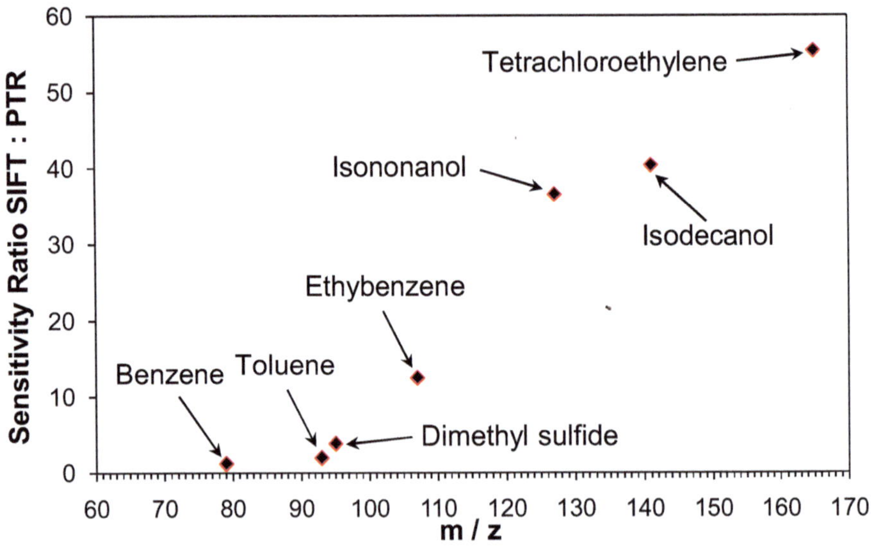

Fig. 8.13 The ratio of ion intensities obtained by a Syft Voice200 SIFT-MS and a standard PTR-MS measuring VOC samples from the same sample stream with the H_3O^+ reagent ion in both instruments. (Courtesy Syft Technologies Ltd)

Because there is no mass-selection filter in the PTR-MS for the reagent ion, it is usually stated that PTR-MS is more sensitive than SIFT-MS as the reagent ion number density in the PTR-MS drift tube is greater than in SIFT-MS. Although this is true in some cases, the difference is not as great as might be imagined as there are other factors that need to be included. In a direct comparison of a PTR-MS quadrupole instrument and a Syft Voice200 SIFT-MS on the same sample stream divided between instruments, the SIFT-MS instrument was found to have greater sensitivity than the PTR-MS at m/z values greater than 80—see Fig. 8.13.

PTR-TOF-MS Recently, some significant gains have been made in sensitivity with PTR-TOF-MS. A new instrument that utilizes a quadrupole ion guide designated PTR-QiTOF has been designed and released [225]. The mass resolution of this instrument is up to 10,000 m/Δm. With a drift tube voltage of 1000 V and an increased pressure of 2.85 Torr in the drift tube, a sensitivity of up to 4700 cps/ppbv for dichlorobenzene at m/z 147 has been reported. This compares with 1900 cps/ppbv on a commercial Voice200*Ultra* SIFT-MS quadrupole instrument operating at a flow tube pressure of 0.6 Torr with a nitrogen carrier gas in which the analyte was included in the carrier gas. For this measurement, the analyte mesitylene was generated from a permeation tube with the oven at 80 °C and the mesitylene concentration in the nitrogen stream was at 225 ppbv.

The advantage of an instrument like the PTR-QiTOF is not only its sensitivity but also the mass resolution enabling separation of some isobaric compounds. It should also be remembered that in a TOF-based instrument, the entire mass spectrum in the specified mass range is recorded on the multichannel plate detector in every TOF pulse, allowing subsequent analysis if necessary at a later time for unidentified analytes. However, the benefits of TOF-based instruments need to be considered against the much simpler operating systems of the quadrupole-based instruments, the isobaric separation that they give from different reagent ion chemistry, their significantly lower cost, and their much more user-friendly software.

Acknowledgments The author gratefully acknowledges Vaughan Langford who made the comparative study of SIFT-MS and PTR-MS; also, the help of John Gray, Daniel Milligan, Yan Li, and Diandree Padayachee, who assisted with various parts of the manuscript, is recognized.

References

1. James AT, Martin AJP. Gas-liquid partition chromatography. Biochem J. 1952;50:679–80.
2. Gohlke RS, McLafferty FW. Early gas chromatography/mass spectrometry. J Am Soc Mass Spectrom. 1993;4:367–71.
3. Watson JT. A historical perspective and commentary on pioneering developments in gas chromatography/mass spectrometry at MIT. J Mass Spectrom. 1998;33:103–8.
4. Horning EC, Horning MG, Carroll DI, et al. New pictogram detection system based on a mass spectrometer with an external ionization source at atmospheric pressure. Anal Chem. 1973;45:936–43.
5. Carroll DI, Dzidic I, Stillwell RN, et al. Atmospheric pressure ionization mass spectrometry: corona discharge ion source for use in liquid chromatograph-mass spectrometer—computer analytical systems. Anal Chem. 1975;47:2369–73.
6. French JB, Thomson BA, Davidson WR, et al. Atmospheric pressure chemical ionization mass spectrometry. In: Karasek FW, Hutzinger O, editor. Mass spectrometry in environmental sciences. New York: Plenum Press; 1985. pp. 101–21.
7. Bruins AP. Mass spectrometry with ion sources operating at atmospheric pressure. Mass Spec Rev. 1991;10:53–7.
8. Carroll DI, Dzidic I, Stillwell RN, et al. Subpicogram detection system for gas phase analysis based on atmospheric pressure ionization (API) mass spectrometry. Anal Chem. 1974;46:706–10.
9. Siegel MW, Fite WL. Terminal ions in weak atmospheric plasmas. Applications of atmospheric pressure ionization to trace impurity analysis in gases. J Phys Chem. 1976;80:2871–81.
10. Sunner J, Ikonomou MG, Kebarle P. Sensitivity enhancements obtained at high temperatures in atmospheric pressure ionization mass spectrometry. Anal Chem. 1988;60:1308–13.
11. Thomson BA. Micro and nano-electrospray ionization techniques. In: Gross M, Caprioli R, editors. Encyclopedia of mass spectrometry Vol6. Ionization methods. Oxford: Elsevier; 2007. pp. 435–44.
12. Huang MZ, Yuan CH, Cheng SC, et al. Ambient ionization mass spectrometry. Annu Rev Anal Chem. 2010;3:43–65.
13. Bruins AP. Atmospheric pressure-ionization mass spectrometry. Trends Anal Chem. 1994;13:37–43.
14. Badjagbo K, Moore S, Sauve S. Real-time continuous monitoring methods for airborne VOCs. Trends Anal Chem. 2007;26:931–40.
15. Lane DA, Thomson BA, Lovett AM, et al. Real time tracking of industrial emissions through populated areas using mobile APCI mass spectrometers. Adv Mass Spectrom. 1980;8B:1480–9.
16. www.epa.gov/region6/6lab/taga.htm. Accessed 16 May 2014.

17. Badjagbo K, Picard P, Moore S, et al. Direct atmospheric pressure chemical ionization-tandem mass spectrometry for the continuous real-time trace analysis of benzene, toluene, ethylbenzene and xylenes in ambient air. J Am Soc Mass Spectrom. 2009;20:829–36.
18. Huang G, Gao L, Duncan J, et al. Direct detection of benzene, toluene and ethyl benzene at trace levels by atmospheric pressure ionization using a hand held mass spectrometer. J Am Soc Mass Spectrom. 2010;21:132–5.
19. Auld J, Hastie DR. Tandem mass spectrometry and multiple reaction monitoring using an atmospheric pressure chemical ionization triple quadrupole mass spectrometer for product identification in atmospherically important reactions. Int J Mass Spectrom. 2009;282:91–8.
20. Warscheid B, Hoffman T. On-line measurement of α-pinene ozonolysis products using an atmospheric pressure chemical ionisation ion-trap mass spectrometer. Atmos Environ. 2001;35:2927–40.
21. Eiceman GA, Bergloff JF, Rodriguez JE, et al. Atmospheric pressure chemical ionization of fluorinated phenols in atmospheric pressure chemical ionization mass spectrometry, tandem mass spectrometry and ion mobility mass spectrometry. J Am Soc Mass Spectrom. 1999;10:1157–65.
22. Taylor AJ, Linforth RST, Harvey BA, et al. Atmospheric pressure chemical ionization mass spectrometry for in vivo analysis of volatile flavour release. Food Chem. 2000;71:327–38.
23. Taylor AJ, Linforth RST, editors. Food flavour technology. 2nd edn. Oxford: Wiley-Blackwell; 2010.
24. Gkatzionis K, Linforth RST, Dodd CER. Volatile profile of Stilton cheeses: difference between zones within a cheese and dairies. Food Chem. 2009;113:506–12.
25. Jublot L, Linforth RST, Taylor AJ. Direct atmospheric pressure chemical ionization ion-trap mass spectrometry for aroma analysis, speed, sensitivity and resolution of isobaric compounds. Int J Mass Spectrom. 2005;243:269–77.
26. Gan HH, Soukoulis C, Fisk I. Atmospheric pressure chemical ionisation mass spectrometry analysis linked with chemometrics for food classification—a case study: geographical provenance and cultivar classification of monovarietal clarified apple juices. Food Chem. 2014;146:149–56.
27. Scholz B, Barnsteiner A, Feist K, et al. Analysis of phytostanyl fatty acids in enriched foods via UHPLC-APCI-MS. J Agric Chem. 2014;62:4268–75.
28. Prelle A, Spadaro D, Garibaldi A, et al. A new method for detection of five Alternaria toxins in food materials based on LC-APCI-MS. Food Chem. 2013;140:161–7.
29. Richon G, Paris C, Girardin M, et al. Highly sensitive, quick and simple quantification method for mono and disaccharides in aqueous media using liquid chromatography-atmospheric pressure chemical ionization-mass spectrometry (LC-APCI-MS). J Chromat B. 2011;879:1529–36.
30. Capriotti AL, Foglia P, Gubbiotti R, et al. Development and validation of a liquid chromatography/ atmospheric pressure chemical photoionization-tandem mass spectrometric method for the analysis of mycotoxins subjected to commission regulation (EC) No. 1881/2006 in cereals. J Chromat A. 2010;1217:6044–51.
31. Wenzel M, Seuss-Baum I, Schliche E. Influence of pasteurization, spray- and freeze-drying and storage on the carotenoid content in egg yolk. J Agric Food Chem. 2010;58:1726–31.
32. Gnolt FM, Davidson WR, Lovett AM, et al. Breath analysis by atmospheric pressure chemical ionization mass spectrometry. Anal Chem. 1983;55:805–7.
33. Schneider C, Sablier M, Desmazieres B. Characterization by mass spectrometry of an unknown polysiloxane sample used under uncontrolled medical conditions for cosmetic surgery. Rapid Commun Mass Spectrom. 2008;22:3353–61.
34. Klys M, Konopka T, Scislowski M, et al. Fatality involving vinblastine overdose as a result of complex medical error. Cancer Chemo Pharmacol. 2007;59:89–95.
35. Ferguson EE, Fehesenfeld FC, Schmeltekopf AL. Flowing afterglow measurements of ion-molecule reactions. Adv Atomic Mol Phys. 1969;5:1–56.
36. Mohler O, Reiner T, Arnold F. The formation of SO_5^- by gas phase ion-molecule reactions. J Chem Phys. 1992;97:8233–9.

37. Viidanoja J, Reiner T, Arnold F. Laboratory investigations of negative ion-molecule reactions of formic and acetic acids: implications for atmospheric measurements by ion-molecule reaction mass spectrometry. Int J Mass Spectrom. 1998;181:31–41.
38. www.vandf.com. Accessed 10/5/2014.
39. Schubert H, Guntow U, Hofmann K, et al. Performance and application potentials of ion-molecule reaction mass spectrometry (IMR-MS) in the analysis of complex gas mixtures. Fresenius J Anal Chem. 1996;56:127–37.
40. Hornuss C, Praun S, Villinger J, et al. Real-time monitoring of Propofol in expired air in humans undergoing total intravenous anesthesia. Anesthesiol. 2007;1006:665–74.
41. Dearth MA. Evaluation of a commercial mass spectrometer for its potential to measure auto exhaust constituents in real time. Ind Eng Chem Res. 1999;38:2203–9.
42. Wanke T, Vehlow J. IMR-MS on-line measurements in the exhaust gas of a municipal solid waste incineration pilot plant (Tamara). Chemosphere. 1997;34:345–55.
43. Wang H, Xie W, Chen M, et al. Determination of hazardous volatile organic compounds in the Hoffman list by ion-molecule reaction mass spectrometry. Rapid Commun Mass Spectrom. 2012;26:1841–8.
44. Hornuss C, Wiepcke D, Praun S, et al. Time course of expiratory propofol after bolus injection as measured by ion-molecule reaction mass spectrometry. Anal Bioanal Chem. 2012;403:555–61.
45. Dolch ME, Frey L, Hornuss C, et al. Molecular breath-gas analysis by online mass spectrometry in mechanically ventilated patients: a new software-based method of CO_2-controlled alveolar gas monitoring. J Breath Res. 2008;2:037010:10.
46. Netzer M, Millonig G, Osl M, et al. A new ensemble-based algorithm for identifying breath gas marker candidates in liver disease using ion molecule reaction mass spectrometry. Bioinformatics. 2009;25:941–7.
47. Dolch ME, Hornuss C, Klocke C, et al. Volatile organic compound analysis by ion-molecule reaction mass spectrometry for Gram-positive bacteria differentiation. Eur J Clin Microbiol Infect Dis. 2012;31:3007–13.
48. Hansel A, Jordan A, Holzinger R, et al. Proton transfer reaction mass spectrometry: online trace gas analysis at the ppb level. Int J Mass Spectrom Ion Proc. 1995;149/150:609–19.
49. Bouchoux G, Salpin JY, Le Blanc D. A relationship between the kinetic and the thermochemistry of proton transfer reactions in the gas phase. Int J Mass Spectrom Ion Proc. 1996;153:37–48.
50. Ellis AM, Mayhew CA. Proton transfer mass spectrometry. UK:Wiley; 2014.
51. Anicich VG. An index of the literature for bimolecular gas phase cation-molecule reaction kinetics JPL Publication 03-19. NASA, JPL; 2003.
52. Marcus RK, editor. Glow discharge spectroscopies. New York:Plenum Publishing; 1993.
53. Ellis HW, Thackston MG, McDaniel EW, et al. Transport properties of gaseous ions over a wide energy range. Part III. At Data Nucl Data Tables. 1984;31:113–51.
54. Dotan I, Albritton DL, Lindinger W, et al. Mobilities of CO_2^+, N_2H^+, H_3O^+, $H_3O^+.H_2O$ and $H_3O^+.(H_2O)_2$ ions in nitrogen. J Chem Phys. 1976;65:5012–30.
55. March RE. Quadrupole ion trap mass spectrometry: theory, simulation, recent developments and applications. Rapid Commun Mass Spectrom. 1998;12:1543–54.
56. Prazeller P, Palmer PT, Boscaini E. Proton transfer reaction ion trap mass spectrometer. Rapid Commun Mass Spectrom. 2003;17:1593–9.
57. Graus M, Mueller M, Hansel A. High resolution PTR-TOF: quantification and formula quantification of VOC in real time. J Amer Soc Mass Spectrom. 2010;21:1037–44.
58. Jordan A, Haidacher S, Hanel G, et al. A high resolution and high sensitivity proton-transfer-reaction time-of-flight mass spectrometer (PTR-TOF-MS). Int J Mass Spectrom. 2009;286:122–8.
59. Wyche KP, Blake RS, Willis KA, et al. Differentiation of isobaric compounds using chemical ionization mass spectrometry. Rapid Commun Mass Spectrom. 2005;19:3356–62.

60. Blake RS, Wyche KP, Ellis AM, et al. Chemical ionization reaction time-of-flight mass spectrometry: multi-reagent analysis for determination of trace gas composition. Int J Mass Spectrom. 2006;254:85–93.
61. Jordan A, Haidacher S, Hanel G, et al. An on-line ultra-high sensitivity proton transfer reaction mass spectrometer combined with switchable reagent ion capability (PTR + SRI-MS). Int J Mass Spectrom. 2009;286:32–8.
62. www.ionicon.com/publications
63. Hewitt CN, Hayward S, Tani A. The application of proton transfer reaction mass spectrometry (PTR-MS) to monitoring and analysis of volatile organic compounds in the atmosphere. J Environ Monit. 2003;5:1–7.
64. de Gouw JA, Warneke C. Measurements of volatile organic compounds in the Earth's atmosphere using proton-transfer mass spectrometry. Mass Spectrom Rev. 2007;26:223–57.
65. Blake RS, Monks PS, Ellis AM. Proton transfer reaction mass spectrometry. Chem Rev. 2009;109:861–96.
66. Cappellin L, Loreto F, Aprea E, et al. PTR-MS in Italy: a multipurpose sensor with applications in environmental, agri-food and health science. Sensors (Basil). 2013;13:11923–55.
67. Davidson B, Taipale R, Langford B, et al. Concentrations and fluxes of biogenic volatile organic compounds above a Mediterranean macchie ecosystem in western Italy. Biogeosc. 2009;6:1655–70.
68. Vesin A, Bouchoux G, Quivet E, et al. Use of the HS-PTR-MS for online measurements of pyrethroids during indoor insecticide treatment. Anal Bioanal Chem. 2012;403:1907–21.
69. Rogers TM, Grimsrud EP, Herndon SC, et al. On-road measurements of volatile organic compounds in the Mexico City metropolitan area using proton transfer mass spectrometry. Int J Mass Spectrom. 2006;252:26–37.
70. Tanimoto H, Kameyama S, Iwata T, et al. Measurement of air-sea exchange of dimethyl sulphide and acetone by PTR-MS coupled with gradient flux technique. Environ Sci Technol. 2014;48:526–33.
71. Yuan B, Warneke C, Shao M, et al. Interpretation of volatile organic compound measurements by proton transfer- mass spectrometry over the deep water horizon spill. Int J Mass Spectrom. 2014;358:43–8.
72. Shilling JE, Zaveri RA, Fast JD, et al. Enhanced SOA formation from mixed anthropogenic and biogenic emissions during the CARES campaign. Atmos Chem Phys. 2013;13:2091–113.
73. Warneke C, Veres P, Holloway JS, et al. Airborne formaldehyde measurements using PTR-MS calibration, humidity dependence, inter-comparison and initial results. Atmos Meas Tech. 2011;4:2345–58.
74. Herbig J, Gutmann R, Winkler K, et al. Real-time monitoring of trace gas concentrations in syngas. Oil Gas Sci Technol. 2014;69:363–72.
75. Warneke C, de Gouw JA, Kuster WC, et al. Validation of atmospheric VOC measurements by proton transfer reaction- mass spectrometry using a gas chromatographic preparation method. Environ Sci Technol. 2003;37:2494–501.
76. Hansen MJ, Liu D, Guldberg LB, et al. Application of proton transfer reaction-mass spectrometry to the assessment of odorant removal in a biological air cleaner for pig production. J Agric Chem. 2012;60:2599–606.
77. Penuelas J, Llusia J. BVOCs: plant defence against climate warming? Trends Plant Sci. 2003;8:105–9.
78. Williams J, Fischer H, Harris GW, et al. Variability—lifetime relationship for organic trace gases; a novel aid to compound identification and estimation of HO concentrations. J Geophys Res. 2000;105:D20473–D86.
79. Eerdekens G, Ganzeveld L, Vila-Guerau de Arellano J, et al. Flux estimates of isoprene, methanol and acetone from airborne PTR-MS measurements over the tropical rainforest during the Gabriel 2005 campaign. Atmos Chem Phys. 2009;9:4207–27.
80. Biasioli F, Gasperi F, Yeretzian C, et al. PTR-MS monitoring of VOCs and BVOCs in food science and technology. Trends Anal Chem. 2011;30:968–77.

81. Zardin E, Tyapkova O, Buettner A, et al. Performance assessment of proton transfer reaction-time-of-flight mass spectrometry (PTR-TOF-MS) for analysis of isobaric compounds in food-flavour applications. Food Sci Technol. 2014;56:153–60.
82. Romano A, Cappellin L, Ting V, et al. Nosespace analysis of PTR-TOF-MS for the characterization of food and tasters: the case study of coffee. Int J Mass Spectrom. 4;365–366:20–7.
83. Gloess AN, Vietri A, Wieland F, et al. Evidence of flavour formation dynamics by roasting coffee from different origins: online analysis with PTR-TOF-MS. Int J Mass Spectrom. 2014;365–366:324–37.
84. Costa F, Cappellin L, Farneti B, et al. Advances in QTL mapping for ethylene production in apple (Malus x domestica Borkh). Post Harvest Biol Tech. 2014;87:126–132.
85. Soukoulis C, Cappellin L, Aprea E, et al. PTR-TOF-MS, a novel, rapid, high sensitivity and non-invasive tool to monitor volatile compound release during fruit post-harvest storage: the case study of apple ripening. Food Bioprocess Tech. 2013;6:2831–43.
86. Tsevdou M, Soukoulis C, Cappellin L, et al. Monitoring the effect of high pressure and transglutaminase treatment of milk on the evolution of flavour compounds during lactic acid fermentation using PTR-TOF-MS. Food Chem. 2013;138:2159–67.
87. Sanchez del Pulgar J, Soukoulis C, Carrapiso AI et al. Effect of the pig rearing system on the final volatile profile of Iberian dry-cured ham as detected by PTR-TOF-MS. Meat Sci. 2013;93:420–8.
88. Morisco F, Aprea E, Lembo V, et al. Rapid "breath print" of liver cirrhosis by proton transfer time-of-flight mass spectrometry. A pilot study. PLoS One. 2013;8:e59658.
89. Kohl I, Beauchamp J, Cakar-Beck F, et al. First observation of a potential non-invasive breath gas biomarker for kidney function. J Breath Res. 2013;7(017110):1–12.
90. Aprea E, Cappellin L, Gasperi F, et al. Application of PTR-TOF-MS to investigate metabolites in exhaled breath of patients affected by coeliac disease under gluten free diet. J Chromatog B. 2014;966:208–13.
91. Righettoni M, Schmid A, Amann A, et al. Correlations between blood glucose and breath components from portable gas sensors and PTR-TOF-MS. J Breath Res. 2013;7(037110): 1–9.
92. Trefz P, Schmidt M, Oertel P, et al. Continuous real time breath gas monitoring in the clinical environment by proton-transfer-reaction-time-of-flight-mass spectrometry. Anal Chem. 2013;2013:10321–9.
93. Veres PR, Faber P, Drewnick F, et al. Anthropogenic sources of VOC in a football stadium: assessing human emissions in the atmosphere. Atmos Environ. 2013;77:1052–9.
94. Spanel P, Smith D. Selected ion flow tube: a technique for quantitative trace gas analysis of air and breath. Med Biol Eng Comput. 1996;34:409–19.
95. Adams NG, Smith D. Selected Ion Flow Tube (SIFT)—a technique for studying ion-neutral reactions. Int J Mass Spectrom Ion Proc. 1976;21:349–59.
96. Ferguson EE, Fehsenfeld FC, Schmeltekopf AL. Flowing afterglow measurements of ion-neutral reactions. Adv Atom Mol Phys. 1969;5:1–54.
97. Smith D, Spanel P. Selected ion flow tube-mass spectrometry (SIFT-MS) for on-line trace analysis. Mass Spec Rev. 2005;24:661–700.
98. Smith D, Spanel P. The novel selected ion flow tube approach to trace gas analysis of air and breath. Rapid Comm Mass Spectrom. 1996;10:1183–98.
99. Freeman CG, McEwan MJ. Rapid analysis of trace gases in complex mixtures using selected ion flow tube-mass spectrometry. Aust J Chem. 2002;55:491–4.
100. Spanel P, Smith D. Progress in SIFT-MS: breath analysis and other applications. Mass Spectrom Rev. 2011;30:236–67.
101. www.syft.com. Accessed 30 May 2014.
102. Milligan DB, Fairley DA, Freeman CG, et al. A flowing afterglow-selected ion flow tube (FA/SIFT) comparison of SIFT injector flanges and H_3^+ + N revisited. Int J Mass Spectrom. 2000;202:351–61.

103. Bohme D. In Ausloos P, editor. Interactions between ions and molecules. New York: Plenum Press; 1975.p. 489.
104. Smith D, Spanel P. Ambient analysis of trace compounds in gaseous media by SIFT-MS. Analyst. 2011;136:2009–32.
105. Wilson PF, Milligan DB, Liew WL, et al. Reactions of $CH_3OCH_2^+$ with hydrocarbons and O, N and S compounds: applications for chemical ionisation in selected ion flow tube studies. J Amer Soc Mass Spectrom. 2002;13:1028–33.
106. Francis GJ, Wilson PF, Milligan DB, et al. A SIFT-MS method for the analysis of small linear hydrocarbons of relevance to oil exploration. Int J Mass Spectrom. 2007;268:38–46.
107. Spanel P, Smith D. Selected ion flow tube studies of the reaction of H_3O^+, NO^+ and O_2^+ with several aromatic and aliphatic hydrocarbons. Int J Mass Spectrom. 1998;181:1–10.
108. Smith D, Spanel P, Dabill D, et al. On-line analysis of diesel engine exhaust gases by selected ion flow tube mass spectrometry. Rapid Commun Mass Spectrom. 2004;18:2830–8.
109. Hastie DR, Gray J, Langford VS, et al. Real-time measurement of peroxyacetyl nitrate using selected ion flow tube-mass spectrometry. Rapid Commun Mass Spectrom. 2010;24:343–8.
110. Prince BJ, Milligan DB, McEwan MJ. Applications of selected ion flow tube-mass spectrometry to real-time atmospheric monitoring. Rapid Commun Mass Spectrom. 2010;24:1763–9.
111. Sovova K, Shestiviska V, Spanel P. Real-time quantification of trace biogenic volatile selenium compounds in humid air by selected ion flow tube mass spectrometry. Anal Chem. 2011;84:4979–83.
112. Francis GJ, Langford VS, Milligan DB, et al. Real-time monitoring of hazardous air pollutants. Anal Chem. 2009;81:1595–9.
113. Francis GJ, Milligan DB, McEwan MJ. Detection and quantification of chemical warfare agent precursors and surrogates by selected ion flow tube-mass spectrometry. Anal Chem. 2009;81:8892–9.
114. Langford VS, Gray JDC, McEwan MJ. Selected ion flow tube studies of several siloxanes. Rapid Commun Mass Spectrom. 2013;27:700–6.
115. Langford VS, Gray JDC, Maclagan RGAR, et al. Detection of siloxanes in landfill gas and biogas using SIFT-MS. Current Anal Chem. 2013;9:558–64.
116. Heynderickx PM, Van Huffel K, Dewulf J, et al. SIFT-MS for livestock emission characterization: application of similarity coefficients. Chem Eng Trans. 2012;30:157–62.
117. Heynderickx PM, Van Huffel K, Dewulf J, et al. Application of similarity coefficients to SIFT-MS data for livestock emission characterization. Biosyst Eng. 2013;114:44–54.
118. Sumonsiri N, Barringer SA. Application of SIFT-MS in monitoring volatile compounds in fruits and vegetables. Curr Anal Chem. 2013;9:631–41.
119. Spanel P, Smith D. Selected ion flow tube-mass spectrometry: detection and real-time monitoring of flavours released from food products. Rapid Commun Mass Spectrom. 1999;13:585–96.
120. Davis BM, Senthilmohan ST, Wilson PF, et al. Major volatile compounds in the headspace above olive oil analysed by selected ion flow tube-mass spectrometry. Rapid Commun Mass Spectrom. 2005;19:2272–8.
121. Davis BM, Senthilmohan ST, McEwan MJ. Direct determination of antixoidants in whole olive oil using the SIFT-MS-TOSC assay. J Am Oil Chem Soc. 2011;88:785–92.
122. Xu Y, Barringer S. Effect of temperature on lipid-related volatile production in tomato puree. J Agric Food Chem. 2009;57:9108–13.
123. Xu Y, Barringer S. Comparison of tomatillo and tomato volatile compounds in the headspace by selected ion flow tube mass spectrometry (SIFT-MS). J Food Sci. 2010;75:C268–C73.
124. Xu Y, Barringer S. Comparison of volatile release in tomatillo and different varieties of tomato during chewing. J Food Sci. 2010;75:C352–C8.
125. Ties P, Barringer S. Influence of lipid content and lipoxygenase on flavour volatiles in the tomato peel and flesh. J Food Sci. 2013;77:C830–C7.

126. Duan H, Barringer S. Changes in furan and other volatile compounds in sliced carrot during air-drying. J Food Proc Preserv. 2012;36:46–54.
127. Bowman T, Barringer S. Analysis of factors affecting volatile compound formation in roasted pumpkin seeds with selected ion flow tube mass spectrometry (SIFT-MS). J Food Chem. 2012;77:C51–C60.
128. Azcarate C, Barringer SA. Effect of enzyme activity and frozen storage on jalapeno pepper volatiles by selected ion flow tube-mass spectrometry. J Food Sci. 2010;75:C710–C21.
129. Wampler B, Barringer SA. Volatile generation in bell peppers during frozen storage and thawing using selected ion flow tube-mass spectrometry (SIFT-MS). J Food Sci. 2012;77:C677–C83.
130. Pothakos V, Nyambi C, Zhang B-Y, et al. Spoilage potential of psychrotrophic lactic acid bacteria (LAB) species: *Leuconostoc gelidum* sbsp. *gasicomitatum* and *lactococcus piscium*, on sweet bell pepper (SBP) simulation medium under different gas compositions. Int J Food Microbiol. 2014;178:120–9.
131. Ozcan G, Barringer S. Effect of enzymes on strawberry volatiles during storage, at different ripeness level and in different cultivars and during eating. J Food Sci. 2011;76:C324–C33.
132. Hansanugrum A, Barringer SA. Effect of milk on the deodorization of malodorous breath after garlic ingestion. J Food Sci. 2010;75:C549–C58.
133. Munch R, Barringer SA. Deodorization of garlic breath volatiles by food and food components. J Food Sci. 2014;79:C526–C33.
134. Huang Y, Baringer SA. Alkyl pyrazines and other volatiles in cocoa liquors at pH 5 to pH 8, by selected ion flow tube mass-spectrometry (SIFT-MS). J Food Sci. 2010;75:C121–C7.
135. Huang Y, Barringer SA. Monitoring of cocoa volatiles produced during roasting by selected ion flow tube-mass spectrometry (SIFT-MS). J Food Sci. 2011;76:C279–C86.
136. Agila A, Barringer S. Effect of roasting conditions on color and volatile profile including HMF level in sweet almonds (*Prunus dulcis*). J Food Sci. 2012;77:C461–C8.
137. Agila A, Barringer S. Volatile profile of cashews (*Anacardium occidentale L.*) from different geographical origins during roasting. J Food Sci. 2011;76:C768–C74.
138. Agila A, Barringer S. Application of selected ion flow tube mass spectrometry coupled with chemometrics to study the effect of location and botanical origin on volatile profile of unifloral American honeys. J Food Chem. 2012;77:C1103–C8.
139. Agila A, Barringer S. Effect of adulteration versus storage on volatiles in unifloral honeys from different floral sources and locations. J Food Sci. 2013;78:C184–C91.
140. Langford V, Gray J, Foulkes B, et al. Application of selected ion flow tube-mass spectrometry to the characterization of monofloral New Zealand honeys. Agric Food Chem. 2012;60:6806–15.
141. Olivares A, Dryahina K, Navarro J, et al. Selected ion flow tube-mass spectrometry for absolute quantification of aroma compounds in the headspace of dry fermented sausages. Anal Chem. 2010;82:5819–29.
142. Flores M, Olivares A, Dryahina K, et al. Real time detection of aroma compounds in meat and meat products by SIFT-MS and comparison to conventional techniques (SPME-GC-MS). Current Anal Chem. 2013;9:622–30.
143. Olivares A, Dryahina K, Spanel P. Rapid detection of lipid oxidation in beef muscle packed under modified atmosphere by measuring volatile organic compounds using SIFT-MS. Food Chem. 2012;135:1801–8.
144. Noseda B, Ragaest P, Pauwels D, et al. Validation of selected ion flow tube mass spectrometry for fast quantification of volatile bases produced on Atlantic cod. J Agric Food Chem. 2010;58:5213–9.
145. Noseda B, Dewulf J, Goethals J, et al. Effect of food matrix and pH on the volatilization of bases in packed north Atlantic gray shrimp (*Crangon crangon*). Volatile bases in MAP fishery products. J Agric Food Chem. 2010;58:11864–9.
146. Noseda B, Islam MdT, Eriksson M et al. Microbiological spoilage of vacuum and modified atmospheric packaged Vietnamese *Pangasius hypophthalmus* fillets. Food Micobiol. 2012;30:408–19.

147. Noseda B, Goethals J, De Smedt L, et al. Effect of O_2-CO_2 enriched atmosphere on microbiological growth and volatile metabolite production in packaged cooked packed gray shrimp (*Crangon crangon*). Int J Food Microbiol. 2012;160:65–75.
148. Broekaert K, Noseda B, Heyndrickx M, et al. Volatile compounds associated with *Psychrobacter* spp. and *Pseudoalteromonas* spp. the dominant microbiota of brown shrimp (*Crangon crangon)* during aerobic storage. Int J Food Microbiol. 2013;166:487–93.
149. West R, Seetharaman K, Duizer LM. Whole grain macaroni: flavour interactions with sodium-reduced cheese sauce. Food Res Int. 2013;53:149–55.
150. Langford VS, Read CJ, Milligan DB, et al. Headspace analysis of Italian and New Zealand parmesan cheese. J Food Sci. 2012;77:C719–C26.
151. Taylor K, Wick C, Castada H, et al. Discrimination of Swiss cheese from 5 different factories by high impact volatile organic compound profiles determined by odor activity values using selected ion flow tube-mass spectrometry and odor threshold. J Food Sci. 2013;78:C1509–C15.
152. Castada HZ, Wick C, Taylor K, et al. Analysis of selected volatile organic compounds in split and non-split Swiss cheese samples using selected ion flow tube mass spectrometry (SIFT-MS). J Food Sci. 2014;79:C489–C98.
153. Spanel P, Smith D. Progress of SIFT-MS: breath analysis and other applications. Mass Spectrom Rev. 2011;30:236–67.
154. Spanel P, Smith D. Recent SIFT-MS studies of volatile compounds in physiology, medicine and cell biology. In: Amman A, Smith D, editors. Volatile biomarkers: non-invasive diagnosis in physiology and medicine. Oxford:Elsevier; 2013. p. 49–76.
155. Turner C. VOC analysis by SIFT-MS, GC-MS, and electronic nose for diagnosing disease. In: Amman A, Smith D, editors. Volatile biomarkers: non-invasive diagnosis in physiology and medicine. Oxford:Elsevier; 2013. p. 343–57.
156. Smith D, Spanel P, Herbig J, et al. Mass spectrometry for real-time quantitative breath analysis. J Breath Res. 2014;8:027101.
157. Smith D, Chippendale TWE, Dryahina K, et al. SIFT-MS analysis of nose-exhaled breath: mouth contamination and the influence of exercise. Curr Anal Chem. 2013;9:565–75.
158. Spanel P, Dryahina K, Smith D. The concentration distributions of some metabolites in the exhaled breath of young adults. J Breath Res. 2007;1:011001.
159. Gilchrist FJ, Bright-Thomas RJ, Jones AM, et al. Hydrogen cyanide concentrations in the breath of adult cystic fibrosis patients with and without *Pseudomonas aeruginosa* infection. J Breath res. 2013;7:026010.
160. Kumar S, Huang J, Abbassi-Ghadi N, et al. Selected ion flow tube mass spectrometry analysis of exhaled breath for volatile organic compound profiling of esophago-gastric cancer. Anal Chem. 2013;85:6121–8.
161. Huang J, Kumar S, Singanayagam A, et al. SIFT-MS analysis of volatile metabolites in urine headspace for the profiling of gastro-esophageal cancer. Anal Chem. 2013;85: 3409–16.
162. Dryahina K, Spanel P, Pospisilova V, et al. Quantification of pentane in exhaled breath, a potential biomarker of bowel disease, using selected ion flow tube mass spectrometry. Rapid Commun Mass Spectrom. 2013;27:1983–92.
163. Davies SJ, Spanel P, Smith D. Breath analysis of ammonia, volatile organic compounds and deuterated water vapour in chronic kidney disease during dialysis. Bioanalysis. 2014;6:843–57.
164. Sturney SC, Storer MK, Shaw GM, et al. Off-line breath acetone analysis in critical illness. J Breath Res. 2013;7:037102.
165. Hanouneh IA, Zein NN, Cikach F, et al. The breath prints in patients with liver disease identify novel biomarkers in alcoholic hepatitis. Clin Gastroenterol Hepatol. 2013;12:516–23.
166. Zeft A, Costanzo D, Alkhouri N, et al. Metabolomic analysis of breath volatile organic compounds reveals unique breathprints in children with juvenile idiopathic arthritis. Arthritis Rheumatol. 2014;66:S159.

167. Alkhouri N, Eng K, Cikach F, et al. Breathprints of childhood obesity: changes in volatile organic compounds in obese children compared with lean controls. Pediat Obesity. 2015;10:23–9
168. Cikach FS, Tonelli AR, Barnes J, et al. Breath analysis in pulmonary arterial hypertension. Chest. 2014;145:551–8.
169. Navaneethan U, Parsi MA, Gutierrez NG, et al. Volatile organic compounds in bile can diagnose malignant biliary strictures in the setting of pancreatic cancer: a preliminary observation. Gastrointest Endosc. 2014;80:1038–45.
170. Cody RB, Laramee JA. Versatile new ion source for the analysis of materials in open air under ambient conditions. Anal Chem. 2005;77:2297–302.
171. Green FM, Salter TL, Stokes P, et al. Ambient mass spectrometry: advances and applications in forensics. Surf Interface Anal. 2009;42:347–57.
172. Alberici RM, Simas RC, Sanvido GB, et al. Ambient mass spectrometry: bringing MS into the "real world". Anal Bioanal Chem. 2010;398:265–94.
173. Gross JH. Direct analysis in real time–a critical review on DART-MS. Anal Bioanal Chem. 2014;406:63–80.
174. Klee S, Derpmann V, Wisdorf W, et al. Are clusters important in understanding the mechanisms in atmospheric pressure ionization? Part 1: reagent generation and chemical control of ion populations. J Am Soc Mass Spectrom. 2014;25:1310–21.
175. Vaclavik L, Cajka T, Hrbek V. Ambient mass spectrometry employing direct analysis in real time (DART) ion source for olive oil quality and authenticity assessment. Anal Chim Acta. 2009;645:56–63.
176. Vaclavik L, Belkova B, Reblova Z, et al. Rapid monitoring of heat-accelerated reactions in vegetable oils using direct analysis in real time ionization coupled with high resolution mass spectrometry. Food Chem. 2013;138:2312–20.
177. Haefliger OP, Jeckelmann N. Direct mass spectrometric analysis of flavors and fragrances in real applications using DART. Rapid Commun Mass Spectrom. 2007;21:1361–6.
178. Zachariasova M, Cajka T, Godula M, et al. Analysis of multiple mycotoxins in beer employing (ultra)-high-resolution mass spectrometry. Rapid Commun Mass Spectrom. 2010;24:3357–67.
179. Block E, Dane AJ, Thomas S, et al. Applications of direct analysis in real time mass spectrometry (DART-MS) in allium chemistry. 2-propenesulfenic and 2-propenesufinic acids, diallyl trisulfane S-oxide, and other reactive sulfur compounds from crushed garlic and other alliums. J Agric Food Chem. 2010;58:4617–25.
180. Vaclavik L, Hrbek V, Cajka T, et al. Authentication of animal fats using direct analysis in real time (DART) ionization-mass spectrometry and chemometric tools. J Agric Food Chem. 2011;59:5919–26.
181. Hrbek V, Vaclavik L, Elich O, et al. Authentication of milk and milk-based foods by direct analysis in real time ionization-high resolution mass spectrometry (DART-HRMS) technique: a critical assessment. Food Control. 2014;36:138–45.
182. Wang Y, Liu L, Liu S. Identification of saccharides by using direct analysis in real time (DART) mass spectrometry. Int J Mass Spectrom. 2014;357:51–7.
183. Bai Y, Zhang J, Hiu H. Direct analysis in real time mass spectrometry combined with single-drop liquid-liquid-microextraction for rapid analysis of multiple phytohormones in fruit juice. Anal Bioanal Chem. 2012;403:2307–14.
184. Krtkova V, Schulzova V, Lacina O, et al. Analytical strategies for controlling polysorbate-based nanomicelles in fruit juice. Anal Bioanal Chem. 2014;406:3909–18.
185. Busman M, Liu J, Zhong H, et al. Determination of the aflatoxin AFB1 from corn by direct analysis in real time-mass spectrometry (DART-MS). Food Addit Contam A. 2014;31:932–9.
186. Hajslova J, Cajka T, Vaclavik L. Challenging applications offered by direct analysis in real time (DART) in food-quality and safety analysis. Trends Anal Chem. 2011;30:204–18.
187. DART book of food abstracts. www.ionsense.com/pdfs/DART_Food_Book2014_web.pdf.

188. Li Y. Applications of a confined DART (direct analysis in real time) ion source for online *in vivo* analysis of human breath. Anal Methods. 2013;5:6933–40.
189. Jones RW, Cody RB, McClelland JF. Differentiating writing inks using direct analysis in real time mass spectrometry. J Forensic Sci. 2006;51:915–8.
190. Sisco E, Dake J, Bridge C. Screening for trace explosives by AccuTOF-DART: an in-depth validation study. Forensic Sci Int. 2013;232:160–8.
191. Fernandez FM, Cody RB, Green MD, et al. Characterization of solid counterfeit drug samples by desorption electrospray ionization and direct-analysis-in-real-time coupled to time-of-flight mass spectrometry. ChemMedChem. 2006;1:702–5.
192. Cody RB. Observation of molecular ions and analysis of nonpolar compounds with the direct analysis in real time ion source. Anal Chem. 2009;81:1101–7.
193. Paseiro-Cerrato R, Noonan GO, Begley TH. Development of a rapid screening method to determine primary aromatic amines in kitchen utensils using direct analysis in real time mass spectrometry (DART-MS). Food Addit Contam A. 2014;31:537–45.
194. Manova RK, Joshi S, Debrassi A, et al. Ambient surface analysis of organic monolayers using direct analysis in real time orbitrap mass spectrometry. Anal Chem. 2014;86:2403–11.
195. Takats Z, Wiseman JM, Gologan B, et al. Mass spectrometer sampling under ambient conditions with desorption electrospray ionization. Science. 2004;306:471–3.
196. Cooks GR, Ouyan Z, Takats Z, et al. Ambient mass spectrometry. Science. 2006;311:1566–70.
197. Venter A, Sojka PE, Cooks RG. Droplet dynamics and ionization mechanisms in desorption electrospray ionization mass spectrometry. Anal Chem. 2006;78:8549–55.
198. Wu C, Dill AD, Eberlin LS, et al. (2013) Mass spectrometry imaging under ambient conditions. Mass Spec Rev 32:218–43.
199. Bennet RV, Gamage CM, Galhena AS, et al. Contrast-enhanced differential mobility-desorption electrospray ionization-mass spectrometry imaging of biological tissue. Anal Chem. 2014;86:3756–63.
200. Hemalatha RG, Pradeep T. Understanding the molecular signatures in leaves and flowers by desorption electrospray ionization mass spectrometry (DESI-MS) imaging. J Agric Food Chem. 2013;61:7477–87.
201. Lanekoff I, Burnum-Johnson K, Thomas M, et al. High-speed tandem mass spectrometric in situ imaging by nanospray desorption electrospray ionization mass spectrometry. Anal Chem. 2013;85:9596–603.
202. Lanekoff I, Thomas M, Carson JP, et al. Imaging nicotine in rat brain tissue by use of nanospray desorption electrospray ionization mass spectrometry. Anal Chem. 2013;85:882–9.
203. Cotle-Rodriguez I, Takats Z, Talaty N, et al. Desorption electrospray ionization of explosives on surfaces: sensitivity and selectivity enhancement by reactive desorption electrospray ionization. Anal Chem. 2005;77:6755–64.
204. Cotle-Rodriguez I, Hernandez-Soto HC, Chen H, et al. In situ trace detection of peroxide explosives by desorption electrospray ionization and desorption atmospheric pressure chemical ionization. Anal Chem. 2008;80:1512–9.
205. Harper JD, Charipar NA, Mulligan CC, et al. Low-temperature plasma probe for ambient desorption ionization. Anal Chem. 2008;80:9097–104.
206. Ifa DR, Gumaelius LM, Eberlin LS, et al. (2007) Forensic analysis of inks by imaging desorption electrospray ionization (DESI) mass spectrometry. Analyst 132:461–7.
207. Ifa DR, Manicke NE, Dill AL, et al. Latent fingerprint chemical imaging by mass spectrometry. Science. 2008;321:805–5.
208. Nyadong L, Green MD, De Jesus VY, et al. Reactive desorption electrospray ionization linear ion trap mass spectrometry of latest-generation counterfeit antimalarials via noncovalent complex formation. Anal Chem. 2007;79:2150–7.
209. Ferreira CR, Wu L, Vogt G, et al. Fiducial markers for distribution of drug and excipient on tablet surfaces by multimodel desorption electrospray ionization-mass spectrometry (DESI-MS) imaging. Anal Lett. 2014;47:91–101.

210. Nielen MWF, Hooijerink H, Zomer P, et al. Desorption electrospray ionization mass spectrometry in the analysis of chemical food contaminants in food. Trends Anal Chem. 2011;30:165–80.
211. Dill AL, Eberlin LS, Costa AB, et al. (2011) Multivariate statistical identification of human bladder carcinomas using ambient ionization imaging mass spectrometry. Chem Eur J 17:2897–902.
212. Eberlin LS, Tibshirani RJ, Jhang J, et al. (2014) Molecular assessment of surgical-resection margins of gastric cancer by mass-spectrometric imaging. Proc Nat Acad Sci 111:2436–41.
213. D'Alvise J, Mortenson R, Hansen SH, et al. Detection of follicular transport of lidocaine and metabolism in adipose tissue in pig ear skin by DESI mass spectrometry imaging. Anal Bioanal Chem. 2014;406:3735–42.
214. Gonzalez-Serrano AF, Felipe A, Pirro V, et al. Desorption electrospray ionization mass spectrometry reveals lipid metabolism of individual oocytes and embryos. PLoS One. 2013;8:e74981.
215. Biasoli F, Yeretzian C, Mark TD, et al. Direct injection mass spectrometry adds the time-dimension to (B)VOC analysis. Trends Anal Chem. 2011;30:1003–17.
216. Pavon JL, Sanchez MdN, Pinto CG, et al. Strategies for qualitative and quantitative analyses with mass spectrometry-based electronic noses. Trends Anal Chem. 2006;25:257–66.
217. Cozzolino D, Smyth HE, Cynka W, et al. Use of direct head-space mass spectrometry coupled with chemometrics to predict aroma properties in Australian Riesling wine. Anal Chim Acta. 2008;621:2–7.
218. Van Leeuwen S, Hendriksen L, Karst U. Determination of aldehydes and ketones using derivatization with 2,4-dinitrophenylhydrazine and liquid chromatography-atmospheric pressure photoionization-mass spectrometry. J Chromatogr A. 2004;1058:107–12.
219. Wang H, Xie W, Chen M, et al. Determination of hazardous volatile organic compounds in the Hoffman list by ion-molecule reaction mass spectrometry. Rapid Commun Mass Spectrom. 2012;26:1841–8.
220. Smith D, Spanel P. Mass spectrometry for real-time quantitative breath sampling. J Breath Res. 2014;8:23 (027101).
221. Langford VS, Gray J, Maclagan RGAR, et al. Real-time measurements of nitrosamines in air. Int J Mass Spectrom. 2015;377:490–5.
222. Blake RS, Wyche KP, Ellis AM, et al. Chemical ionization reaction time-of -flight mass spectrometry: multi-reagent analysis for determination of trace gas composition. Int J Mass Spectrom. 2006;254:85–93.
223. Spanel P, Smith D. SIFT studies of the reactions of H_3O^+, $NO^+$and $O_2^+$with a series of alcohols and ketones. Int J Mass Spectrom Ion Proc. 1997;165/166:25–37.
224. Spanel P, Smith D. SIFT studies of the reactions of H_3O^+, $NO^+$and $O_2^+$with a series of volatile carboxylic acids and esters. Int J Mass Spectrom Ion Proc. 1998;172:137–47.
225. Sulzer P, Hartungen E, Hanel G, et al. A proton transfer reaction-quadrupole interface time-of-flight mass spectrometer (PTR-QiTOF): high speed due to extreme sensitivity. Int J Mass Spectrom. 2014; Dx.doi.org/10.1016/j/ijms.2014.05.004.

Chapter 9
Summary and Perspectives

W. M. A. Niessen

9.1 Introduction

From a chemistry point of view, the two most interesting processes in mass spectrometry (MS) involve the analyte ionization and the fragmentation of analyte ions. Analyte ionization essentially enables the mass analysis, which is based on the separation of analyte ions according to their *m/z* and subsequent detection. The wide variety of MS ionization techniques can be classified in methods that achieve analyte ionization by abstraction (or attachment) of electrons, e.g., electron ionization (EI) and electron-capture negative ionization, and methods that achieve analyte ionization by attachment (or abstraction) of ions to (or from) the neutral analyte molecules. Analyte ionization by EI generally results in excited molecular ions $M^{+\bullet}$ (odd-electron ions), which undergo rapid in-source fragmentation. The H^+-ion-attachment reaction is widely used in the ionization methods that involve protonation (proton attachment) to generate protonated molecules $[M+H]^+$, In negative-ion mode, a similar process involves deprotonation (proton abstraction) to generate deprotonated molecules $[M–H]^-$. Both the protonated and the deprotonated molecules are even-electron ions. This is observed for soft-ionization techniques such as chemical ionization (CI), field ionization (FI), field desorption (FDI), electrospray ionization (ESI), atmospheric-pressure chemical ionization (APCI), matrix-assisted laser desorption ionization (MALDI), as well as techniques that have become obsolete, such as fast-atom bombardment (FAB) and thermospray ionization (TSI).

This book explores an alternative analyte ionization strategy involving attachment of alkali-metal ions such as Li^+, Na^+, and K^+ as well as, to some extent, other metal ions. Theoretical, instrumental, and practical aspects of ion-attachment mass spectrometry (IAMS) have been thoroughly discussed in the previous chapters. Fujii [1–3] introduced the attachment of alkali-metal ions as a soft ionization technique to overcome problems that may arise due to the extensive fragmentation of some

W. M. A. Niessen (✉)
Hyphen MassSpec, de Wetstraat 8, 2332 XT Leiden, The Netherlands
e-mail: mail@hyphenms.nl

T. Fujii (ed.), *Ion/Molecule Attachment Reactions: Mass Spectrometry*,
DOI 10.1007/978-1-4899-7588-1_9

analytes in EI. Unlike in EI, analyte ionization by IAMS is nondissociative, thus readily enabling determination of the molecular mass of labile analyte molecules. Subsequently, by the use of ion-dissociation techniques and tandem mass spectrometry (MS–MS), these ion-attachment products, such as $[M+Li]^+$ or $[M+Na]^+$, can be fragmented and structural information can be obtained. Quite often, this even results in different fragmentation pathways compared to the radical-site directed fragmentations observed in EI and/or the fragmentation of (de)protonated molecules, thus providing additional tools in structure elucidation.

Chapter 1 provides a general introduction in and overview of IAMS and indicates typical application areas of IAMS. It demonstrates the important role of IAMS in fundamental studies of chemical reactions in the gas phase of a mass spectrometer, in terms of both ionization and fragmentation. It also highlights the relevance of IAMS in solving analytical problems. The next two chapters are mainly devoted to theoretical aspects. Chapter 2 discusses the typical termolecular reactions involved in ion attachment in the gas phase as well as other types of ion attachment reactions. It also pays attention to the influence ion attachment has on fragmentation of ions and discusses ion-dissociation techniques in use in MS. Chapter 3 addresses ion attachment reactions from a theoretical point of view. It introduces theoretical models and computational methods in order to study aspects of ion attachment related to site specificity, influence on ion structures, etc. Chapter 4 provides a concise overview of MS technology, describing general instrumentation available, with some emphasis on commercially available instruments. IAMS can be achieved via different ionization techniques, which can be classified according to the physical state of the sample. Chapter 5 discusses analyte ionization by gas-phase ion-attachment reactions. Attention is paid to both fundamental and instrumental aspects, while an important focus is on the applications of gas-phase IAMS in various areas of interest. Chapter 6 is devoted to dedicated instrumental setups for the study of gas-phase ion-attachment reactions, where IAMS is mostly applied in combination with other sample-introduction technologies, such as aerosol generators, pyrolysis, as well as various types of ion-trapping devices such as ion-traps and ion cyclotron resonance (ICR) cells. Chapter 7 addresses ion-attachment phenomena in analyte ionization in the condensed phase (liquids or solids). By far, the majority of MS users are familiar with this type of ion attachment reactions, as these involve the formation of the sodiated and potassiated molecules, that is: $[M+Na]^+$ and $[M+K]^+$, frequently observed in the currently most widely used ionization techniques in the MS analysis of polar molecules, i.e., ESI and MALDI. Some of these MS users experience the condensed-phase ion-attachment reactions as disadvantageous, as they result in a, sometimes sample-dependent, distribution of charge over several analyte-related ionic species, which may compromise efficient and sensitive detection using targeted selected reaction monitoring (SRM) strategies. Other users, especially those involved in structure elucidation rather than routine quantitative analysis, appraise the benefits of the ion-attachment reactions in helping them to solve their analytical problems. For some classes of analytes, ion-attachment reactions present the most important way to actually achieve analyte ionization. In still other cases, ion-attachment reactions with specific metal ions opens specific fragmentation pathways to strengthen the structure elucidation of these compound classes, where difficulties are experienced, for instance, due to isomerism issues. In recent years, technologies have been developed to perform direct MS analysis of samples, that is, without extensive sample pretreatment and/or prior chromatographic separation. Important developments in this area are discussed in Chap. 8, with considerable emphasis on these technologies that gained wide acceptance and application in specific application areas.

Next to the discussion on principles, mechanisms and fundamental aspects, the previous chapters provide information on the applications of IAMS in their specific areas of interest. Chapter 9 is intended to provide a brief perspective from an application point of view and to indicate the IAMS methods and technologies that are especially useful in selected application areas. As such, it addresses the question that many application-oriented MS users may have: what is the usefulness of IAMS in solving the analytical challenges of today and what role IAMS may play in the future.

Perhaps, one of the clearest facts demonstrating the importance of IAMS in real-life applications is the recent introduction of a commercial instrument for Li^+-IAMS, the so-called IA-Lab [4]. The instrument features an M-400-QA-M quadrupole mass spectrometer (Canon Anelva Corporation, Tokyo, Japan) with a model L-241G-IA Li^+-cationization source as well as a conventional EI source for complementary fragment information.

9.2 Studying Dynamic Chemical Systems

IAMS and its associated technologies can be used to study gas-phase processes, e.g., to determine and characterize molecular constituents as well as intermediate free radicals in dynamic chemical systems such as flames, combustion, plasmas, and discharges. Over the years, this has been especially challenging to MS, because remote sampling without interfering or destroying the intrinsic nature of such dynamic systems is quite challenging. The advantages of detecting free radicals by IAMS comprise the formation of intact molecule-related ions as a guide to radical identity. The ion–molecule reactions involved actually provide high sensitivity. In contrast to EI, IAMS is readily applied in high-pressure conditions, and is thus more compatible to the atmospheric-pressure conditions of the dynamic chemical systems studied. It enables direct and continuous measurements of any species in such a dynamic system. As such, IAMS provides means to characterize free radicals formed in plasmas, flames, and combustion processes as well as radicals produced during the volatilization and pyrolysis, e.g., flash pyrolysis coupled with supersonic free jet technique. Examples of this are discussed in detail in Chaps. 5 and 6. Interesting results have been obtained in the study of the microwave discharge in gas systems such as CH_4, C_2H_4, C_2H_2, CH_4/H_2, CH_4/O_2, or CH_4/N_2. It enables qualitative analysis of the compositions of hydrocarbon radicals in discharges. In a microwave discharge with CH_4/O_2, for instance, species with molecular formulae such as $C_nH_{2n}O$ and $C_nH_{2n}O_2$, thus presumably aldehydes and/or ketones, and $C_nH_{2n+2}O$, presumably alcohols, are observed with *n* up to 8 and even higher [5]. The compounds involved in various chemical vapor deposition (CVD) techniques have also been studied in more detail using IAMS, such as Diamond film growth. Li^+-IAMS allows the detection of the thermally labile copper precursor known as CupraSelect, a complex of Cu with hexafluoroacetylacetonate and trimethylvinylsilane, which is used in the CVD of Cu in the production of integrated circuit boards [6].

Similarly, gas-phase processes in the atmosphere, ionosphere, or interstellar environments may be studied, thus providing contributions to an understanding of important issues such as acid rain, smog, and the chemistry of the upper atmosphere. Characterization of incineration products of polyethylene glycols (PEGs) may serve as an example. PEGs are widely used in a wide variety of industrial products, such as detergents, surfactants, packaging bags, cosmetics, ointments, and high-energy-density batteries. The thermal decomposition of PEGs results in, for instance, highly reactive organic peroxides, such as CH3OOH and HOCH2OOH [7]. Li^+-IAMS provides characteristic profiles, showing both the individual components, such as formaldehyde and organic peroxides, and the distribution of the pyrolysis products. The formation of peroxides in PEG incineration may have important environmental implications [8]. The EU-Directive on the Restriction of Hazardous Substances (RoHS) has stimulated the analysis of flame-retarding compounds such as polybrominated biphenyls (PBB) and polybrominated diphenyl ethers (PBDEs). Li^+-IAMS in combination with a direct inlet probe has been applied to analyze PBBs and PBDEs in plastic materials, even up to deca-BDE. The compounds are readily identified from their *m/z* and isotope pattern [9]. To some extent, this type of studies again involves the study of ion–molecule reactions, which may even be extended towards the upper regions of the Earth's and other planets' atmospheres. An attractive example is the use of IAMS in simulation of the atmospheric chemistry of Titan, a satellite of Saturn, using microwave discharges of a 90% N_2 in CH_4 gas mixture at 27 mbar [10].

In quite a number of these applications, IAMS actually broadens the potential application of MS, enabling the detection of intermediary free radical species, of atmospheric and interstellar species, of environmentally important species, and the identification of unfamiliar or unstable species. Ample examples of this have been summarized in Chap. 5.

9.3 Analysis of Drugs

An early account on IAMS for compounds with biological relevance is the thermal desorption ion source described by Röllgen [11]. Electrically heated metal surfaces doped with alkali-cation salts and compounds such as crown ether, sugars, drugs, peptides, steroids, and nucleosides were used to generate $[M+Na]^+$ at surface temperatures way below the threshold temperatures for thermionic emission.

Alkali$^+$-cationization has subsequently been observed for a wide range of compounds studied with gas-phase as well as condensed-phase (liquids or solids) ionization techniques such as FDI, FAB, ESI, and MALDI. Thus, next to protonation, ion-attachment processes with sample-related Na^+- or K^+-ions may contribute to the ionization of polar analytes in in these condensed-phase ionization techniques. Residual alkali-metal ions in solvents (typically < 1 mM) may also play a role. The mechanisms behind the formation of Alkali$^+$-cationized molecules is still unclear, both condensed-phase and gas-phase processes may actually contributed to their formation. Sodiated and potassiated molecules, often indicated as so-called adduct ions, are especially observed for compounds containing multiple vicinal hydroxyl and/or carboxyl groups. As such, $[M+Na]^+$ is readily observed for, for instance, sugar-containing analytes, but for many other small-molecule analytes as well, including lipids, (oligo)nucleotides, and peptides. One should be aware that apparent $[M+Na]^+$-ions may also be observed if compounds with acidic functional groups, e.g., carboxylic acids, sulfonates, sulfates, or phosphates, are analyzed. In that case, exchange of the acidic hydrogen for sodium may occur, thus resulting in the formation of protonated sodium salts, e.g., $[(M–H+Na)+H]^+$, $[(M–2H+2Na)+H]^+$, or sodiated sodium salts, e.g., $[(M–H+Na)+Na]^+$. From their *m/z* values, these H^+/Na^+-exchange products cannot be discriminated from Na^+-adduct ions $[M+Na]^+$. This behavior is frequently observed for peptides and oligonucleotides, carrying multiple carboxylic acid and phosphate groups, respectively. Ion-attachment processes, both adduct-ion formation and H^+/Alkali$^+$-exchange, occur in all condensed-phase ionization techniques, including in ESI-MS and MALDI-MS, which are often based on the formation of preformed ions in the condensed phase [12, 13]. In negative-ion mode, ion-attachment processes and/or adduct-ion formation with solvent constituents may occur as well, resulting in the observation of ions such as $[M+HCOO]^-$, $[M+CH_3COO]^-$, or $[M+Cl]^-$, depending on the additives used.

Gas-phase Li^+-IAMS was used in a temperature-programmed heating probe for evolved gas analysis (EGA) for rapid monitoring of the anticancer drug $Ti(C_5H_5)_2Cl_2$, used in the treatment of leukemia, in a hospital environment [14]. Na^+- and/or K^+-cationization is frequently observed for many classes of drugs, such as simvastatin and related substances, paclitaxel, gingkolides, antibiotics and coccidiostats, steroids, carotenoids, to quote some of the examples discussed in Chap. 7. In most cases, one tries to avoid Na^+- and/or K^+-cationization, especially in routine quantitative analysis involving SRM on tandem-quadrupole instrument.

9.4 Biochemical and Biological Applications: Biomacromolecules

From a fundamental point of view, the study of gas-phase and condensed-phase interactions between alkali-metal ions and biologically active molecules is of great interest, as these interactions relate to chemical and biological processes occurring in the heterogeneous regions. This can help to study processes such as ion solvation, catalysis, transport through membranes, affinity of active compounds toward receptors, and antibiotic activity. Alkali-metal ions are involved in biological processes such as enzyme regulation, transmission of cellular signals (Na^+ or K^+ ion channels), and transport through transmembrane channels. They also play key roles in enzymatic activity, cellular metabolism, and structural stabilization, e.g., protein folding (see Chap. 3).

Na^+-cationization plays an important role in the analysis of glycosides, peptides, proteins, nucleotides, lipids, and oligosaccharides. Quite often, it can be demonstrated that different fragmentation behavior can be observed for $Alkali^+$-cationized molecules compared to protonated molecules. This has been frequently exploited. A good examples is the possibility to induce charge-remote fragmentation in $[(M–H+Li)+Li]^+$-ions of fatty acids, enabling determination of double-bond position [15]. Subsequently, Li^+-cationization has become an important tool in the characterization of many classes of lipids, including for instance phospholipids [16]. Different ionization strategies have been applied in the characterization of oligosaccharides and glycans, nowadays involving mainly MALDI [17] and ESI [18]. Amino-acid sequencing of peptides is mainly based on the fragmentation of (multiple-charge) protonated peptides, leading to characteristic protonated oxazolone derivatives and protonated peptides as fragment ions [19]. However, *C*-terminal amino-acid residue losses are observed upon fragmentation of Li^+- or Na^+-cationized peptides [20].

9.5 Environmental Analysis of Atmospheric and Water Pollution

As discussed in Sect. 9.2, IAMS has been applied in the study of atmospheric processes. As such, IAMS can contribute to an understanding of acid rain, smog, and other air contamination processes in the atmosphere. Important additional tools in this respects are the aerosol mass spectrometer (Sect. 6.2) and the infrared image furnace (IIF, Sect. 6.4.3). The latter system was involved in studying combustion of polycarbonates, which may result in the emission of the endocrine disruptor bisphenol A. Li^+-IAMS enabled the detection of the bisphenol A biradical [21]. Similarly, EGA-IAMS has been used to study the pyrolysis of polyacrylamide, which among others results in the suspected carcinogenic monomeric acrylamide [22].

The occurrence of highly-reactive and labile peroxides and hydroperoxides in the atmosphere is another issue of concern in environmental analysis. Peroxides are used in polymer industry as a source of free radicals. Conventional EI-MS of peroxides results in extensive fragmentation, whereas IAMS, e.g., in EGA–IAMS (Sect. 6.4.2) enables molecular profiling of peroxides. Proton-transfer reaction MS (PTR-MS) technique has also been used for kinetics studies of organic peroxy radicals such as $CH_3O_2^•$, $CH_3CH_2O_2^•$, and $C_3H_7O_2^•$ [23, 24].

Portable MS systems based on IAMS have been developed. Such a system enables the real-time analysis in applications such as emission control and monitoring of volatile organic compounds (VOCs) in urban and rural environments, emissions from construction material and furniture, and emissions from industrial facilities and industrial fermentation and food production processes [25]. Further uses also include the monitoring of catalytic processes, cigarette smoke, and breath. Application in continuous process monitoring and in relation to homeland security issues can also be readily foreseen.

9.6 Food-Related Applications

The feature of IAMS to generate intact molecule ionic species is of benefit in any MS applications, especially involving VOCs, where conventional EI-MS results in extensive fragmentation and often loss of the intact molecule information. In food analysis, this certainly applies to the analysis of flavors, fragrances, and aroma compounds. GC–IAMS can be a powerful tool in that respect. Several examples have been discussed in Chap. 8, involving proton-transfer MS (PTR-MS) and selected-ion flow tube MS (SIFT-MS).

Most widely applied in the study of food-related compounds, is the Li^+- or Na^+-cationization in the IAMS analysis of lipids and related substances, which is discussed in considerable detail in Sect. 7.5.4. It seems that in this area Li^+- or Na^+-cationization is especially important in detailed characterization of lipids, e.g., with respect to determination of double-bond positions in fatty acids, and given the fact that often complementary structural information is obtained compared to the information obtained from the fragmentation of protonated or deprotonated molecules. This is clearly demonstrated for phospholipids [16].

9.7 Polymer Characterization

In previous sections of this chapter, the potential of IAMS in the characterization of polymeric materials has already been discussed, although mainly attention was given to monitoring of low-molecular-weight compounds resulting from the combustion or pyrolysis of polymers. Over the years, IAMS has also been frequently used to enable the MS analysis of intact synthetic oligomers and polymers, e.g., by addition of Li^+ or Na^+ to condensed-phase sample preparation to be analyzed using ionization techniques such as FDI, FAB, ^{252}Cf plasma desorption (PDI), and secondary ion mass spectrometry (SIMS). More recently, alkali-metal adduct ion formation is applied in the MALDI-MS [26] and ESI-MS analysis of synthetic polymers [27, 28]. ESI-MS is not as extensively used as MALDI-MS, because of the more complex spectra obtained in ESI-MS, where charge state distributions may overlap chain length distributions.

Another interesting development in the field of synthetic polymer characterization is the emergence of ambient-pressure desorption ionization techniques such as desorption electrospray ionization mass spectrometry (DESI-MS) and many related techniques [29], involving either liquid extraction, thermal desorption, or laser ablation as the important desorption/extraction step from the condensed phase, which takes place in conjunction with the analyte ionization, either by protonation or alkali-metal ion attachment. This field has recently been reviewed [30]. The ultrahigh resolution of an Orbitrap mass spectrometer may help in the data interpretation of ion envelopes of multiple-charge ions generated in DESI-MS of synthetic polymers [31].

9.8 Miscellaneous Applications

IAMS can also be used in the nonmainstream application areas. Examples have been provided in the previous chapters of the book. However, an interesting development in condensed-phase IAMS has been given little attention, and that is a technique called "coordination ionspray" (CIS-MS) [32]. In this technique, cationization agents are applied in the ESI-MS analysis of nonpolar analytes to induce coordination and thereby charging. Characteristic early examples involve Ag^+ addition to achieve ionization of arenes, (poly)olefins, and carotenoids, and Pd^{2+}-addition for the ionization of estrogens and fat-soluble vitamins (A, D, and E). Similarly, CIS-MS was applied in the analysis of, for instance, peroxidation products of cholesterol linoleate and cholesterol arachidonate [33], and triglycerides in vegetable oils, both using Ag^+-IAMS [34].

9.9 Future Perspectives

IAMS is generally not considered as a separate technique in MS. Quite often, one considers Alkali$^+$-cationization as a byproduct of what is actually aimed at in analyte ionization: the formation of a protonated molecule. However, the chapters of this book demonstrate that there in fact is more to it. Even in ionization techniques such as ESI and MALDI, where one most tries to avoid Alkali$^+$-cationization as much as possible, one must admit that some especially identification problems are better solved by fragmenting the Alkali$^+$-cationized molecule rather than the protonated. It is certainly true that fragmentation of the Alkali$^+$-cationized molecules often provides additional information, that cannot be obtained from fragmenting the protonated molecule. In addition, analyte cationization by other metal ions can show additional features as well. In this context, it might actually be worth investigating the other (instrumental) ways the ionization can be directed towards either protonated or for instance lithiated molecules, either by liquid-phase or gas-phase processes.

Apart from the fragmentation and structure elucidation point of view, the chapters of the book readily demonstrate that Alkali$^+$-cationization in IAMS enables the ionization of various classes of molecules that cannot be ionized via (de)protonation, or the ionization by Alkali$^+$-cationization is (much) more efficient than by protonation. This is not only true for the condensed-phase ionization techniques, which most researchers apply most frequently today, but also true for gas-phase ionization, e.g., in combination with gas chromatography (GC) or other gas-phase analyte introduction techniques. Alkali$^+$-cationization of constituents of flames, discharges, etc., be it molecules or radicals, is a clear example of the potential of IAMS in a wide variety of important application areas of MS.

The current rapid introduction of tandem mass spectrometers for GC–MS applications in a wide variety of application areas come with an additional promise. The commercial available Li^+-cationization source may be fitted onto these instrument. As in this way, intact $[M+Li]^+$-ions could be generated for a wide variety of analyte classes and with excellent ionization efficiency, the combination of GC–IAMS–MS–MS might hold great promise for both routine targeted quantitative analysis using SRM and advanced structure elucidation of unknowns. This is certainly worth paying far more attention to.

This book provides an extensive overview of the potential of IAMS. The interesting technology just awaits more interest from the scientific community to deepen the application.

References

1. Fujii T, Ogura M, Jimba H. Chemical ionization mass spectrometry with lithium ion attachment to the molecule. Anal Chem. 1989;61:1026–9.
2. Fujii T. Alkali-metal ion/molecule association reactions and their applications to mass spectrometry. Mass Spectrom Rev. 2000;19:111–38.
3. Sablier M, Fujii T. Mass spectrometry of free radicals. Chem Rev. 2002;102:2855–924.
4. C & V Technix Co., Ltd. http://c-vtechnix.com/. Accessed 14 April 2015.
5. Fujii T, Syouji K. Mass spectrometric studies of the neutral and ionic products in a CH_4/O_2 microwave discharge plasma. J Phys Chem. 1993;97:11380–4.
6. Fujii T, Arulmozhiraja S, Nakamura M, Shiokawa Y. Chemistry of Cu deposition by Cu(hfac)(tmvs) monitored by Li^+ ion attachment mass spectrometry. J Appl Phys. 2006;100:084912.
7. Kitahara Y, Takahashi S, Fujii T. Thermal analysis of polyethylene glycol: evolved gas analysis with ion attachment mass spectrometry. Chemosphere. 2012;88:663–9.
8. Hua W, Chen ZM, Jie CY, Kondo Y, Hofzumahaus A, Takegawa N, Chang CC, Lu KD, Miyazaki Y, Kita K, Wang HL, Zhang YH, Hu M. Atmospheric hydrogen peroxide and organic hydroperoxides during PRIDE-PRD'06, China: their concentration, formation mechanism and contribution to secondary aerosols. Atmos Chem Phys. 2008;8:6755–73.
9. Sato Y, Oki M, Kondo A, Takenaka M, Satake H. Rapid analysis of polybrominated diphenyl ethers by ion attachment mass spectrometry. Anal Methods. 2010;2:701–6.
10. Fujii T, Arai N. Analysis of N-containing hydrocarbon species produced by CH_4/N_2 microwave discharge. Astrophys J. 1999;519:858–63.
11. Stoll R, Rollgen FW. Thermal desorption of quasimolecular ions. Org Mass Spectrom. 1981;16:72–5.
12. Rollgen FW, Borchers F, Giessmann U, Levsen K. Collisional activation of ions formed by $[Li]^+$ ion attachment. Org Mass Spectrom. 1977;12:541–3.
13. Vestal ML. Ionization techniques for nonvolatile molecules. Mass Spectrom Rev. 1983;2:447–80.

14. Takahashi S, Suga T, Kitahara Y, Fujii T. Evolved gas analysis of $Ti(C_5H_5)_2Cl_2$ by means of Li^+ ion attachment mass spectrometry. J Phys Chem A. 2012;116:865–9.
15. Adams J, Gross ML. Energy requirement for remote charge site ion decompositions and structural information from collisional activation of alkali metal cationized fatty alcohols. J Am Chem Soc. 1986;108:6915–21.
16. Hsu FF, Turk J. Electrospray ionization with low-energy collisionally activated dissociation tandem mass spectrometry of glycerophospholipids: mechanisms of fragmentation and structural characterization. J Chromatogr B. 2009;877:2673–95.
17. Harvey DJ. Analysis of carbohydrates and glycoconjugates by matrix-assisted laser desorption/ionization mass spectrometry: an update for 2009–2010. Mass Spectrom Rev. 2015;34. in press. doi: 10.1002/mas.21411.
18. Zaia J. Mass Spectrometry of oligosaccharides. Mass Spectrom Rev. 2004;23:161–227.
19. Paizs B, Suhai S. Fragmentation pathways of protonated peptides. Mass Spectrom Rev. 2005;24:508–48.
20. Lin T, Glish GL. C-terminal peptide sequencing via multistage mass spectrometry. Anal Chem. 1998;70:5162–5.
21. Kitahara Y, Takahashi S, Tsukagoshi M, Fujii T. Formation of bisphenol A by thermal degradation of poly(bisphenol A carbonate). Chemosphere. 2010;80:1281–4.
22. Kitahara Y, Okuyama K, Ozawa K, Suga T, Takahashi S, Fujii T. Thermal decomposition of acrylamide from polyacrylamide: time-resolved pyrolysis with ion-attachment mass spectrometry. J Therm Anal Calorim. 2012;110:423–9.
23. Hanson D, Orlando J, Noziere B, Kosciuch E. Proton transfer mass spectrometry studies of peroxy radicals. Int J Mass Spectrom. 2004;239:147–59.
24. Zhang X, Kato S, Bierbaum VM, Nimlos MR, Ellison GB. Use of a flowing afterglow SIFT apparatus to study the reactions of ions with organic radicals. J Phys Chem A. 2004;108:9733–41.
25. Takahashi S, Nakamura M, Fujii T. Design and performance of a compact Li^+ ion attachment mass spectrometry system with an atmospheric sampling device. J Am Soc Mass Spectrom. 2012;23:547–52.
26. Bahr U, Deppe A, Karas M, Hillenkamp F, Giessmann U. Mass spectrometry of synthetic polymers by UV-matrix-assisted laser desorption/ionization. Anal Chem. 1992;64:2866–9.
27. Rizzarelli P, Carroccio S. Modern mass spectrometry in the characterization and degradation of biodegradable polymers. Anal Chim Acta. 2014;808:18–43.
28. Altuntaş E, Schubert US. "Polymeromics": mass spectrometry based strategies in polymer science toward complete sequencing approaches: a review. Anal Chim Acta. 2014;808:56–69.
29. Van Berkel GJ, Pasilis SP, Ovchinnikova O. Established and emerging atmospheric pressure surface sampling/ionization techniques for mass spectrometry. J Mass Spectrom. 2008;43:1161–80.
30. Paine MR, Barker PJ, Blanksby SJ. Ambient ionisation mass spectrometry for the characterisation of polymers and polymer additives: a review. Anal Chim Acta. 2014;808:70–82.
31. Friia M, Legros V, Tortajada J, Buchmann W. Desorption electrospray ionization—orbitrap mass spectrometry of synthetic polymers and copolymers. J Mass Spectrom. 2012;47:1023–33.
32. Bayer E, Gfrörer P, Rentel C. Coordination-Ionspray-MS (CIS-MS), a universal detection and characterization method for direct coupling with separation techniques. Angew Chem Int Ed. 1997;38:992–5.
33. Havrilla CM, Hachey DL, Porter NA. Coordination (Ag^+) ion spray-mass spectrometry of peroxidation products of cholesterol linoleate and cholesterol arachidonate: High-performance liquid chromatography-mass spectrometry analysis of peroxide products from polyunsaturated lipid autoxidation. J Am Chem Soc. 2000;122:8042–55.
34. Sandra P, Medvedovici A, Zhao Y, David F. Characterization of triglycerides in vegetable oils by silver-ion packed-column supercritical fluid chromatography coupled to mass spectroscopy with atmospheric pressure chemical ionization and coordination ion spray. J Chromatogr A. 2002;974:231–41.

Index

T. Fujii (ed.), *Ion/Molecule Attachment Reactions: Mass Spectrometry*,
DOI 10.1007/978-1-4899-7588-1

MIX
Papier aus verantwortungsvollen Quellen
Paper from responsible sources
FSC® C105338

If you have any concerns about our products,
you can contact us on
ProductSafety@springernature.com

In case Publisher is established outside the EU,
the EU authorized representative is:
Springer Nature Customer Service Center GmbH
Europaplatz 3, 69115 Heidelberg, Germany

Printed by Libri Plureos GmbH
in Hamburg, Germany